THE CHEMICAL EVOLUTION OF PHOSPHORUS

An Interdisciplinary Approach to Astrobiology

THE CHEMICAL EVOLUTION OF PHOSPHORUS

An Interdisciplinary Approach to Astrobiology

Enrique Maciá-Barber

AAP | APPLE ACADEMIC PRESS

Apple Academic Press Inc.
4164 Lakeshore Road
Burlington ON L7L 1A4
Canada

Apple Academic Press Inc.
1265 Goldenrod Circle NE
Palm Bay, Florida 32905
USA

Exclusive worldwide distribution by CRC Press, a member of Taylor & Francis Group

International Standard Book Number-13: 978-1-77188-804-2 (Hardcover)
International Standard Book Number-13: 978-0-42926-513-6 (eBook)

Library and Archives Canada Cataloguing in Publication

Title: The chemical evolution of phosphorus : an interdisciplinary approach to astrobiology / Enrique Maciá-Barber.

Names: Barber, Enrique Maciá, author.

Description: Includes bibliographical references and index.

Identifiers: Canadiana (print) 20190164123 | Canadiana (ebook) 20190164158 | ISBN 9781771888042 (hardcover) | ISBN 9780429265136 (ebook)

Subjects: LCSH: Phosphorus.

Classification: LCC QD181.P1 B37 2020 | DDC 546/.712—dc23

CIP data on file with US Library of Congress

Apple Academic Press also publishes its books in a variety of electronic formats. Some content that appears in print may not be available in electronic format. For information about Apple Academic Press products, visit our website at **www.appleacademicpress.com** and the CRC Press website at **www.crcpress.com**

Printed and bound by CPI Group (UK) Ltd, Croydon, CR0 4YY

About the Author

Enrique Maciá-Barber, PhD
Professor of Condensed Matter Physics, Universidad Complutense de Madrid, Spain

Enrique Maciá-Barber, PhD, is currently a full professor of condensed matter physics at the Universidad Complutense de Madrid. His research interests include the thermoelectric properties of quasicrystals and DNA biophysics. He is also the author of several monographs and the books *Aperiodic Structures in Condensed Matter: Fundamentals and Applications* (CRC Press, Boca-Raton, 2009) and *Thermoelectric Materials: Fundamentals and Applications* (Pan Stanford Publishing, Singapore, 2015). Prof. Maciá-Barber holds a PhD in Physical Sciences from the Complutense University of Madrid (UCM); he was the winner of the Extraordinary Doctorate Award for his thesis on Elementary Excitations in Aperiodics Systems. He received his MSc degree in astrophysics in 1987.

Contents

Abbreviations

AGB	asymptotic giant branch
ATP	adenosine triphosphate
AU	astronomical unit
CH_3CONH_2	acetamide
CO	carbon monoxide
CO_2	carbon dioxide
COOH	carboxylic group
COSAC	cometary sampling and composition
CS	carbon monosulfide
DEPA	diethylphosphorodithioic acid
DIPA	diisopropylphosphorodithioic acid
DMPA	dimethylphosphorodithioic acid
DNA	deoxyribonucleic acid
FAD	flavine-adenine dinucleotide
GCR	galactic cosmic rays
HC_3N	cyanoacetylene
HCN	hydrogen cyanide
HR	Hertzsprung–Russell
HST	Hubble Space Telescope
HZ	habitable zone
IDPs	interplanetary dust particles
IGM	intergalactic medium
IP6	inositol hexaphosphate
IR	infrared
ISM	interstellar medium
KREEP	potassium, rare-earth, and phosphorus
LED	light emission device
LTE	local thermodynamical equilibrium
LUCA	last universal common ancestor
MACHOs	massive compact halo objects
MPO_4	phosphate minerals
NADPH	nicotine-adenine dinucleotide
PAH	polycyclic aromatic hydrocarbon

PN	protoplanetary nebulae
PO	phosphorus monoxide
PO_4^{3-}	phosphate ion
QSO	quasi-stellar object
RNA	ribonucleic acid
ROSINA	Rosetta orbiter spectrometer for ion and neutral analysis
SEP	solar energetic particles
SN	supernova
SNR	supernova remnant
UV	ultraviolet
WHO	World Health Organization
WIMPs	weakly interacting massive particles

Foreword

Since the pioneering work on stellar nucleosynthesis by Fred Hoyle and A.G.W. Cameron in the 1950s, astronomers have been aware that most of the chemical elements in the universe are produced in stars. Chemical elements are made by nuclear reactions deep in the interior of stars, carried up to the surface, ejected into the interstellar medium, and distributed throughout the Milky Way galaxy. The primordial solar nebula, out of which the Sun and the Earth were formed, was enriched by the ejection of the previous generation of stars.

Since living organisms evolved from ingredients of the early Earth, it is therefore not an exaggeration to say that every atom in our body was once inside a star and humans are made of stellar material. The most common elements in our bodies are hydrogen, oxygen, carbon, and nitrogen, which are also among the most common elements found in stars and gaseous nebulae in the galaxy.

The only exception is phosphorus. Phosphorus is the fifth most common element in human bodies but only ranks 18th among the most abundant chemical elements in our parent star, the Sun, and in the interstellar medium of our galaxy. Phosphorus is a major constituent of nucleic acids and cell membranes and plays a major role in storage and transmission of genetic information as well as metabolism. Why is phosphorus so heavily concentrated and utilized in our bodies is one of the most interesting unanswered questions in the origin of life. Among the primordial ingredients available, why did terrestrial biochemistry preferentially select phosphorus to create the prebiotic molecules that eventually led to life?

These are the questions that Enrique Maciá-Barber tries to answer in this book. Prof. Maciá-Barber is a world-renowned expert in the astrobiological study of phosphorus. This book beautifully traces the stellar origin of the element phosphorus, its chemical properties, and the observations of phosphorus-based molecules and minerals in the interstellar medium and in the solar system. He then connects the astronomical studies with the role that phosphorus plays in living organisms, presenting the biochemistry of biomolecules that incorporates phosphorus, and the roles that these molecules play in the origin of life on Earth.

This book presents a comprehensive summary of our current understanding of the astrochemical and astrobiological significance of phosphorus. It is invaluable for researchers and students who are interested in the question of the origin of life and the search for extraterrestrial life.

Is the importance of phosphorus in life only confined to terrestrial biochemistry? Can we imagine other biochemical pathways and structures in an unknown alien life where phosphorus plays a different role? These are fascinating questions for future researchers to explore.

— **Sun Kwok**
President, International Astronomical Union Commission on Astrobiology (2015–2018), University of British Columbia, Vancouver, Canada

Acknowledgments

This book is dedicated to the memory of Professor John Oró with my heartfelt gratitude for the deep interest, valuable advice, and motivating support he provided me during the time I had the fortune to share with him. I warmly thank Marcelino Agúndez, Lou Allamandola, Martin Asplund, Sun Kwok, and Lucy Ziurys for their interest in this research, as well as for fruitful correspondence and useful comments during the last two decades. I am also indebted to Rafael Bachiller, Gabriele Cescutti, Fred Goesmann, Chiaki Kobayashi, Terence Kee, Katharina Lodders, Daniel Murphy, Matthew Pasek, Ilka Peterman, Cameron Pritekel, Víctor M. Rivilla, Caleb A. Scharf, Xiaoping Sun, Josep María Trigo-Rodríguez, Takashi Tsuji, Channon Visscher, and Ian P. Wright, for sharing useful materials and relevant information. Last, but not least, I thank the Apple Academic Press Vice President, Sandra Sickels, for making this book possible, and to Victoria Hernández for her continued interest in my long-standing research project, and for her collaboration and assistance with the manuscript.

Preface

Phosphorus (light-bearer in Greek) was discovered in 1669 by German physician and alchemist Hennig Brand by distilling human urine, a biological waste product. Indeed, phosphorus belongs to the exclusive set of those chemical elements, namely, H, O, C, N, S, and P, which are present in large enough amounts in all living beings, and for that reason, they are referred to as main biogenic elements. Nevertheless, at variance with the remaining bioelements, which are also among the most abundant atoms in the universe, phosphorus is remarkably scarce on a cosmic scale, occupying the eighteenth position in the cosmic elemental abundance ranking.

The scarcity of phosphorus atoms in the universe sharply contrasts with the high abundance of this species in the elemental composition of living cells and tissues as well as in vertebrates' bones and teeth. This feature is quite intriguing, especially when one realizes that phosphorus compounds exist in a great variety in living systems, where they perform many fundamental biochemical tasks. Thus, the esters of phosphoric acid, including sugar phosphates and nucleotides, play a leading role in most biochemical processes, such as glycolysis and nucleic acid metabolism. These esters also determine the structural stability of DNA and RNA nucleic acids. Adenosine triphosphate (ATP) is the best-known conveyer of chemical energy in most metabolic routes, and this molecule also acts as an inorganic phosphate carrier in many important enzymatic reactions. In addition, some cyclic nucleotide derivatives play a significant role in the biochemical activity of diverse hormones, in the synaptic transmission of the nervous system, in cellular division regulation, and even in immune and inflammation responses. Definitively, phosphorus is everywhere in the living world!

Therefore, how did phosphorus atoms, which are produced inside the inner cores of a handful of huge stars, concentrate in relatively high proportions, mainly in the form of PO_4 phosphate groups, in the organisms composing Earth's biosphere to rise up from its 18th cosmic abundance place to the fifth or sixth position in the elemental abundance ranking of biomass representatives so diverse as ancient archeobacteria and modern human beings?

And, closely related to the question above, how did these phosphate derivatives manage to be included in such a great variety of organic molecules playing essential biochemical roles in all known life forms?

These two questions define the essence of what I refer to as the phosphorus enigma, and this book is devoted to describe the very nature of this puzzle, and to provide some hints towards suitable answers to both queries. To this end, within the grand panorama of cosmos evolution, we will grab the limelight onto this chemical element as the main character of our story, and will follow its evolutionary path from massive stars inner cores, where it is formed ruled by the strong nuclear forces, all the way long until its incorporation in the minerals dispersed all over the solar system, ending up with the macromolecular biopolymers encoding the genetic information and the metabolic molecules setting the rhythms of life inside the living beings that populate the planet we inhabit.

We will start this scientific exploration journey by first considering thermonuclear reactions involving atomic nuclei which interact at a subatomic domain, to progressively zoom out the spatial scale by considering simple phosphorus-bearing molecules flowing through circumstellar shells around aged stars to dilute in the cold interstellar medium, and then we will further increase the magnifying-glass power up to the size of schreibersite or apatite mineral crystals readily visible to the naked eye. During this bottom-up voyage across our hierarchically structured cosmos, we will also pay attention to the different energy intervals where these phosphorus-containing structures can exist, ranging in temperature from the thousand million Kelvin degrees required to ignite the nuclear fuels leading to the synthesis of ^{31}P nuclei, to the much lower room-temperature values necessary to preserve the structural integrity of the sugar-phosphate backbone of DNA macromolecules and the stability of the fragile organic compounds moving around inside living cells alike.

The contents of this book are arranged according to three main conceptual stages, which are explicitly distributed in 10 chapters. Due to the interdisciplinary nature of the problem to be addressed, the first stage aims to introduce the fundamental concepts and notions of physics, chemistry, and biology within the astronomical framework that one needs to be able to achieve a proper understanding of the topics discussed in the subsequent stages. In this way, Chapter 1 provides a brief introductory overview describing the main goals of the book. Afterward, some fundamentals of astrophysics, chemistry, and biology disciplines are reviewed in the Chapters 2 and 3, closing the first conceptual stage with a detailed description of the main physical and chemical properties of phosphorus compounds of interest, which is given in Chapter 4.

Equipped with this basic knowledge, the second stage of the trip focuses on the presence and distribution of phosphorus, and its compounds as the universe evolves following the arrow of time. To this end, through the Chapters 5–8, we will discuss the relevance of phosphorus among the main biogenic elements by considering its crucial role in most essential biochemical functions as well as its peculiar chemistry under different physicochemical conditions. In doing so, we will review the phosphorus compounds which have been found in different astrophysical objects, such as planets and moons, interplanetary dust particles, asteroids and comets, stars of different sorts, and the interstellar medium. In this way, we realize that this main biogenic element is both scarce and ubiquitous in the universe. These features can be related to the complex nucleosynthesis of phosphorus nuclei in the cores of massive stars under explosive conditions favoring a wide distribution of P atoms throughout the interstellar medium, where they would be ready to react with other available atoms. The tendency towards oxidized or reduced phosphorus compounds will be scrutinized as chemical evolution proceeds from circumstellar and interstellar grains covered with icy mantles to full-fledged minerals resulting from condensation and aggregation into planetesimal bodies within protoplanetary disks ultimately leading to the formation of planetary systems. In the light of these results, we also will discuss some possible routes allowing for the incorporation of phosphorus compounds of prebiotic interest during the earlier stages of solar system formation and the emergence of life on Earth.

Then, in Chapter 9, we will narrow down our perspective from the workings of Nature all over the universe to focus on the workshops of men laboring on the Earth, and its close planetary neighborhood, by considering industrial applications of manmade phosphorus compounds in current areas of research of solid-state physics, materials engineering, nanotechnology or medicine. Finally, some preliminary answers regarding the three main queries related to the phosphorus enigma, namely, the phosphorus nucleosynthesis problem, the phosphorus chemistry puzzle, and the prebiotic phosphate conundrum, will be given in Chapter 10, along with some suggestions for future research work which could be carried out in order to further clarify the cosmic history of phosphorus in the years to come.

Previous books covering the broad topic of chemical evolution have paid little attention to two main aspects which are considered in detail in this book, namely: the significant role of energy fluxes as a parallel development to the increasing structural complexity in the physicochemical processes eventually leading to the possible emergence of life, and the fundamental importance of entropy as a driving force in cosmic evolution, which complements the broadly considered role of minimum energy principle in the appearance of complex enough stable structures as time goes on.

The book contains 16 proposed exercises accompanied by their detailed solutions, along with an Appendix including the values of some important physical constants, astronomical quantities, and conversion factors. I have prepared the exercises mainly from results published and discussed in regular research papers during the last decade, in order to provide a glimpse into the main current trends in the field. Although the exercises and their solutions are given at the end of the book for convenience, it must be understood that they are an integral part of the presentation, either motivating or illustrating the different concepts and notions introduced in the main text. Accordingly, it is highly recommended to the reader that he/she tries to solve the exercises in the sequence they appear in the text, then check his/her obtained result with those provided at the end of the book, and only then to resume the reading of the corresponding chapter. In this way, the readers (who are intended to be both graduate students as well as senior scientists approaching this topic from other research fields) will be able to extract the maximum benefit from the materials contained in this book in the shortest time. For the sake of completeness, the most relevant technical terms introduced through the book are compiled in a detailed Glossary, which is included in the front matter of the book.

The book can be used as a reader-friendly textbook for undergraduate, graduate, or postgraduate students, senior scientists and researchers coming from diverse related fields of physics, chemistry, astrophysics, biology or geology and approaching this topic from other research fields. Indeed, most of the contents covered in this book are included in the curricula of different science courses. Accordingly, I confidently hope this book may provide a motivating unifying topic, the chemical evolution of phosphorus compounds, within a transversal framework which could be fruitfully used by both students and teachers in order to gain a broader perspective on the intertwined workings of Nature as the universe unfolds.

—Enrique Maciá Barber, PhD
Madrid, September 2019

CHAPTER 1

The Phosphorus Enigma: An Overview

"Phosphorus is of interest because little is known about its gas-phase chemistry, although it plays a fundamental role in biological systems."

(Lucy Ziurys, 1987)

"The relative abundance of phosphorus in the human body is several orders of magnitude greater than in Solar System, where it is only the seventeenth most common element. So, how did phosphorus concentrate on earth, ultimately becoming part of us?"

(Sun Kwok, 2006)

1.1 THE ORIGIN OF CHEMICAL ELEMENTS

By the late 1960s, astrophysicists already knew that hydrogen, a significant fraction of helium and some traces of lithium atomic nuclei, were made during the primordial nucleosynthesis episode, which took place under the extremely high temperature and density conditions prevailing in the first few minutes of our early expanding universe. Most of the remaining heavier chemical elements present in the periodic table were made a long time after this stage, through an ordered sequence of nuclear reactions occurring in the cores of stars (Burbidge et al., 1957; Fowler, 1984). Astronomers had also discovered that not all stars equally contribute to the final inventory of chemical elements in the galaxy they inhabit, but less massive stars yield relatively lighter atoms, while more massive stars produce heavier elements. Since less massive stars significantly outnumber more massive ones in typical galaxies, the resulting averaged elemental abundance distribution systematically decreases as the atomic number (Z) increases, according to a nearly exponential decline until $Z \sim 42$, thereafter decreasing more gradually, as can be seen in Figure 1.1. Along with this general main trend, we can also appreciate some remarkable dips and peaks in the elemental distribution curve, namely, the beryllium, fluorine, and scandium relative dips, on the one hand, and the oxygen, iron, and lead relative peaks, on the other hand.

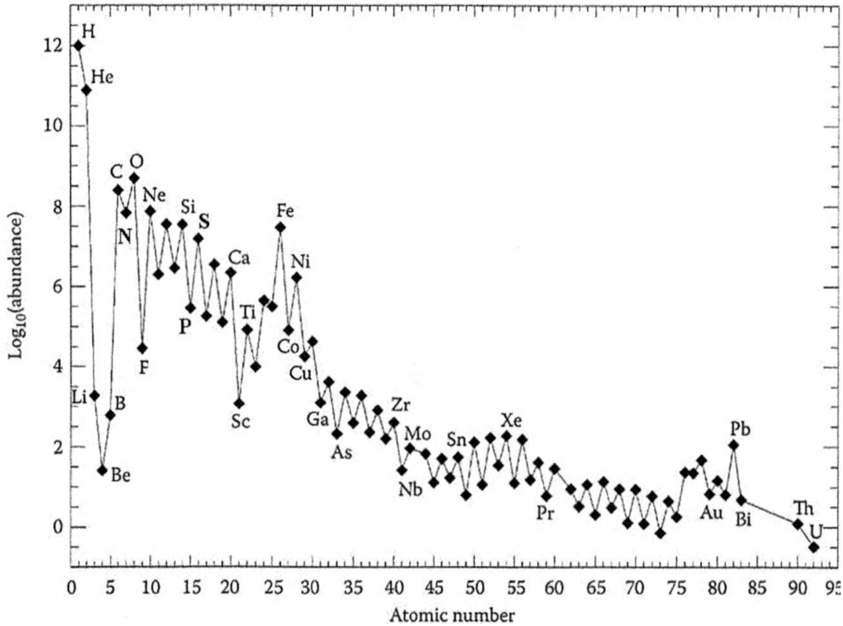

FIGURE 1.1 Plot showing the average cosmic abundance of elements as a function of their atomic number Z. This curve combines data retrieved from spectroscopic observations of the Sun, the Stars, and the interstellar medium (ISM), as well as from cosmic-ray particles and direct chemical analysis of samples collected from Earth, the Moon, Mars, meteorites, asteroids, cometary nuclei and interplanetary dust particles, hence providing the so-called universal (or cosmic) abundances of the elements. Although the obtained chemical distribution curves sometimes differ in detail for particular elements depending on the considered sources, they rarely do so by more than a factor of three on a scale that spans more than 12 orders of magnitude.

Source: Greenwood and Earnshaw, 1986; Asplund et al., 2009; Data taken from Lodders, 2003.

1.2 THE ROLE OF BIOGENIC ELEMENTS IN CHEMICAL EVOLUTION

Living cells consist of a large variety of different biopolymers, namely, proteins, sugars, lipids, ribonucleic acids (RNA), and deoxyribonucleic acids (DNA). Together with smaller molecules, such as water, phosphates, sulfates, and a few metallic ions, these components give the molecular content of the so-called biomass. Elemental analyses of these biochemical compounds have revealed that H, O, C, N, P, and S atoms are needed in large quantities to make living organisms, and for this reason, these elements are referred to as the main biogenic elements. For instance, by properly averaging the main biomass constituents of a typical yeast cell, its elemental composition (normalized to the carbon content) can be expressed by the stoichiometric formula $H_{1.748}CO_{0.596}N_{0.148}P_{0.009}S_{0.0019}M_{0.0018}$, where M stands for metal atoms belonging to the so-called oligo-elements set, including K, Na, Mg, Ca, Fe, Mn, Cu, and Zn, which are required in minor quantities only (Lange and Heijnen, 2001).

By inspecting the elemental abundance curve depicted in Figure 1.1, and comparing it with the elemental composition of living beings, derived from detailed biochemical analysis, some scientists realized that the four most abundant elements in the universe, with the exception of the noble gases helium and neon, are hydrogen, oxygen, carbon, and nitrogen, which are also precisely the four major constituent elements of organic compounds and of living matter (Greenstein, 1961; Oró, 1963). This is illustrated in Figure 1.2, where we can also appreciate that C and P elements are enhanced in biomass as compared to their cosmic abundances. In fact, albeit the cosmic elemental abundance of S is about two orders of magnitude larger than that of P, the latter is more abundant in living beings by a factor of four at least.

Indeed, phosphorus compounds exist in a great variety in living systems, where they perform many fundamental biochemical functions involving storage and transfer of information (nucleic acids), energy transfer (adenine and guanine nucleotides), membrane structure (phospholipids), and signal transduction (cyclic nucleotides). Accordingly, the elemental abundance of phosphorus ranks at the fifth (sixth) position in the chemical inventory of unicellular (pluricellular) organisms, respectively. However, by inspecting Figure 1.1, we realize that phosphorus, occupying the eighteenth position in the cosmic elemental abundance ranking, is the less abundant species among the third-row elements of the periodic table, as well as in the group of the main biogenic elements. Thus, the only biogenic element present in biological tissues at a concentration substantially above its solar abundance is phosphorus (Whittet and Chiar, 1993). This suggests the probable existence of physical and chemical processes favoring a differential enhancement of organic matter in general, and phosphorus bearing compounds in particular, at certain astrophysical environments, stemming from *chemical evolution* in the Galaxy, that is, a progressive and general tendency of matter to go from simpler to more complex atomic and molecular arrangements at certain astrophysical places as times goes by (Mason, 1992; Rauchfuss, 2008; Kwok, 2013).

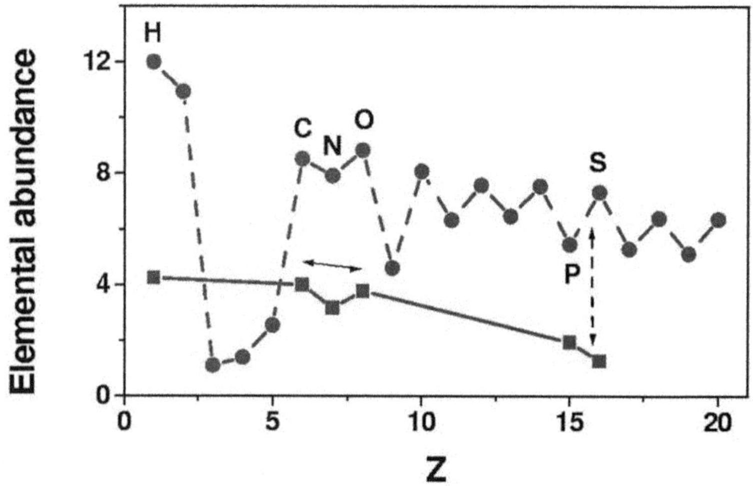

FIGURE 1.2 Elemental abundance comparison between the dry biomass of representative yeast with stoichiometric composition $H_{1.748}CO_{0.596}N_{0.148}P_{0.009}S_{0.0019}M_{0.0018}$ (squares) and the cosmic elemental distribution curve shown in Figure 1.1 (circles). The abundances are given on a logarithmic scale with H = 12 for cosmic abundances and C = 4 for biomass abundances. The elements are arranged according to their atomic number Z, and the main biogenic elements are explicitly labeled. Note that the relative importance of C and O peaks is reversed in biomass and cosmic curves, respectively, and that the P/S ratio in the biomass curve is remarkably enhanced as compared to the observed cosmic ratio.
Source: Maciá, 2005; with permission from the Royal Society of Chemistry.

It should be emphasized, however, that this trend towards a higher chemical complexity is not the result of a continuous process occurring everywhere, but rather the final outcome of a lot of intertwined processes, which take place at different scales of space and time, in such a way that synthesis and destruction events alternate each other along the arrow of time. For instance, as we will describe it detail in Chapter 5, chemical elements heavier than helium must be synthesized by means of thermonuclear reactions requiring very high temperatures, well above those we find in our ever cooling expanding universe once the primeval nucleosynthesis episode ended. The appearance of stars able to reach the required high temperatures in their inner cores then becomes a crucial event for the origin of most of the periodic table elements and their related chemistry. Stellar nucleosynthesis illustrates an important synthesis mechanism which started when the universe was about 200 million years old and will be enduring for a long period in the future universe's history. Now, the elements formed inside stellar cores must be subsequently liberated to the ISM in order to undergo further chemical processing. Such a release takes place through mass loss processes during the life-cycle of stars, first by means of stellar winds, eventually followed by either the formation of planetary nebulae (in the case of low mass stars) or via supernova (SN) explosions (in the case of high mass stars), all of which being essentially disruptive processes by themselves, ultimately leading to the destruction of the original stellar structure (Trimble, 1982, 1983; Kwok, 2000, 2013). Once in the ISM the atoms delivered from stars can undergo chemical reactions among them, promoting the formation of polyatomic molecules, either on the surfaces of minute dust grains (previously condensed in the circumstellar envelopes around aging stars) or in the gas-phase among the stars (Zeng et al., 2018), as it is illustrated in Figure 1.3.

As we see, in the course of these chemical reactions some molecules are formed, while others are destroyed according to complex entangled networks driven by energetic ultraviolet (UV) photons emitted by young stars or fast cosmic rays: wandering particles spiraling throughout the Galaxy as material echoes of ancient SN blasts.

In this way, we realize that notwithstanding the important role certain astrophysical objects may play in their due time, different material structures are progressively emerging and fading away, as the cosmic clock ticks on, ruled by energy optimization and entropy production criteria. Keeping this in mind we realize that the emergence of life, which requires the existence of very complex and fragile biopolymer systems, continuously exchanging matter and energy with their surroundings under conditions very far from thermodynamical equilibrium, can only take place in a mature and cold enough evolved universe.

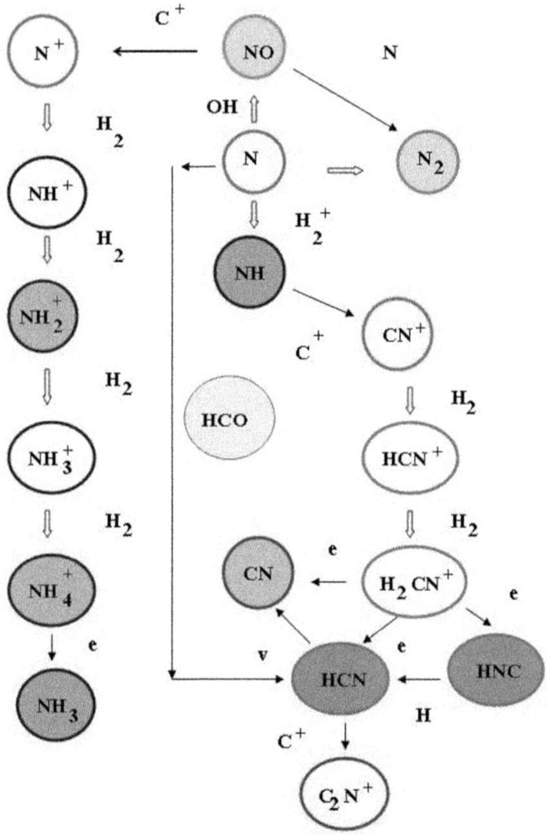

FIGURE 1.3 Some important chemical reactions in the ISM gas-phase nitrogen chemistry. The initiating N^+ and C^+ ions are produced by the intense UV radiation field of nearby stars and cosmic rays ionizing radiation (dissociative processes). The presence of the molecules highlighted in the filled circles has been spectroscopically confirmed in the ISM.
Source: Maciá, 2005; with permission from the Royal Society of Chemistry.

1.3 THE ROLE OF PHOSPHATES IN LIVING SYSTEMS

Phosphorus, mainly in the form of phosphate derivatives, is a universal constituent of cells protoplasm and is required for growth, health, and reproduction in all forms of animals, plants, and bacteria. Accordingly, this element belongs to the selected group of the main biogenic elements, which along with H, C, O, N, and S are present in all known life beings. In fact, phosphorus compounds profusely appear in living systems where they perform many fundamental biochemical functions. Thus, the esters of phosphoric acid, including sugar phosphates and nucleotides, play a leading role in most biochemical processes, such as glycolysis and nucleic acid metabolism, and these esters also determine the structural stability of DNA and RNA nucleic acids. Almost all coenzymes of photosynthesis, fermentation, respiration, and biosynthesis, such as nicotine-adenine dinucleotide (NADPH), flavine-adenine dinucleotide (FAD), or coenzyme A contain phosphoric acid derivatives as an essential component. Adenosine triphosphate (ATP) is the best-known conveyer of chemical energy in most metabolic routes, and this molecule also acts as an inorganic phosphate carrier in many important enzymatic reactions. In addition, some cyclic nucleotide derivatives play a significant role in the biochemical activity of diverse hormones, in the synaptic transmission of the nervous system, in cellular division regulation, and even in immune and inflammation response. On the other hand, the ion HPO_4^{2-} plays a crucial role in tasks ranging from active carrier transport through cellular and mitochondrial membranes to bone metabolism. The $H_2PO_4^-$-HPO_4^{2-} system is also an important intracellular buffer. The main biochemical roles played by phosphorus compounds are summarized in Table 1.1.

By inspecting this table, we realize that different phosphoric acid moieties cover a broad spectrum of biochemical activities involving storage and transfer of information, energy transfer, membrane structure, or signal transduction. Thus, phosphate esters and anhydrides dominate the living world (Westheimer, 1987). This properly illustrates the chemical unity of phosphorus compounds in living matter, expressed by the fact that such diverse and fundamental biological tasks are related to a unique basic chemical motive, namely, the orthophosphoric acid molecule H_3PO_4.

TABLE 1.1 Biochemical Roles of Compounds Containing Phosphorus*

Compound	Biochemical Role
Nucleic acids	Storage and transmission of genetic information
Nucleotides	Coenzymes; carriers of P; precursors in DNA and RNA synthesis
	Chemical energy transfer (ATP)
Phospholipids	Main characteristic components of cellular membranes
Sugar phosphates	Intermediate molecules in carbohydrates metabolism
HPO_4^{2-}	Intracellular buffer; ionic carrier; bone metabolism

*Main biochemical roles of different phosphorus-bearing compounds in living systems.

1.4 THE PHOSPHORUS ENIGMA

So, how did phosphorus atoms produced inside giant stars concentrate as phosphate derivatives in the organisms composing earth's biosphere to rise up from its 18th cosmic abundance place to the fifth or sixth position in the elemental abundance ranking of biomass representatives, going from archeobacteria to human beings? And, closely related to the question above, how did phosphorus atoms belonging to PO_4 phosphate groups manage to be included in such a great variety of molecules playing essential biochemical roles in all currently existing life forms?

The two questions above define the essence of what we refer to as the phosphorus enigma.

The first step towards its elucidation will be presented in Chapter 5, where we compare the different nucleosynthesis routes yielding the main biogenic elements C, O, N, S, and P. In doing so, we will realize that the case of phosphorus is quite remarkable. In the first place, the nucleosynthesis of the ^{31}P nucleus can only take place in the minor subset of stars which are massive enough to ignite the previously synthesized C and Ne fuels under explosive conditions. In the second place, its synthesis proceeds through an involved nuclear reactions network, rendering a very low overall yield of phosphorus. These facts account for the scarcity of this element in the cosmic elemental abundance inventory, hence highlighting the importance of differential abundance enhancement mechanisms occurring during the condensation and aggregation episodes giving rise to the synthesis and degradation of different chemical compounds in the course followed by atoms from dying stars' external atmospheres (Chapter 6) to newborn planetary systems around new generation stars (Chapter 7). Indeed, once the atoms are formed inside the stars, the next step in chemical evolution is the synthesis of molecular compounds joining them. During the past 80 years astronomers have detected many molecules containing biogenic elements in different astrophysical environments beyond our solar system, such as extended stellar atmospheres, circumstellar shells, diffuse nebulae and dense clouds interspersed in our Galaxy, as well as in other galaxies far away, and in the intergalactic medium (IGM) between them. For the sake of illustration, a detailed account of the ISM gas-phase molecular inventory in our Galaxy is given in Figures 1.4 and 1.5.

Quite interestingly, the number of carbon-containing compounds observed in diverse galactic sources amounts to about 75% of the 221

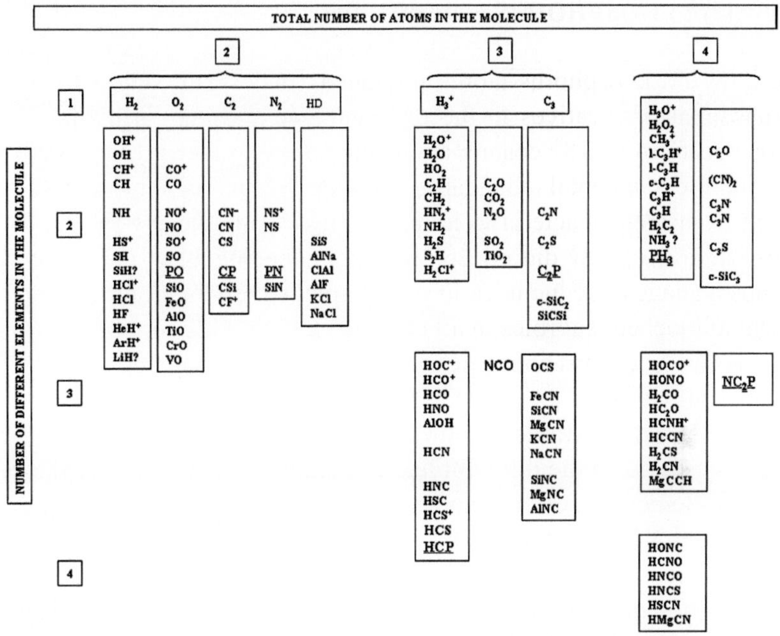

FIGURE 1.4 Molecules containing up to four atoms detected in different astrophysical environments of the ISM (including diffuse and dense clouds, circumstellar envelopes around aged stars, and planetary nebulae) are arranged according to their chemical complexity, measured in terms of the number of different elements present in the considered molecule (in ordinates) and the total number of atoms they contain (in abscissas). In each box, the molecules are listed attending to the elemental abundance rank of the atoms they contain. The seven P-bearing compounds observed to date are highlighted.
Source: Tielens, 2013; Agúndez et al., 2014b, 2018, Ziurys et al., 2015. http://www.astrochymist.org/astrochymist_ism.html; http://www.astro.uni-koeln.de/cdms/molecules/; https://en.wikipedia.org/wiki/List_of_interstellar_and_circumstellar_molecules.

molecules identified up to now (28 August, 2019), so that one may properly state that the chemistry of the universe as a whole is mainly organic chemistry (Oró, 1963). Nitrogen and oxygen-bearing compounds are also profusely found throughout the Galaxy, each one accounting for about 34% of the molecules listed in Figures 1.4 and 1.5, whereas the 21 sulfur-bearing molecules only represent a 10%. The number of molecules containing phosphorus is even smaller, since just seven representatives have been reported to date in the ISM. Suitable information regarding these molecules is provided in Tables 1.2 and 1.3. Most molecules listed in Table 1.2 are notable in that all compounds but PH_3 contain a strong

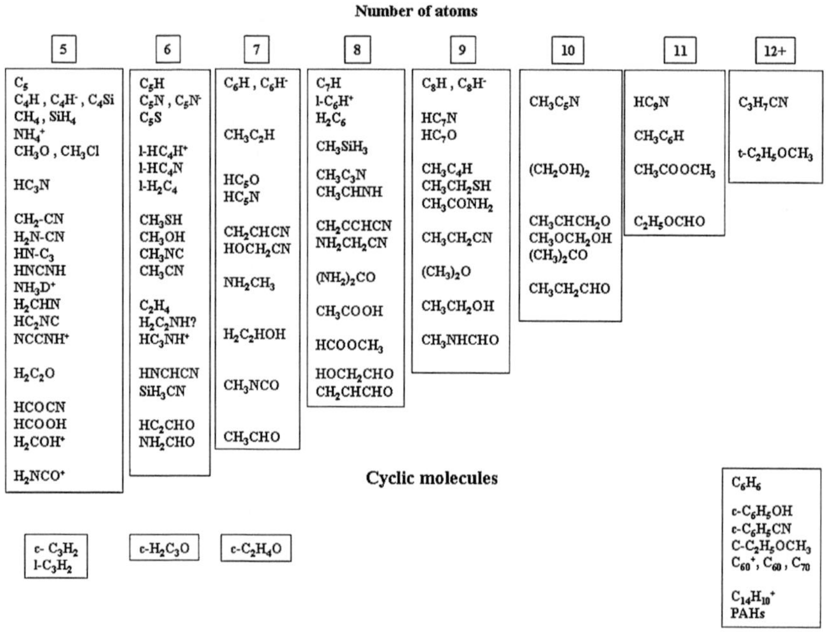

FIGURE 1.5 Molecules containing more than four atoms detected in different astrophysical environments of the ISM (including diffuse and dense clouds, circumstellar envelopes around aged stars, and planetary nebulae) are listed according to their chemical complexity measured in terms of the total number of atoms present in the considered molecule. No P-bearing compounds containing more than four atoms have been detected to date.
Source: Tielens, 2013; Ziurys et al., 2015. http://www.astrochymist.org/astrochymist_ism.html; http://www.astro.uni-koeln.de/cdms/molecules/; https://en.wikipedia.org/wiki/List_of_interstellar_and_circumstellar_molecules.

double or triple bond amounting to energies between 600 and 760 kJ mol^{-1} (6.2–7.8 eV, see Table 4.2). In contrast, phosphine only contains P – H single bonds with a bond energy of just 343 kJ mol^{-1} (3.5 eV). It is also interesting to note that analogs to each of the compounds listed in Table 1.2 have been discovered in the ISM with P substituted by its isovalent element N, namely, N_2, CN, NO, HCN, CCN, and NH_3 (see Figure 1.4).

In this regard, it is worthy to note that, albeit both N and P atoms belong to the same group in the periodic table, phosphorus chemistry is quite peculiar as compared to that of nitrogen (Cummins, 2014). A significant difference between N and P chemistries is properly illustrated by the fact that, on the grounds of energetic considerations, earlier theoretical calculations predicted that the production of the PO molecule should be highly favored in the extreme conditions (low temperature, very low densities) prevailing in the ISM (Thorne et al., 1984).

However, although the extensive search was performed, the detection of this molecule in the ISM has remained very elusive, and it was first observed in the circumstellar regions around oxygen-rich evolved stars, in agreement with previous suggestions (Maciá et al., 1997). Similarly, while nitrogen hydrides are relatively abundant in the ISM (see Figures 1.4 and 1.5), and phosphine has been long known to be present in the atmospheres of the giant planets Jupiter and Saturn (Sánchez-Lavega, 2011), no hydrides of phosphorus have been detected in the ISM yet, although PH_3 has been observed in both a circumstellar shell and a protoplanetary nebula to date (see Tables 1.2 and 1.3).

TABLE 1.2 Phosphorus Bearing Molecules Detected in the ISM Arranged in Chronological Order

Compound	Chemical Formula		Source		References
Phosphorus mononitride	PN	P≡N	Ori KL	SFR	Ziurys, 1987
					Turner and Bally, 1987
			W51M	SFR	
					Ziurys, 1987
			Sgr B2	SFR	Turner and Bally, 1987
					Ziurys, 1987
			M17SW	SFR	Turner and Bally, 1987
			DR 21OH	SFR	
			NGC 7538	SFR	Turner et al., 1990
			IRC +10216	C-rich CSE	Turner et al., 1990
			IRC +10216	C-rich CSE	Turner et al., 1990
			VY CMa	O-rich CSE	Guélin et al., 2000
			CRL 2688	PPN	Agúndez et al., 2007
			L1157 B1	SFR	Ziurys et al., 2007
			IK Tau	O-rich CSE	Milam et al., 2008
					Yamaguchi et al., 2011
					De Beck et al. 2013

TABLE 1.2 *(Continued)*

Compound	Chemical Formula		Source		References
			W51, W3(OH)	SFR	Fontani et al., 2016
			TX Cam	O-rich CSE	Ziurys et al., 2018
			R Cas	O-rich CSE	Ziurys et al., 2018
			NLM Cyg	O-rich CSE	Ziurys et al., 2018
Carbon monophosphide	CP	•C≡P	IRC +10216	C-rich CSE	Guélin et al., 1990
			VY CMa	O-rich CSE	Milam et al., 2008
Phosphorus monoxide	PO	•P=O	VY CMa	O-rich CSE	Tenenbaum et al., 2007
			IK Tau	O-rich CSE	De Beck et al. 2013
			TX Cam	O-rich CSE	De Beck et al. 2013
			R Cas	O-rich CSE	De Beck et al. 2013
			L1157	SFR	Lefloch et al., 2016
			W51, W3(OH)	SFR	Rivilla et al., 2016
			G+0.693–0.03	DMC	Rivilla et al., 2018
			NLM Cyg	O-rich CSE	Ziurys et al., 2018
Phosphaethyne	HCP	H–C≡P	IRC +10216	C-rich CSE	Agúndez et al., 2007
			CRL 2688	PPN	Milam et al., 2008
Dicarbon phosphide	CCP	•C–C≡P ↕ C=C=P•	IRC +10216	C-rich CSE	Halfen et al., 2008
Phosphine	PH_3	H | H–P–H	CRL 2688	PPN	Tenenbaum and Ziurys, 2008
			IRC +10216	C-rich CSE	Agúndez et al., 2014a
Cyano phosphaethyne	NCCP	N≡C–C≡P	IRC +10216	C-rich CSE	Agúndez et al., 2014b

Keys: DMC (dense molecular cloud), CSE (circumstellar shell envelope), PPN (protoplanetary nebula), SFR (star forming region). Among the detected P-bearing molecules, PN, and PO are the only ones that have been reported in star-forming regions.

TABLE 1.3 Abundances of Phosphorus-Bearing Molecules (Relative to Molecular Hydrogen) Observed in Several Circumstellar Envelopes and the Protoplanetary Nebula CRL 2688*

Source	Molecule	Abundance $f(X/H_2)$	References
VY CMa	PO	$(5 \pm 3) \times 10^{-8}$	Ziurys et al., 2018
	PN	$(7 \pm 3) \times 10^{-9}$	Ziurys et al., 2018
IK Tau	PO	$(4.5 \pm 2.5) \times 10^{-8}$	Ziurys et al., 2018
	PN	$(1.0 \pm 0.2) \times 10^{-8}$	Ziurys et al., 2018
TX Cam	PO	$(5.5 \pm 2.5) \times 10^{-8}$	Ziurys et al., 2018
	PN	$(1.0 \pm 0.3) \times 10^{-8}$	Ziurys et al., 2018
NLM Cyg	PO	$(7 \pm 3) \times 10^{-8}$	Ziurys et al., 2018
	PN	$(3 \pm 1) \times 10^{-9}$	Ziurys et al., 2018
R Cas	PO	$(1.0 \pm 0.3) \times 10^{-7}$	Ziurys et al., 2018
	PN	$(2.0 \pm 0.5) \times 10^{-8}$	Ziurys et al., 2018
CRL 2688	HCP	2×10^{-7}	Milam et al., 2008
	PN	$(3–5) \times 10^{-9}$	Milam et al., 2008
IRC +10216	HCP	3×10^{-8}	Milam et al., 2008
	CP	1×10^{-8}	Milam et al., 2008
	PH_3	1×10^{-8}	Agúndez et al., 2014a
	CCP	1×10^{-9}	Halfen et al., 2008
	PN	3×10^{-10}	Milam et al., 2008
	NCCP	$< 4\times10^{-10}$	Agúndez et al., 2014b

*The model parameters employed in the analysis of the O-rich stars IK Tau, TX Cam, R Cas, and NLM Cyg are given in Tables 6.3 and 6.4 in Section 6.2.2. Those corresponding to IRC +10216 and VY CMa stars are given in Sections 2.4.4 and 2.4.5, respectively, and those of the source CRL 2688 are: stellar radius $R_* \sim 9 \times 10^{12}$ cm, photosphere temperature $T_e \sim 3000$ K, distance of 1,000 pc, and a mass-loss rate of 1.7×10^{-4} $M_\odot$ yr^{-1}.

By inspecting the data listed in Table 1.3, we see that the fractional abundances (relative to molecular hydrogen) of the P-bearing compounds given in Table 1.2 fall in the range $f = 10^{-10}–10^{-7}$. The relative abundance of PN molecules observed in dense molecular clouds is even lower, ranging from $f = 10^{-12}$ to 10^{-10}, relative to H_2 (Ziurys, 2008). Keeping in mind the cosmic abundance value [P] = 2.6×10^{-7} relative to hydrogen we conclude that only a minor fraction of the available phosphorus is in the gas phase in the considered sources. For instance, it has been estimated that HCP and PH_3 molecules account for 5% and 2% of the total P budget around

IRC +10216, respectively (Agúndez et al., 2014a). In addition, we see that the fractional abundance of PO is more abundant than PN by a factor of five in the circumstellar envelopes around the evolved O-rich stars TX Cam, IK Tau and R Cas, and by a factor of ~ 20 in the envelope of the supergiant star NLM Cyg. On the other hand, it can be reasonably assumed that the organophosphorus HCP moiety may likely be the precursor species for the two organophosphorus compounds CCP and CP in the envelope around C-rich star IRC +10216. Accordingly, the amount of phosphorus contained in the scarce molecules so far detected in the gas phase is clearly insufficient to account for the quantity of phosphorus-bearing compounds one may expect from its elemental cosmic abundance, strongly suggesting that most phosphorus inventory could be stored in suitable condensed forms in solid dust particles (Guélin et al., 1990; Turner et al., 1990; Maciá, 2005). Indeed, according to recent observations of PN and PO molecules towards seven molecular clouds located at the Galactic Center, these molecules are probably formed in the gas-phase after the shock-induced sputtering of the dust grain mantles, while they are efficiently destroyed in those regions dominated by intense UV, X-ray, and cosmic ray radiation fields. Indeed, PN was detected in five out of seven sources, whose chemistry is thought to be shock dominated. The two sources where PN was not detected, on the other hand, correspond to clouds exposed to intense radiation fields. PO molecule was only detected towards the cloud G+0.693–0.03, with a PO/PN abundance ratio of ~1.5 (Rivilla et al., 2018).

It is interesting to note that phosphorus compounds exhibiting different oxidation states are found in diverse astrophysical environments, as it is depicted in Figure 1.6, where we observe the presence of:

1. Several phosphate minerals (MPO_4) in rocky planets and satellites (Earth, Mars, Moon), stony meteorites, and interplanetary dust particles (IDPs);
2. Phosphorus monoxide (PO) in O-rich circumstellar shells;
3. Organophosphorus compounds such as phosphonic acids in carbonaceous chondrites, and the molecules CP, CCP, HCP, and NCCP found in circumstellar regions;
4. The ubiquitous PN molecule detected in eight star forming regions, six circumstellar envelopes, and a protoplanetary nebula; and

5. Reduced phosphorus compounds like schreibersite (Fe_3P) in iron meteorites, or PH_3 in a carbon-rich circumstellar shell, a protoplanetary nebula, and the atmospheres of Jupiter and Saturn.

As we see, phosphates are the predominant form in terrestrial planets, stony meteorites, and IDPs of possible cometary origin. Reduced moieties predominate in the atmospheres of giant planets and iron-rich meteorites. Finally, intermediate oxidation states are observed in some chondrites and the ISM.

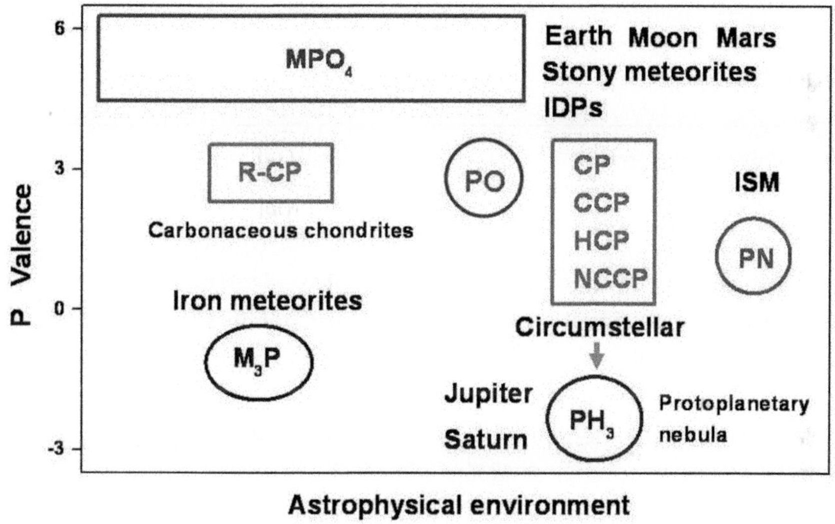

FIGURE 1.6 Distribution of phosphorus compounds in several astrophysical environments arranged according to their oxidation degree.
Source: Maciá, 2005; with permission from the Royal Society of Chemistry.

Quite remarkably, the fact that most phosphorus on earth's surface is in the form of phosphate leads to an important problem in prebiotic chemistry arising from the difficulty of the spontaneous phosphorylation of organic compounds by minerals likely present on the early earth's crust (Pasek, 2015b). In fact, a major open question in evolutionary biochemistry concerns both the role of phosphorus compounds in the chemical evolution which preceded the emergence of life on earth and the primary sources of such phosphorus compounds (Chapter 8).

In summary, keeping in mind the queries we have enumerated in this introductory chapter, the phosphorus enigma can then be briefly posed as follows: Why should such a relatively scanty element be so important for biological systems? In order to provide a plausible answer, in the following chapters we will adopt an interdisciplinary approach by addressing key physical, chemical, and biological aspects of phosphorus atoms journey from their very formation in the cores of massive stars to their incorporation into the early planet earth, or similar exoplanets orbiting around other stars, and then to the possible emergence of the first metabolic pathways inside the first living cell on our planet… and elsewhere in the universe?

KEYWORDS

- **biogenic elements**
- **biomass**
- **biosphere**
- **chemical evolution**
- **circumstellar shell**
- **cosmic rays**
- **galaxies**
- **intergalactic medium**
- **interstellar chemistry**
- **interstellar medium**
- **nucleosynthesis**
- **oligo-elements**
- **organophosphorus compounds**
- **prebiotic chemistry**
- **supernova event**
- **thermodynamical equilibrium**

CHAPTER 2

The Unfolding Universe

"*Wisdom is one thing. It is to know the thought by which all things are steered through all things (ἓν τὸ σοφόν, ἐπίστασθαι γνώμην, ὁτέη ἐκυβέρνησε πάντα διὰ πάντων).*"

Heraclitus (535–475 BC)

2.1 THE COSMOS AS AN ORDERED EVOLVING WHOLE

We inhabit a structured universe which ancient Greeks named *cosmos* (κόσμος). If we observe it with care and attention, we can identify a broad palette of different basic structures, such as galaxies, stars, planets, minerals, living cells, molecules or atoms, to name but the most relevant ones, each one displaying a characteristic range of typical sizes at different spatial scales. Their characteristic scales are determined by a precise balance involving at least one of the fundamental interactions of Nature, namely, gravitation, electric, and magnetic forces, weak nuclear interaction, and strong nuclear force. These fundamental interactions have different ranges of action: gravitational and electromagnetic interactions are indefinitely long-ranged, whereas nuclear forces are extremely short-ranged instead, and they have quite different relative strengths too. For instance, the gravitational force is about 40 orders of magnitude weaker than electromagnetic interactions (Exercise 1). In turn, the strong nuclear force is about 10^2 and 10^5 times stronger than the electromagnetic and weak nuclear forces, respectively. As a consequence, one can usually see remarkable scale jumps among the different basic structures listed above. For example, galaxies contain stars, which are much smaller than galaxies (e.g., the Sun to Galaxy size ratio is about 9×10^{-13}, see the Appendix), while stars themselves are made of atoms, which are many orders of magnitude smaller than stars (e.g., the hydrogen atom to Sun size ratio is about 7 × 10^{-20}). On the other hand, galaxies associate in larger clusters which, in turn, group themselves in even larger superclusters hundreds of millions of light-years across.

In fact, *clustering* is a common trend in most of the structural patterns we observe in the cosmos. Thus, dust particles ranging in size from nanometer (10^{-9} m) to micrometer (10^{-6} m) scales condense from certain atoms and molecules present in circumstellar shells formed around aged stars (see Section 6.2.2). These dust particles then cluster to each other in the ISM to form millimeter-sized fluffy interstellar grains that subsequently stick together in protostar and preplanetary nebulae, giving rise to little icy rocks which are the seeds of cometary nuclei and the planetesimals out of which full-fledged planets are born (Bentley et al., 2016).

This sequence of events properly illustrates the trend of matter to go from small to progressively larger aggregates, through processes driven by electrostatic forces at the smaller scales, which are superseded by gravitational attraction at the larger ones. This transition from electrostatic to gravitational dominated structures is conveyed by the natural tendency of charges of opposite sign to attract to each others, hence progressively canceling their electrical force field around them. As the assembly of charged particles grows larger and larger, the resulting gravitational attraction field systematically grows in strength, while the electrostatic influence progressively disappears in the neighboring space.

Empirical observation also shows us that all structures we see in the universe are transitory, lasting for a longer or shorter characteristic lapse of time. For instance, beryllium nuclei formed during the primordial nucleosynthesis episode were quickly decomposed into two α particles (He nuclei) almost as soon as they were formed, whereas some stars in globular clusters may be older than 11 billion years. However, even these long-living stars will eventually fade away forever as their inner cores finally exhaust their continuously decreasing energy budget. In this way, we can chronologically catalog the different structures present at any given time in the cosmic history attending to the epoch when they first appeared, thereby defining a suitable sort of cosmic timetable. Following this criterion, we can enumerate the following sequence of events: the prime entities to appear were photons (light particles), elementary particles (neutrinos, electrons, positrons, protons, neutrons), and low mass nuclei (deuterium, ^{2}H, tritium, ^{3}H, helium, ^{4}He, and lithium, ^{6}Li), formed during the first three minutes of the cosmic history (Weinberg, 1977).

They were then followed by neutral atoms, formed after the so-called first recombination era, which occurred about 380,000 years after the very beginning of time. Subsequently, the emergence of globular clusters and halo structures surrounding protogalaxies took place, along with the first generation of very massive stars (Population III). Afterwards, the formation of full-fledged galaxies, containing Population II and Population I stars (in chronological order), most of them accompanied by the formation of planetary systems around them (such as our solar system), successively followed (Table 2.1).

In the present epoch, we can observe many different nested structures distributed throughout the universe, spanning more than forty orders of magnitude in size (Figure 2.1). Going from larger to smaller ones we have superclusters and clusters of galaxies, galaxies, star clusters, stars, planetary systems, planets, rocks, mineral crystals, pluricellular living beings, organic tissues, cells, bacteria, viruses, biological macromolecules, atomic clusters, simple molecules, atoms, nuclei, protons, neutrons, and elementary particles. Since our expanding universe systematically increases its size as the arrow of time goes on, we realize that larger structures were preceded by the smaller ones, although this does not prevent the possible disruption of some relatively large structures into smaller ones when it is required by the forces driving matter and energy interactions.

TABLE 2.1 Chronology of the Main Physical Events During the Universe History*

Time Elapsed Since $t = 0$	Radiation T (K)	Main Events
10^{-43} s	10^{32}	Gravity decouples from the three other forces: strong, weak, and electromagnetic (Planck time)
10^{-35}–10^{-33} s	10^{27}	Inflation. Strong force decouples from electromagnetic one
		Quarks and leptons are no longer exchangeable
10^{-12}–10^{-10} s	10^{15}	Weak and electromagnetic forces separate, neutrons, and protons are formed by photon-photon collisions
10^{-6} s	10^{13}	Quarks bind together to form neutrons and protons

TABLE 2.1 *(Continued)*

Time Elapsed Since $t = 0$	Radiation T (K)	Main Events
10^{-2} s	10^{11}	Electrons and positrons are formed through collisions of photons
1–10 s	1.2×10^{10}	Electrons and positrons annihilation creates a primordial fireball
		The universe becomes transparent to neutrinos
880 s ~ 15 min	9.5×10^{9}	Neutron decay meantime becomes lower than the universe age
10^{2}–10^{3} s	10^{9}–10^{8}	Primordial nucleosynthesis: H, He, and Li nuclei form
24,000 yr	14,000	The radiation-dominated universe becomes a matter-dominated one
380,000 yr	3,000	The universe becomes transparent to radiation: Cosmic background radiation is emitted
		Era of recombination: H atoms form
~ 2–4 × 10^{8} yr	80–30	First stars appear (Population III)
~ 2–8 × 10^{8} yr	80–20	First galaxies form
9 × 10^{9} yr		Solar System formation
1.38 × 10^{10} yr	2.726	Present

*Note the progressive and rapid decrease of the temperature of radiation as the universe grows older. After the universe becomes transparent to radiation, the temperature of matter is no longer the same as the temperature of the background radiation.
Source: Freeman et al., 2014; Barkana, 2006; Shaw, 2006.

Quite remarkably, the clock that regulates the pace of our unfolding universe runs in a different way than that ticking in the timekeepers we generally use to measure the passage of time in our scientific laboratories and sports events alike. Indeed, to this end, we use clocks which are based on the occurrence of *cyclic* processes that we take as reference time scales. For example, atomic vibrations inside quartz crystals determine the atomic clocks ticking in the picoseconds (10^{-12} s) domain, while the beating of our hearts fix the rhythm of human life at the second scale.

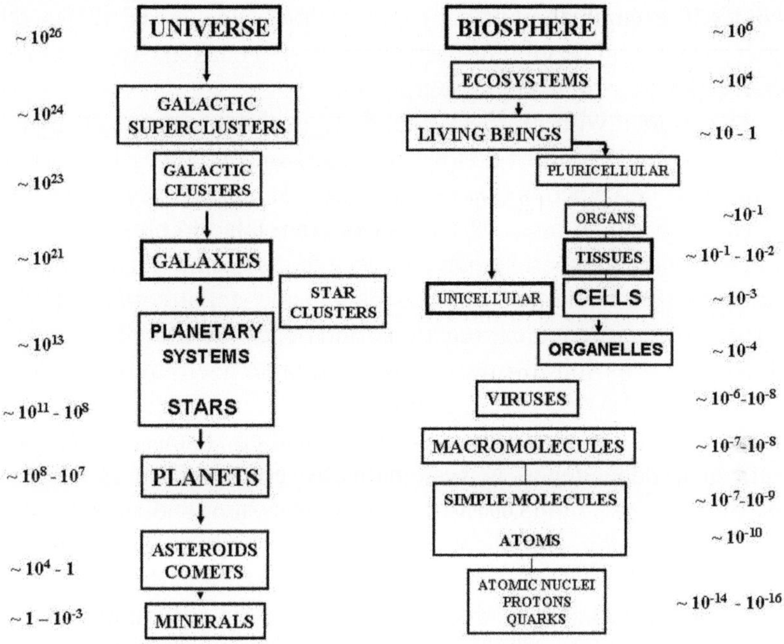

FIGURE 2.1 Typical sizes (in meters) of the main stable structures in the universe. On the left, we have the structures assembled by gravitational interaction, ranging down from the overall universe size to the planetary bodies size. On the upper right, we have the typical structures present in biological systems, ranging from the entire biosphere at the planetary scale down to viruses, biological macromolecules, smaller molecules, and atoms. All these structures are mainly ruled by electromagnetic interactions. Finally, at the bottom right, we encounter the atomic nuclei composed of protons and neutrons, which in turn are made of quarks. These minute structures are governed by nuclear interactions.

Mimicking a heart beating on a stellar size, Cepheid variable stars alternatively expand and contract their external shells in periods ranging from days to weeks, whereas energy pulses coming from quickly spinning neutron stars wipe out our planet within the millisecond to millionth-second rates. Turning the timescale upside down, our Sun completes one revolution around the center of our Galaxy in about 200 million years.

In the same vein, in the course of human history, cyclic features related to the motion of the nearest heavenly bodies have also been used to measure the passage of time since ancient ages. Thus, the spin of the earth's axis determines the length of days, its period of revolution around the Sun determines the length of years, and the synodic period of the Moon provides the length of months. In modern times, mechanical clocks, relying upon springs and pendulum-based devices, have taken over astronomical observations for timekeeping purposes.

At variance with all these cyclically evolving natural or man-made clocks, we also found timekeepers based on processes that are mathematically described by *monotonically decreasing* functions of time instead of periodic ones. These clocks are related to physical processes which spontaneously take place in decaying systems. This is the case of the decreasing content of nuclear fuels inside stellar cores, determining the pace of thermonuclear reactions, and hence the lifetime of every star according to its initial mass content. Another well-known example is provided by the decreasing number of radioactive atoms present in a given piece of matter, which depends on their disintegration rates. In fact, such dependence is systematically employed to determine the age of minerals found on earth or in solar system bodies. On a cosmological scale, the Hubble's expansion rate (nowadays known to be changing itself) determines a cosmic scale of time, directly related to the temperature value of the black-body cosmic radiation background, which progressively decreases in a remarkably non-linear way. Accordingly, different cosmic eras or epochs can be properly characterized by threshold temperature values closely related to the occurrence of specific physical events (Table 2.1).

Thus, the contemplation of our unfolding evolving universe shows us an impressive display of nested structures which successively emerge from the shadows of underlying energy fields permeating the space-time to flourish, shrouded in the plentiful colors of the electromagnetic spectrum, just lasting for a more or less brief time interval, ultimately fading away forever. This cosmic scenery introduces the arrow of time under the semblance of an ordered schedule, which determines the due order of appearance of different structures, stemming from the mutual interactions of all the cosmic players present at a given epoch among themselves. Just in the spirit of the Heraclitus' sentence quoted in this chapter opening.

In order to illustrate the nature of this fundamental cosmic schedule, which ancient Greeks dubbed Kairos (καιρός), let us start from the very beginning by considering the emergence of the fundamental constituents of the so-called hadronic matter, namely, protons (p) and neutrons (n), during the first seconds elapsed in the nascent universe (see Section 5.1). Within this scenario, we find two coupled reactions $p + e^- \leftrightarrow n + \upsilon$, $p + \overline{\upsilon} \leftrightarrow n + e^+$, involving electrons ($e^-$), positrons ($e^+$), neutrinos ($\upsilon$) and antineutrinos ($\overline{\upsilon}$), and resulting in the simultaneous creation and annihilation of protons and neutrons under equilibrium conditions at a given temperature. Now, due to the cosmological space-time expansion, the temperature of the universe was rapidly decreasing, and at a time about 10 seconds after the initial singularity, its value fell below $T = 1.19 \times 10^{10}$ K, the threshold figure required for primordial gamma photons to be able to create the electron-positron pairs necessary to drive the reactions above. Sometime later, the weak interaction governing the neutron decay reaction rate, $n \rightarrow p + e^- + \overline{\upsilon}$, freeze-out below the Hubble parameter expansion rate, so that the meantime for neutron decaying became lower than the age of the universe. As a consequence, the relative number of protons and neutrons was fixed to the ratio p/n ~ 5 (Exercise 2) when the expanding universe reached the critical temperature $T^* = 9.5 \times 10^9$ K (see Table 2.1). Subsequent primordial nucleosynthesis proceeded by the fusion of these protons and neutrons to form deuterons (^2H). Further capture of deuterons, protons, and remaining neutrons led to the production of tritium (^3H), ^3He, and ultimately ^4He nuclei. A minor fraction of ^6Li, ^7Be, ^7Li nuclei was also produced during this primordial nucleosynthesis episode, until by about 15 minutes after the origin of the universe, its temperature had dropped below 4×10^8 K, and further nucleosynthesis was not possible anymore. The key point here is that the neutron mean life controls the neutron-to-proton ratio and directly affects the primordial helium abundance (~ 25% of the current cosmic value), as the required free neutrons must be captured and processed to helium on a time scale short enough compared to their own decay time (about 880s).

Similar episodes of simultaneous synthesis and destruction intertwined processes, acting in a synergic way, took, and currently take place in many other astrophysical scenarios, under ever-changing physical and chemical conditions. They constitute the fundamental engines driving the emergence of progressively more complex structures in our evolving universe, leading to the appearance of arrangements of matter complex enough to sustain a stable genetic code along with robust molecular metabolic networks, while the universe perennially decreasing temperature value ticks the end of old-fashioned systems and the dawn of novel ones. From this viewpoint, our evolving cosmos shares some resemblance with the morphogenetic processes that interweave growth and form at the much smaller embryo's spatial scale (Thomson, 1992), where the passage of time shows up in terms of unfolding hierarchical patterns echoing the essential rhythms of life (Noble, 2008).

2.2 CHEMICAL EVOLUTION: A GALACTIC LANDSCAPE

One of the main assumptions permeating contemporary astrobiology refers to the emergence of life as the natural outcome of a gradual sequence of chemical evolution steps taking place at different sites inside galaxies during their own physical evolution. Accordingly, one would not expect that the biological macromolecules involved in the structure and metabolic processes occurring in current living beings on earth were originally synthesized at once in just a single privileged place of our solar system. Rather, we should assume that these complex organic compounds were progressively assembled at different stages, starting from quite simple moieties interspersed throughout the Galaxy, under the specific physicochemical conditions prevailing in different appropriate astrophysical places, such as those illustrated in the collage displayed in Figure 2.2 and the schematic diagrams shown in Figures 2.3 and 2.4.

As we mentioned in the previous chapter, most chemical elements heavier than helium, including the main biogenic atoms C, O, N, S, and P, are synthesized by means of thermonuclear reactions requiring the very high temperature and density values found in the stars' cores. The elements so formed inside stars must be subsequently released to the ISM, where they undergo further chemical processing via complex reactions networks, which are triggered by ion-molecule interactions (see Figure 1.3). The formation of the necessary ions requires the presence of either UV radiation (in diffuse clouds) or more energetic cosmic rays (in dense clouds), as it is indicated in Figure 2.3. The UV radiation comes from luminous O and B type stars, whose initial burst is generally promoted by the passage of shock waves through the relatively denser regions of the ISM conforming the galaxy's spiral arms. These shock waves, in turn, come from supernova (SN) star blasts echoing the final disruption stage of massive stars.

FIGURE 2.2 (See color insert.) The cycling of matter and energy in the interstellar medium of spiral galaxies (background) involves the injection of material from stars stemming from condensing dense molecular clouds (M42), first in the form of continuous wind (young open cluster M45) and later on during briefer episodes which take place in their final stages. The majority of stars (those with masses comprised in the range 1–8 $M_{\odot}$) transition into planetary nebulae (M57), a stage which lasts about 10,000 years, where the star sheds its outer layers, losing almost all of its original mass and becoming a very hot, UV emitting white dwarf that ionizes its surrounding envelope, rendering it pretty colorful (Kwok, 2010). Higher mass stars undergo explosive events, yielding rapidly expanding supernova remnants (M1). The materials from these nebulae flow into the diffuse ISM, forming relatively dense clouds (ρ Oph) in its due time. These clouds eventually collapse into even denser molecular clouds (M16), which generally become gravitationally unstable and form a new generation of stars along with preplanetary disks ultimately evolving into planetary systems (β-Pic). During this stage, most of the material originally released from the original stars is heavily processed, but some pristine matter from molecular clouds still survives, preserved in planetesimals and the nuclei of comets. (The Messier objects M1, M15, M16, M42, M45, and M57, as well as the background galaxy NGC 1232, are reproduced by courtesy of The Hubble Heritage Team – AURA/STScI/NASA, Baltimore, MD).

Source: The artist's conception showing planet formation around the very young type A star β Pictoris was downloaded from http://imagine.gsfc.nasa.gov/Images/bios/roberge/BetaPictoris.jpg; NASA/FUSE/Lynette Cook.

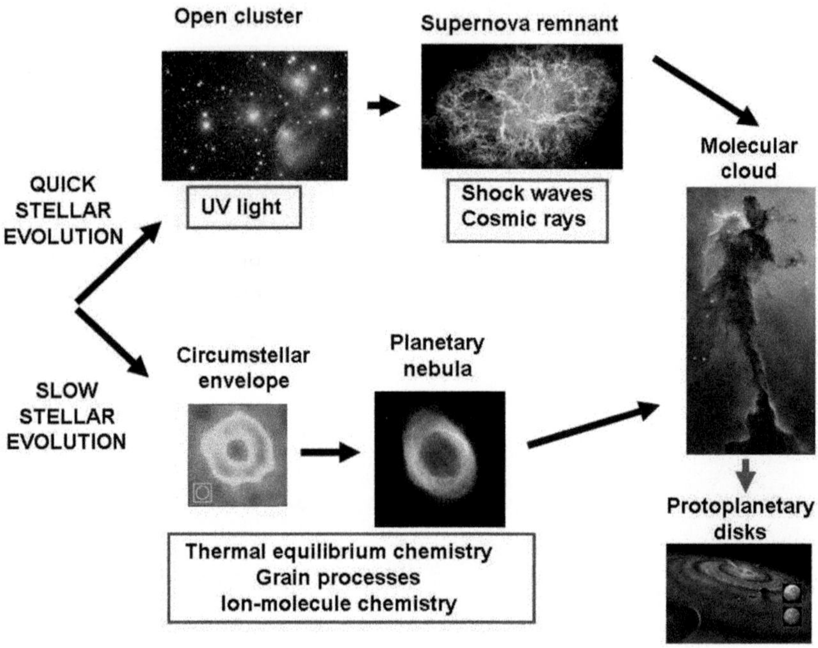

FIGURE 2.3 (See color insert.) This diagram indicates the physicochemical mechanisms that dominate molecule and dust-grain production in several astrophysical objects shown in Figure 2.2. Additional information regarding most astrophysical objects in this diagram is given in Section 2.4.

At the same time, these SN events eject to the galactic medium high energy particles swarms moving at relativistic velocities: the above mentioned cosmic rays. These highly ionizing particles are able to penetrate the shrouded regions of dense molecular clouds to activate a rich chemistry therein. Cosmic rays also collide with heavy atoms in the ISM disrupting them into lighter nuclei: a spallation process that accounts for the very presence of beryllium and boron atoms in the universe. Indeed, although beryllium nuclei are extremely unstable at the high temperatures prevailing at the early universe, and they are readily destroyed by the thermonuclear reactions that occur in stellar interiors, once they are formed by spallation reactions involving heavier nuclei in stellar atmospheres and SN shells, these nuclei are not radioactive and can form stable atoms which are incorporated to molecules and minerals, such as emerald $(SiO_3)_6Al_2Be_3$.

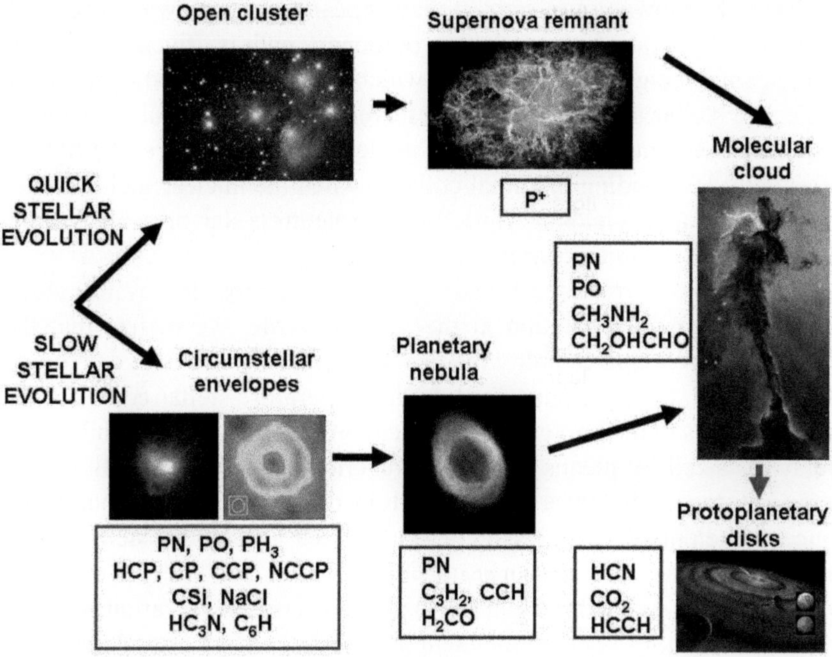

FIGURE 2.4 (See color insert.) This diagram illustrates how molecules and dust grains cycle from circumstellar envelopes and supernova remnants through various phases of interstellar matter in the Galaxy. Some representative molecules appearing in different astrophysical objects during the whole ISM life cycle are indicated (Tielens, 2013; Ziurys et al., 2015), including the seven phosphorus compounds detected to date (see Table 1.2). Additional information regarding most astrophysical objects in this diagram is given in Section 2.4.

In the ISM the material circulates rapidly between diffuse nebulae and molecular clouds (say, with an average time-scale of about 100 Myr, less than a half the rotation period of our Galaxy).

In each stellar generation, the newborn stars can evolve following either a rapid (a few Myr) or a slow (typically a few Gyr) time scales, depending on their initial masses, and they eject much of their gas back into the ISM at their late stages. The length of the cycle between two successive stellar generations is then determined by the slow evolving stars. In our Galaxy, two successive stellar generations have been spectroscopically identified by attending to the chemical composition of their atmospheres, and they are referred to as Population II (old one) and Population I (young one), respectively.

From the viewpoint of stars, it may appear that their own existence is ruled by a fatal fate. Indeed, their evolutionary path is driven by a titanic fight against the gravitational force, which originally created them. Since the entire stellar structure is just preserved by a delicate energy balance requiring the continuous burning of nuclear fuels in its core, the fight is eventually lost, ending in a final collapse when the nuclear fuel becomes exhausted, which yields a white dwarf, a neutron star or a black hole, depending on the initial star mass.

Nevertheless, from the broader perspective, provided by the viewpoint of chemical evolution at the galactic scale, we realize that the stellar life cycle makes pretty much sense, albeit its ultimate defeat. On the one hand, highly compressed, high-temperature stellar cores are the unique places in the universe where elements heavier than He and Li can be synthesized by means of nuclear reactions once the overall cosmic temperature fell below a million Kelvin degrees. On the other hand, since the elements progressively synthesized inside the stellar cores must be released to the world outside in order to allow for chemical compounds to be made (Figure 2.4), a long-term preservation of stars' structural stability is not convenient at all. Thus, when considered from a cosmic perspective, the disruptive death of stars is not a tragedy but rather a convenient event.

In fact, according to the minimum free energy and maximum entropy grand laws ruling the entire universe, the synthesis of progressively more complex nuclei inside stellar cores (a process decreasing entropy) requires the continuous loss of hot matter and high energy photons across their photospheres, thereby increasing the entropy in the needed amount to pay the required thermodynamic bill. In the same vein, when the stellar structure is ultimately destroyed a huge entropy production takes place, affording the subsequent possible formation of highly ordered molecular structures of increasing complexity in the ISM first and during the planetary system formation stage afterwards, which will eventually lead to the formation of the highly ordered DNA macromolecules encoding the required genetic program engraved within the atomic structure of the *aperiodic crystal* of life (Schrödinger, 1944; Maciá-Barber, 2009).

2.3 A HIERARCHY OF NESTED STRUCTURES

2.3.1 *Galaxies and the Intergalactic Medium (IGM)*

Galaxies are the fundamental building blocks of the universe at a cosmic scale. They consist of stars, planets, dust, gas, and dark matter in varying proportions. Galaxies come in a wide range of luminosities, masses, sizes, and shapes, which determine and express their evolutionary stage. Galaxies cover a large range of total luminosities, going from 10^3 to 10^{12} in units of solar luminosity, $L_{\odot}$, giving rise to their relative classification into ultra-faint, intermediate, bright, and ultra-luminous galaxies. They can also have a variety of masses, ranging from 10^5 to 10^{13} in units of the solar mass, $M_{\odot}$, leading to a relative classification into a dwarf, medium, and giant galaxies.

Their shapes are traditionally classified according to the so-called Hubble diagram, where one can distinguish elliptical, spiral, barred spiral and irregular galaxies. Irregular galaxies host stars that follow complex orbits without a well-defined rotation center, and are generally rich in the gas and dust component. Spiral galaxies (Figures 2.2 and 2.5) are also gas-rich systems but, unlike irregular ones, they are rotation supported, resulting in a well-defined set of structural components: an inner bulge, a disk with spiral arms, an ellipsoidal halo and, sometimes, a central bar.

It is believed that the presence or absence of the bar is dependent on the mass available in the galaxy (specially its central mass value) and on the gas plus stars to dark mass ratio, as well as on its interaction history with other galaxies. These two types of galaxies contain clouds of cold gas dense enough to undergo gravitational collapse and keep star-forming bursts. By contrast, elliptical galaxies have exhausted their cold gas and cannot form new generations of stars, whereas the aged stars they contain exhibit metastable orbits around a well-defined center. Irregular and spiral galaxies are typically associated with the intermediate and smaller mass ranges, whereas elliptical galaxies can be observed in all mass ranges (Pila-Díez, 2015).

Different kinds of stars are found in the various components of our Galaxy (Figure 2.5). The globular clusters in the halo are composed of old, metal-poor, Population II stars. The stars in the disk are mostly young, metal-rich, Population I stars like the Sun. The central bulge contains both

Population I and Population II stars, suggesting that some stars are quite ancient whereas others were born more recently.

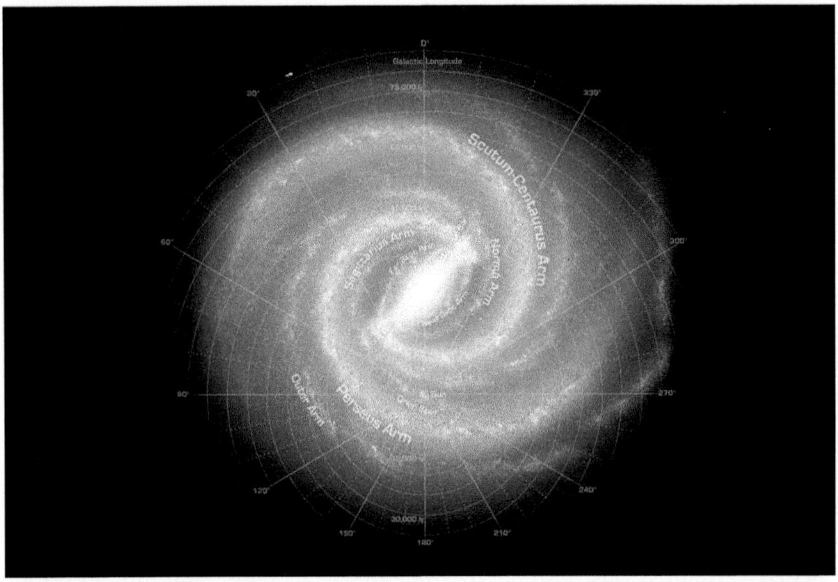

FIGURE 2.5 Graphic view of our Milky Way galaxy. The Milky Way galaxy is a disk-shaped collection of hundreds of billions of stars, organized into spiral arms of giant stars that illuminate interstellar gas and dust. The Sun is in a finger called the Orion Spur. The disk of our Galaxy is about 50 kpc (160,000 ly) in diameter and about 0.6 kpc (2,000 ly) thick. The center of the Galaxy is surrounded by a distribution of stars, called the central bulge that is about 2 kpc (6,400 ly) in diameter. Observations suggest that this bulge is not shaped like a flattened sphere but may be elongated. The halo of the Galaxy contains a spherical distribution of globular clusters, which are spherical aggregates containing roughly a million stars each, closely packed in a volume only a few hundred light-years across. Although these clusters are conspicuous, they contain only about 1% of the total number of stars in the halo. Most halo stars are old, metal-poor, Population II stars in isolation. These ancient stars orbit the Galaxy along paths tilted at random angles to its plane. Overlaid is a graphic of galactic longitude in relation to our Sun.
Credit: NASA/Adler/U. Chicago/Wesleyan/JPL-Caltech. https://www.nasa.gov/mission_pages/sunearth/news/gallery/galaxy-location.html.

Both radio and optical observations reveal that our Galaxy has spiral arms: spiral-shaped concentrations of gas and dust that extend outward from the center in shape reminiscent of a pinwheel. According to these observations, our Galaxy has four major spiral arms and several short arm segments. The Sun is located on a relatively short arm segment called the Orion arm, which includes the Orion nebula (M42) and neighboring sites of intense star formation. Two major spiral arms border either side of the Sun's position. The Sagittarius arm is on the side towards the Galaxy center. This is the arm one sees during the summer months stretching across Sagittarius and Scorpius. During winter, we see the Perseus arm. The remaining two major spiral arms are usually referred to as the Centaurus and the Cygnus arms, respectively.

The spiral arms in the disk of our Galaxy suggest that the stars, gas, and dust are all orbiting the galactic center. The orbit of the Sun around the center of the Galaxy is roughly circular, and it takes about 220 million years to complete one trip at an average speed about 220 km s^{-1}. By carefully measuring the velocity field related to the orbital motions of many other stars a remarkable conclusion was reached: most of the mass of the Galaxy must be in the form of dark matter, a mysterious sort of material that emits no light at all. In fact, according to Kepler's third law, the orbital speed of stars and gas clouds beyond the confines of most of the Galaxy's mass should decrease with increasing distance from the Galaxy's center according to an r^{-1} law. But the measured Galaxy's rotation curve is quite flat instead, indicating uniform orbital speeds well beyond the visible limit of the galactic disk. To explain these almost constant rotational speed astronomers concluded that a large amount of mass must lie outside the Sun's orbit, even if that matter does not show up in any part of the electromagnetic spectrum. One proposal is that such a dark matter is composed, at least in part, of brown and white dwarf stars, too dim to be detected, or black holes with masses between 0.01 $M_\odot$ and 1 $M_\odot$ located in the halo. These are called massive compact halo objects (MACHOs), and observations indicate that they make up no more than 40% of the dark matter. Other possible candidates are neutrinos, which have a tiny amount of mass, and weakly interacting massive particles (WIMPs), whose existence is suggested by certain theories but has not yet been confirmed experimentally.

It is considered that the first galaxies appeared about 2–8 × 10^8 years after the initial singularity (Table 2.1), probably as a consequence of primordial quantum fluctuations in the early universe that were amplified by the inflation event, leading to tiny density variations spread throughout the universe, which grew through gravitational instability. Thus, these density variations were the seeds for current galaxies: as a small over density starts to gravitationally attract matter, the more matter it accretes, becoming an increasingly strong gravitational well. Eventually, these seeds accreted enough gas to form clouds that could (gravitationally) collapse and produce the first stars and galaxies. These protogalaxies, in turn, merged with each other into increasingly massive galaxies (White and Rees, 1978).

The radiation of the earliest generation of stars turned the surrounding atoms into ions, and this ionization eventually pervaded all the space outside and between galaxies, so that few hydrogen atoms remain today between galaxies (Barkana, 2006). Indeed, intergalactic space is filled with a pervasive medium of ionized gas, which is referred to as the intergalactic medium (IGM). To learn about the chemical composition of the IGM gas astronomers study its absorption against distant quasars, the brightest known astrophysical objects. In this way, a residual neutral gas fraction is detected in the spectra of quasars at both low and high redshifts, revealing a highly fluctuating medium. Common elements spectroscopically detected in the IGM are C, N, Si, and Fe, along with O, Mg, Ne, and S (Meiksin, 2009). Phosphorus ions have been detected in the so-called damped Lyman α systems, observed in quasars spectra. These systems span the whole epoch of galaxy formation and are generally considered the progenitors of the present-day galaxies.

2.3.2 Stars and the Interstellar Medium (ISM)

Stars are unique structures where the four fundamental forces of Nature concurrently act in such a way that, in order to preserve their own structure, they generate in their cores many of the elements present in the periodic table. These atoms subsequently move from the central regions towards the outer envelopes where they are able to combine to form relatively simple molecules, which are ultimately expelled into the ISM during the late stages of stellar evolution.

Observation of stars has revealed a wide variety of stellar conditions, expressed in terms of three main physical parameters, namely, mass, luminosity, and surface temperature, which are related among them. The most important physical parameter concerning a star is its mass, usually expressed in terms of the solar mass $M_{\odot} \sim 2 \times 10^{30}$ kg (see the Appendix).

Stars with masses very much lower, or very much higher than the mass of the Sun, beyond the broad interval ranging from 0.01 $M_{\odot}$ up to 100 $M_{\odot}$, are relatively infrequent. Luminosities relative to the solar value, $L_{\odot} = 3.9\times10^{26}$ W, span the range 10^{-4}–10^{6} $L_{\odot}$ and surface temperatures are within the interval T ~ 2,000–50,000 K. The relationship between surface temperature, T, and luminosity, $L = 4\pi R^2\sigma T^4$, where R is the star radius, and σ is the Stephan-Boltzman constant (see the Appendix), plays a very important role in the stability of stars and represents a major step towards an understanding of stellar evolution. This property was first disclosed by E. Hertzsprung and H. N. Russell in 1910 by representing the star's luminosities versus their temperatures in a Cartesian plot, as it is illustrated in Figure 2.6. The location in the diagram provides information about the mass of a star and characterizes its evolutionary phase (see Section 5.4). Most of the stars occupy the region in the diagram along the line called the main sequence. During the stage of their lives in which stars are found on the main sequence line, they are fusing hydrogen in their cores. When hydrogen fuel is almost exhausted the stellar core contracts under its own weight while the residual hydrogen burning zone expands outward. The contraction of the core continues until the temperature, and pressure values in the center are high enough to trigger the helium-burning phase. In this way, these stars enter the so-called red giant (or supergiant) phase undergoing helium fusion in the core and hydrogen burning in a shell surrounding it. The energy released in the core leads to a significant expansion of the outer star layers, thereby increasing the star's luminosity, albeit its surface temperature is decreased, hence shifting the surface color towards red. Accordingly, giant stars concentrate on a region referred to as the horizontal branch of the HR diagram. When all the available helium in the core is converted to carbon and oxygen, energy production in the stellar core relies on the fusion reaction of these two nuclei, followed by a number of subsequent rapid evolution phases that convert these elements, through a number of characteristic thermonuclear processes, referred to as carbon, neon, oxygen, and silicon burnings, to iron and nickel in the inner stellar core (Wiescher and Langanke, 2015). It is suspected that the red supergiant Betelgeuse (near the top right corner in Figure 2.6) is close to entering this final phase.

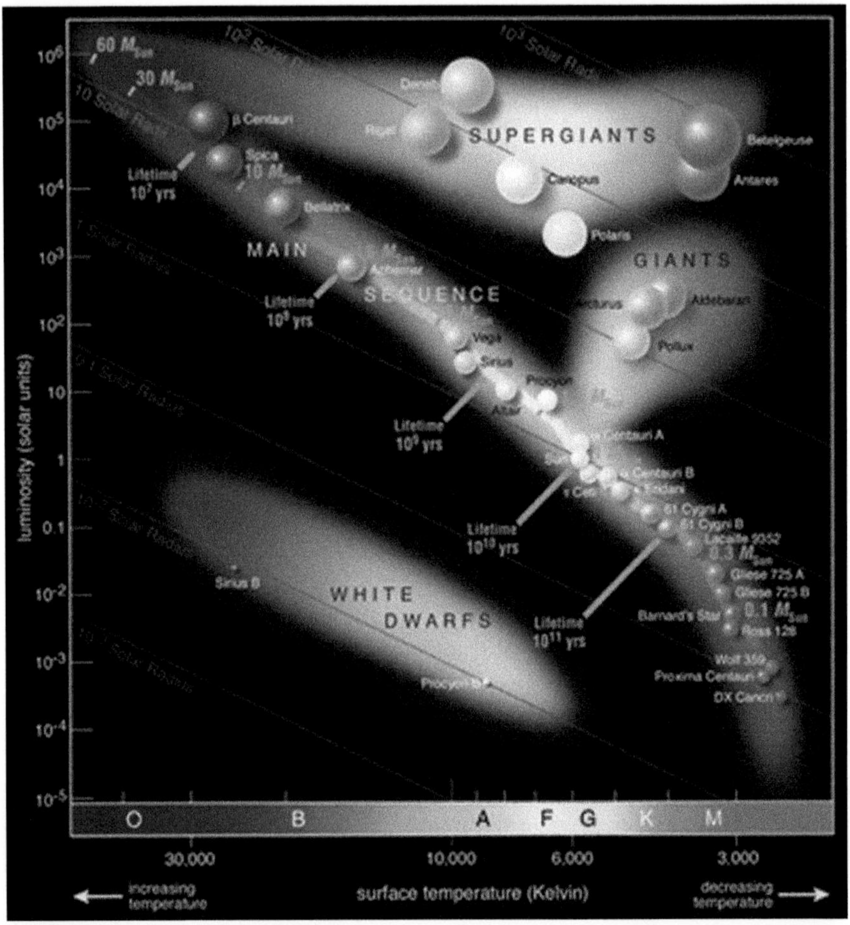

FIGURE 2.6 (See color insert.) The Hertzsprung-Russell (HR) diagram displays the luminosity of a star against its surface temperature (spectral class). Stars of greater luminosity are toward the top of the diagram, and stars with higher surface temperature are toward the left side of the diagram.
By ESO - https://www.eso.org/public/images/eso0728c/.

Under the conditions found in stellar interiors, iron cannot be turned into a heavier element through fusion processes because, unlike fusion of the previous lighter elements, iron fusion requires energy to proceed rather than producing it. Thus, once iron is produced in the star inner core energy production ceases, and since the pressure is what prevents gravitational collapse, this leads to the death of stars. There are two main pathways a dying star can take. For low- to intermediate-mass stars, the core collapses while the outer layers are ejected into the space to become a white dwarf. This low luminosity, dense stars line up along the bottom left corner in the HR diagram. In the case of more massive stars, the collapse triggers an explosion, referred to as SN, that gives rise to the formation of the elements heavier than iron. Depending on the star's initial mass, the remaining core collapses to become either a neutron star or a black hole.

Associated with the rapid energy release in an SN explosion is an enormously high emission of light over the entire electromagnetic spectrum within seconds. The sequences of nuclear reactions that develop during this explosion involve many short-lived nuclei far beyond the limits of nuclear stability. Mapping the sky with γ-telescopes reveals the image of a radioactive universe resulting from the slow radioactive decay processes associated with SN remnants along the galactic plane of the Milky Way. This demonstrates that the synthesis of new elements is an on-going process, continuously changing the abundances and distribution of atoms in the universe as time goes on (Wiescher and Langanke, 2015).

The tenuous medium containing gas laced with microscopic dust particles between stars is referred to as the ISM. The interstellar dust abundance relative to hydrogen can vary both within a given galaxy and from galaxy to galaxy. Nevertheless, in general, the mass of material in dust particles is about 1% of the mass in hydrogen. Dust particles range in size from the molecular domain to sizes of about 0.5 μm and are composed largely of amorphous carbon, graphite, aromatic hydrocarbons, silicates, water, ammonia, methane, and formaldehyde ices, silicon carbide, and possibly iron particles, metallic oxides and sulfides (Hollenbach and Tielens, 1999). The presence of P-bearing compounds (apatite crystals, see Section 4.5.4) in interstellar grains of probably SN origin has been recently reported (see Section 5.4).

Diffuse interstellar clouds are regions of relatively low temperature ($T \sim 80$ K) and density ($n \sim 50$ cm^{-3}), which is about an order of magnitude above the average density of matter in the Galaxy (Exercise 3). Typical sizes and masses of diffuse clouds are ~ 10 pc and ~ 500 $M_{\odot}$, respectively, though they show a broad mass and size distribution which joins quite smoothly into those of more compact molecular clouds. With total hydrogen column densities of $N_H \sim 10^{21}$ cm^{-2}, diffuse clouds are quite transparent to the ambient interstellar radiation field. Hence, UV photons play an important role in physics and chemistry. Diatomic radicals (e.g.,

CH, CH^+, CN) with electronic transitions in the visible were the first molecular species to be observed, back in the late 1930s, but they have now also been observed through millimeter and submillimeter absorption lines. Later on, C_2 and C_3 were added to this list of simple radicals detected through electronic transitions in the visible. CO and OH have been studied in diffuse clouds through both UV electronic and millimeter rotational transitions. The nitrogen-bearing species NH and N_2 have been detected with great difficulty in the UV. Molecular hydrogen (H_2) can be observed directly in diffuse clouds through the Lyman and Werner bands in the UV, and these provide a direct probe of the column density and excitation of this key species. The observations also reveal significant rotational and vibrational excitation of H_2. A variety of polyatomic species have been detected through absorption and emission studies in the millimeter range including HCN, H_2CO, C_2H, C_3H_2, H_2, and HCS^+.

Thus, diffuse clouds show a wide variety of relatively simple molecules, including radicals and ions. While molecular hydrogen is very abundant, most of the oxygen, nitrogen, and carbon are in atomic form, with abundances of 1.6×10^{-4}, 2.5×10^{-5}, and 7×10^{-6} relative to H_2, respectively, in the diffuse cloud located toward the star ζ Oph, a bright O9.5V early-type star at a distance of about 138 pc from the Sun. Furthermore, most of the carbon is in the form of C^+, with an abundance of 2.6×10^{-4} relative to H_2, which reflects the harsh conditions related to the presence of a high UV radiation field. The most abundant molecular species, other than molecular hydrogen, is CO, with an abundance of 5×10^{-6} relative to H_2, which locks up only 1% of the elemental carbon (Tielens, 2013).

On the contrary, carbon monoxide (CO) is the most abundant carbon-bearing molecule in dense molecular clouds, where no simple atomic species have been detected to date. For instance, molecular abundances of 5×10^{-5}, and 3×10^{-5} relative to H_2, have been measured for CO and N_2, respectively, toward the dark cloud core TMC-1 (Tielens, 2013). Quite interestingly, the phosphorus monoxide (PO) molecule has recently been detected in three molecular, star-forming clouds. Molecular clouds are characterized by high particles number density in their centers (~3×10^6 cm^{-3}), low kinetic temperatures (~ 6 K), sizes of the order of 0.1 ly, masses of 1–3 $M_{\odot}$, and features in their molecular spectra indicative of collapse motion. The Sagittarius arm of our Galaxy harbors two hot molecular cores with a rich chemistry, located at a distance of 5.1 kpc, which are associated with compact radio sources known as e1 and e2. Similarly, two massive objects separated by 6," the so-called UC HII region and the chemically rich W3(H2O) hot molecular core, constitute the W3(OH) complex, located at 2.04 kpc. As another example, we can mention the dark molecular clouds labeled L1157, which harbor a low-mass protostar, and L1157 B1, which is a relatively close shocked region formed by interactions between the molecular outflow and ambient gases.

2.3.3 Planetary Systems Around Stars

Planetary systems form in collapsing dense clouds that form new stars. A typical giant molecular cloud will need to have a mass of a few hundred solar masses in order for gravitational collapse to occur. Rather than forming a single massive star, the cloud is fragmented into many low-mass protostellar cores, with masses within the range 0.3–2 $M_{\odot}$. As the material collapses to form a young star, it spins faster and faster due to the conservation of the angular momentum. Thus, most of the material cannot directly fall onto the protostar. Instead, it falls parallel to the rotation axis and forms a thin disk around the protostar. The radial extent of the disk depends on the angular momentum of the material from which it forms, but these disks are typically a hundred to a few hundred AU in size. Instabilities in the disk provide a mechanism for transporting angular momentum outwards, allowing the mass to accrete onto the protostar, hence allowing the star to grow (Rice, 2015). About 1% of the mass in the disk is solid, in the form of micron-sized dust grains. Since the disk temperature decreases with increasing distance from central protostar, at some distance (referred to as the snow line) the temperature becomes cold enough for water, ammonia, and methane ices to condense on the dust grains, so that solid material in the inner part of the disk is composed of dust grains only, while in the outer part of the disk it is composed of dust grains with substantial ice mantles. In a disk around a young solar-like star, the radius beyond which the ices form on the dust grains occurs at about 2.7 AU.

The basic planet formation scenario is that the solid phase grows via collisions to ultimately form kilometer-sized planetesimals. These then coagulate to form planetary size mass bodies. Due to the enhancement in solid materials (dust plus ice), planetary bodies beyond the snow line grow faster than those inside the snow line. As a consequence, the cores of these planets can become massive enough to gravitationally attract a dense gaseous envelope which characterizes the so-called giant gas planets.

The first detection of extrasolar planets took place in two very different places: around a dead pulsar star (Wolszczan and Frail, 1992) and around a solar-type star (Mayor and Queloz, 1995). The latter one was found to have a mass of 0.47 Jupiter masses, but to have an orbital period of only 4.2 days. Hence, this is a gas giant planet that is orbiting its parent star with an orbital radius remarkably smaller than that of Mercury (whose orbital period is 88 days). Due to its close location to the star, its surface temperature must be high, and for this reason, it was referred to as a hot Jupiter planet. Many more hot Jupiter planets have been discovered during the last two decades. Due to its close location to the star, the planet receives a great amount of radiation, so that the planet's effective temperature can reach more than 2,000 K (Sedaghati et al., 2017). The picture that the basic architecture of a planetary system should consist of rocky, small planets close to the star and large, gaseous, and icy planets far from it was rapped as a consequence of these findings, which broke the paradigm of the solar system as an archetype of the formation and structure of a planetary system. In main sequence stars, the basic mechanisms described above are expected to operate as well, but the processes which may be occurring in systems with multiple stars or in evolved stars that contain planets open new possibilities. New discoveries, disclosing the presence of giant massive planets very close to the stars require the action of mechanisms that could redistribute these planets into very different orbits from where they had been originally formed, or could create isolated planets not gravitationally bounded to a star (the so-called rogue planets). Thus, we now know that the orbital configuration of a planetary system basically depends on the formation mechanisms operating during its preplanetary stage, including tidal interactions, and on the subsequent orbital evolution due to the mutual gravitational interaction among its components, leading to planetary migration from an outward to an inward orbit or the expulsion of a planet from the system through scattering (Sánchez-Lavega, 2011).

2.4 ASTROPHYSICAL OBJECTS GALLERY

2.4.1 *Rho (ρ) Ophiuchi Cloud* (Figure 2.7)

Located about 460 light-years away from earth the ρ Ophiuchi cloud complex is a near star-forming region, which allows us to observe finer details than other distant similar sites, such as the M42 Orion nebula

(Figure 2.9). The diverse colors represent different infrared light wavelengths obtained from NASA's Wide-field Infrared Explorer. The central bright white nebula is an emission nebula and glows due to heating from nearby stars, as does most of the gas seen through the image, including the bluish bow spot in the bottom right close to the bright red area which is a reflection nebula: the light from the star in its center – σ Scorpii – is reflected off of the surrounding dust. The interspersed darker areas are absorption nebulae, where cool dense gas pockets block out the background light. The bright pink spots sited left of center are young stellar objects. Many of these newborn stars are still enveloped in compact nebulae, and in visible light, they are completely hidden in the dark nebula around them.

FIGURE 2.7 (See color insert.) Rho (ρ) Ophiuchi cloud
Reproduced by courtesy of NASA/JPL-Caltech/WISE Team; https://commons.wikimedia.org/wiki/File:Rho_Ophiuchi.jpg.

2.4.2 Pleiades Open Cluster (Figure 2.8)

The Pleiades (M45) star cluster is 440 light-years on average from the earth in Taurus constellation. It includes some 500 hot B-type stars, of which a half dozen can be directly seen with the naked eye. The gas that must once have formed a HII region around this cluster has dissipated into interstellar space, leaving only traces of dusty material as a faint reflection nebulosity surrounding the brightest cluster's stars. Nearly all the stars in the Pleiades are on the main sequence stage, so the cluster's age is about 50–100 million years, the time the least massive stars take to finally begin hydrogen burning in their cores. The cluster estimated survival is about another 250 million years, then it will be dispersed by gravitational interactions with its surroundings.

FIGURE 2.8 (See color insert.) Pleiades open cluster
Reproduced by courtesy of The Hubble Heritage Team – AURA/STScI/NASA, Baltimore, MD.

2.4.3 Orion Molecular Cloud (Figure 2.9)

The Orion nebula is situated in Orion constellation, south of Orion's Belt. It is one of the brightest nebulae, visible to the naked eye as the middle star of the three in Orion's constellation sword. M42 is located at about 1,300 light-years and is the massive star formation region closest to earth. Estimations for the M42 nebula are 24 light-years across in size and a mass of about 2,000 times the Sun's. Because this mass is spread over a huge volume, its density is quite low: only a few hundred hydrogen atoms per cubic centimeter. Most of the UV light that causes the nebula glow comes from just five hot, luminous O and B type massive stars at the heart of the nebula, comprising the so-called Trapezium cluster. Over time the

UV light from the inner massive stars will push away the surrounding gas and dust by photoevaporation. Hubble Space Telescope observations have yielded the major discovery of proplyds: more than 150 preplanetary disks around newborn stars within the Orion Nebula, considered to be in the earliest stages of planetary system formation. The nebula is part of a much larger nebula, known as the Orion Molecular Cloud Complex.

FIGURE 2.9 (See color insert.) Orion molecular cloud
Reproduced by courtesy of The Hubble Heritage Team – AURA/STScI/NASA, Baltimore, MD.

2.4.4 IRC+ 10216 Circumstellar Shell

Located at a distance of 130 pc from the Sun, the C9.5 type giant ($R_* = 6.5 \times 10^8$ km ~ 935 $R_\odot$) Mira type variable star (with a period of 630 days), referred to as IRC+10216 (also known as CW Leo), is a prototype circumstellar envelope containing a great deal of atomic carbon and a large number of complex molecular species, including over 80 molecules discovered to date, some of them not found anywhere else (see Figure 6.7). As the central star is at the tip of the asymptotic giant branch phase, the star is expelling matter at a high rate of ~ 2×10^{-5} $M_\odot yr^{-1}$, (Exercise 4a) so that the inner shell is a dust-grain factory, while in its outer layers the photochemistry driven by the penetration of interstellar UV photons produces a wealth of exotic molecules (Figure 2.10c). In fact, close to star's photosphere grain formation is allowed due to the relatively low star surface temperature (T_e = 2320 K) and the presence of the required chemical elements in the photosphere, namely, C, N, O, Si, Mg, Al, P, and

K, which are released to the circumstellar space by strong stellar winds. In fact, the inner region of IRC+10216 shell lies about 40 AU from the star. In this region about 10^{11} atoms occupy every cubic centimeter of space, moving at kinetic temperatures of about 600 K, and a pressure of only a billionth of an atmosphere. Under these conditions, thermal-equilibrium chemistry readily takes place, leading to a steady formation of molecules and grains (Figure 2.10a). Heavier elements such as silicon provide the matrix for building solid grains to which more volatile atoms and molecules can stick. In this way, IRC +10216 becomes a huge molecular factory, with a large fraction of molecules conveniently stuck to grains that will ultimately carry them away into deep space. Because of the carbon-rich character of the star, the molecular inventory of the envelope is dominated by carbon-rich chains such as polyyne and cyanopolyyne radicals (Figure 2.10b), carbon compounds containing sulfur or silicon, molecules containing phosphorus or metals, and negatively charged carbon chains. The extremely cold outer layers of the envelope (at about 10 K) reach a light-year radius, so that the density drops below 30 atoms per cubic centimeter (Exercise 4b). Here further molecule formation takes place, now involving photon-chemistry along with grain-surface reactions. UV photons from distant stars penetrate into the circumstellar outer regions and break up molecules stuck to grains, so that the resulting fragments can rejoin in different ways to form novel species (Verschuur, 1992).

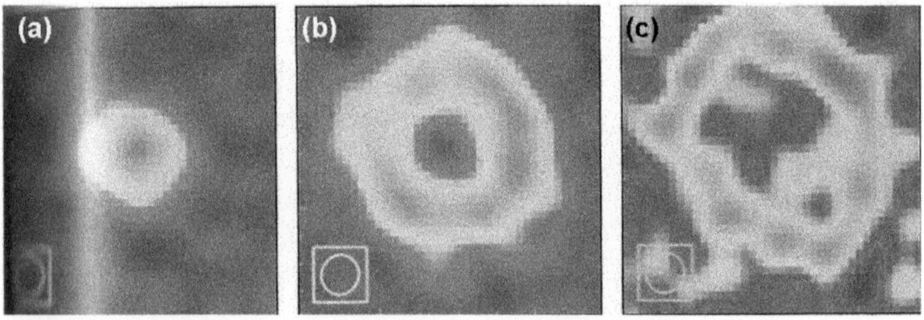

FIGURE 2.10 The layered molecular envelope around the carbon star CW Leo is revealed from these images from the Berkeley-Illinois-Maryland Millimeter array. The angular resolution, indicated by the boxed circle in each image, is 7 arc seconds. (a) Silicon sulfide (SiS) emission is produced in the star's atmosphere, where this molecule rapidly freezes out onto dust grains. (b) Emission from cyanoacetylene (HC_3N) occurs a little further out. This molecule is formed through ion-molecule reactions in the colder, outer zone of the envelope. (c) On the fringes of the envelope (about a light-year from the star) UV starlight leaking in from outside dissociates the HC_3N to produce C_3N. Courtesy the NRAO, John Bieging and N. Q. Rieu (Verschuur, 1992).

2.4.5 *VY Canis Majoris Circumstellar Envelope*

Situated at a distance of 1,140 pc, this M5-type supergiant ($R_* = 1.4\times10^9$ km ~ 2,011 $R_\odot$) star, with a mass of ~ 25 $M_\odot$, T_e = 2,800 K, a luminosity 220,000 times the Sun's, VY CMa is a cool, old star like CW Leo. But it differs in one crucial aspect: its photosphere is rich in oxygen rather than carbon. It exhibits a substantial mass-loss rate of $2\times10^{-4} - 4\times10^{-4}$ $M_\odot$ yr^{-1}, giving rise to a clumpy envelope of arcs, knots, and jets, on a scale of 10" (~15,000 AU) and it is regarded a prototypical example of oxygen-rich related circumstellar envelope (Kwok, 2007) in which a variety of chemical compounds (~70, see Figure 6.7) have been detected to date (August 2019). According to their spectral line profiles, these molecules' origin can be traced back to three distinct regions: a spherical outflow, a blue-shifted expansion envelope, and a red-shifted flow. Certain species (SiO, PN, and NaCl) are only related to the spherical flow, whereas HNC and sulfur-bearing molecules are created in the expanding winds, probably due to shock waves (Ziurys et al., 2007). Because VY CMa is quite massive and it is believed to be on its way to becoming a SN (Richards et al., 1998) it may be currently synthesizing afresh P nuclei in its inner core. If so, some enhancement of phosphorus in its photosphere is possible, depending on the mixing degree of the stellar interior due to convective dredge-up (Figure 2.11).

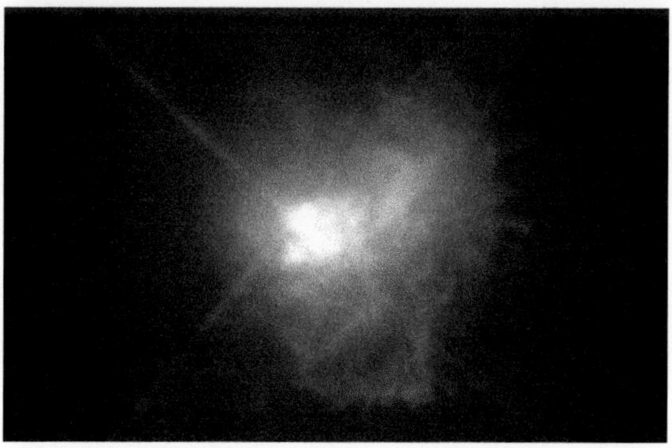

FIGURE 2.11 (See color insert.) This color composite image of VY Canis Majoris consists of a mixture of HST Wide Field and Planetary Camera 2 images, taken in blue, green, red, and near-infrared light. The image reveals a complex pattern of circumstellar ejecta, with arcs, filaments, and knots of material formed by the massive outflows stemming from several outbursts. (Composite image created from HST data taken by R. Humphreys. Reproduced by courtesy of The Hubble Heritage Team – AURA/STScI/NASA, Baltimore, MD.

2.4.6 The Ring and Helix Planetary Nebulae

The planetary nebula stage, which lasts about 10,000 years, follows the asymptotic giant branch phase in the life-cycle of low to intermediate-mass stars. Ultraviolet (UV) radiation from the now white dwarf star ionizes the remnant shell material as it flows into the ISM. The white dwarf typical mass is about 0.5 $M_{\odot}$, so that between 0.5 and 7.5 $M_{\odot}$ is ejected into the diffuse ISM by the planetary nebula progenitor star (Kwok, 2000, 2013). Chemical models predicted that the UV radiation field from the central star is sufficiently intense that the overall molecular content of planetary nebulae will decrease steadily with time, so that at the end of the planetary nebula phase the molecular inventory is expected to be negligible around it. Recent astronomical observations, however, suggest that molecular abundances do not significantly vary over the typical lifetime of planetary nebulae (Ziurys et al., 2015).

Since its discovery in 1779, NGC 6720 (M 57), also known as the Ring Nebula, is the best-known example of a planetary nebula (Figure 2.12). Its progenitor was probably a slightly more massive star than the Sun

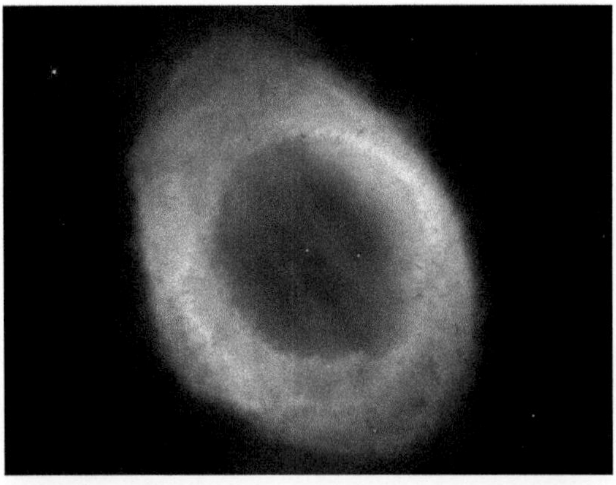

FIGURE 2.12 (See color insert.) NASA's Hubble Space Telescope captured this sharp view of the best known Ring Nebula (M57). The photo, in approximately true colors, shows long dark clumps of material in the gas at the edge of the nebula, and the dying central star floating in a hot gas blue haze. The nebula is about a light-year in diameter and is located some 2,000 light-years from earth towards the constellation Lyra. Going outwards from the very hot central star, blue shows emission from helium atoms, green represents ionized oxygen, and red shows ionized nitrogen, radiated from the coolest, farthest gas. The ultraviolet radiation from the remnant central star, with a white-hot 120,000 K surface temperature, gives rise to the gradations of color in the glowing gas.
Reproduced by courtesy of The Hubble Heritage Team – AURA/STScI/NASA, Baltimore, MD.

which exhausted its core hydrogen and helium and shed its outer layers thus producing the surrounding nebula. The spectra reveal the presence of nitrogen, oxygen, helium, and other gases.

NGC 7293 or the Helix nebula is a planetary nebula originated by an intermediate to low-mass star, which shed its outer layers while approaching the end of its evolution (Figure 2.13). Its remnant stellar core is bound to become a white dwarf star. The energetic glow of the central star causes the bright fluorescence of the previously expelled gases. This is the oldest known planetary nebula, estimated to be 12,000 years old by a measured expansion rate of about 31 km s^{-1} for the inner disk (O'Dell, 2002). The excitation temperature of the atoms varies across the Helix nebula, ranging from 1,800 K in its inner region (at about 2.5′ from the central nucleus) to about 900 K in its outer region at the distance of 5.6′ (Matsuura et al., 2007). This object has a large spatial distribution on the sky such that detailed images of a given atomic or molecular species can be made. Many atomic lines have been studied across the face of the nebula, and also emission from H_2 and CO molecules. The list of molecules detected to date includes HCN, HNC, CN, HCO^+, CCH, c-C_3H_2, and H_2CO (see Table 6.5).

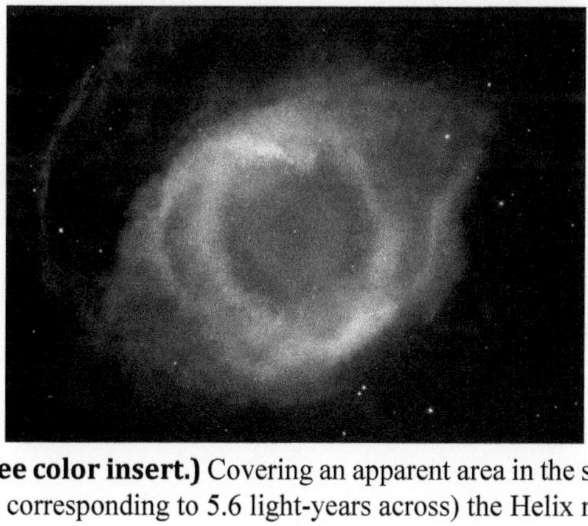

FIGURE 2.13 (See color insert.) Covering an apparent area in the sky equal to half the full moon (actually corresponding to 5.6 light-years across) the Helix nebula (NGC 7293) is the closest of all known planetary nebulae, and it lies 140 parsecs (456 light-years) away from Earth in the constellation Aquarius. The star that ejected these gases is at the center of the glowing shell. The bluish-green color comes from ionized oxygen, the pink and red from ionized nitrogen and hydrogen. The small radial blobs in the red shell (each one is about 150 AU across) give the object its alternate name, the Sunflower Nebula. The ejected shell is actually nearly spherical; it looks like a ring because of the substantial thickness of the shell when we look near the shell's rim.
Reproduced by courtesy of The Hubble Heritage Team – AURA/STScI/NASA, Baltimore, MD.

2.4.7 M1 Supernova (SN) Remnant (Figure 2.14)

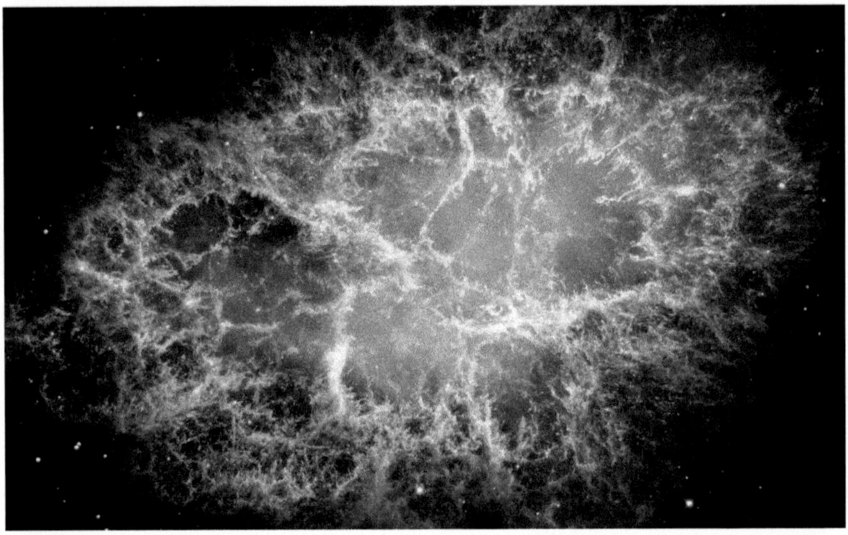

FIGURE 2.14 (See color insert.) The Crab Nebula, named for the arm like appearance of its filamentary structure, is the remnant of a supernova observed on July 4, 1054 A. D. in the constellation of Taurus by the imperial astronomer to the Chinese court, Yang Wei-T'e. A thousand years after the explosion these gases are still moving outward at about 1,800 km s^{-1}. The distance of the nebula is about 2,000 parsecs (6,520 light-years), so its present angular dimensions of 4x6 arc minutes correspond to linear dimensions of approximately 2 by 3 parsecs (6.5 by 9.8 light-years). At the center of the nebula, there is a rapidly rotating neutron star. This pulsar, the first one discovered, has a period of 0.033s. The cloud of gases shines with a luminosity of 75,000 times the Sun's powered by the synchrotron radiation stemming from electrons whirling along the powerful magnetic field of the neutron star.
Reproduced by courtesy of The Hubble Heritage Team – AURA/STScI/NASA, Baltimore, MD.

2.4.8 Beta Pictoris Protoplanetary Disk

In 1984 Beta Pictoris (β Pic), a nearby star located 63.2 light-years (19.4 pc) away with 1.85 $M_{\odot}$ and an effective temperature of about 8,000 K, was the first star discovered to host a debris disk, which was associated with planet formation processes (Figure 2.15). Since then, β Pic has been the object of intense scrutiny by Hubble and ground-based telescopes. The disk's edge-on angle makes it easy to see, and its large amount of starlight-scattering dust makes it especially bright. In addition, β Pic is closer to earth than most of the approximately two-dozen light-scattering circumstellar disks known to date. Accordingly, β Pic provides a nice example of what a young planetary system looks like.

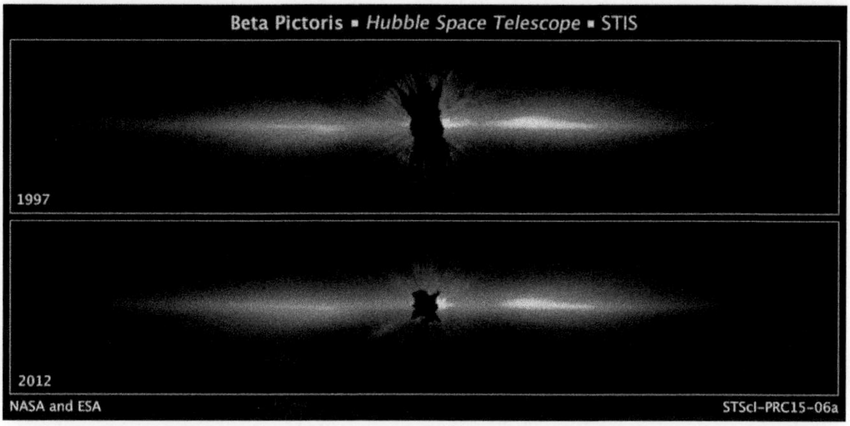

FIGURE 2.15 The 2012 image (bottom) picture of a large, edge-on, gas-and-dust disk encircling the 24 ± 3 million-year-old star Beta Pictoris (spectral class A5 V, approximately 1.8 times more massive than the Sun), is the most detailed to date. The 1997 Hubble image (top) shows the disk's dust distribution has barely changed over 15 years.
Source: https://www.nasa.gov/content/goddard/hubble-gets-best-view-of-circumstellar-debris-disk-distorted-by-planet. Credits: NASA, ESA, University of Arizona.

Nevertheless, β Pic may not be the best representative, since its disk is exceptionally dusty, and carbon is extremely overabundant relative to every other measured element (Roberge et al., 2006). The disk's dusty appearance may be due to recent major inner collisions among an unseen embedded planet and asteroid-sized objects. In fact, β Pic is the only star where astronomers have found an embedded giant planet in a directly-imaged debris disk to date. The planet, discovered in 2009, has a mass 12.9 M_J, a radius 1.46 R_J, and moves on a 20–26-year orbit that is highly inclined with respect to the line of sight from earth and has an effective temperature of about 1,720 K. This relatively short orbital period allows scientists to study how a large planet distorts the mass of gas and dust encircling the star. By comparing the latest 2012 images to those by Hubble in 1997, it is found that the disk's dust distribution has barely changed over 15 years even though the entire structure is orbiting the star like a carousel, thus showing that the disk's structure is smooth and continuous, at least over this interval between the Hubble observations.

2.5 SUMMARY AND REVIEW QUESTIONS

Within the grand panorama of cosmos evolution we will focus on just one chemical element, namely, phosphorus, in order to follow its evolutionary path from massive stars inner cores, where it is formed under the action of the strong nuclear forces to its incorporation in the minerals we found all over the solar system, and the metabolic molecules inside the living beings on earth as well. Following this route, we will start at the small scales of atomic nuclei, and then progressively increasing in size by considering simple P-bearing molecules around circumstellar shells and the ISM all the way long to the sugar-phosphate backbone of DNA macromolecules and apatite crystals readily visible to the naked eye. In our scale increasing voyage along with our hierarchically structured cosmos, we also review the energy scales where these phosphorus-containing structures can exist, ranging in temperature from the 1000 million K required for the thermonuclear synthesis of ^{31}P nuclei to the about 300 K required for the stability of the fragile organic compounds flowing inside bacteria and cells.

KEYWORDS

- **arrow of time**
- **chemical evolution**
- **circumstellar shells**
- **clustering**
- **cosmic evolution**
- **dense clouds**
- **diffuse clouds**
- **fundamental interactions**
- **galaxies classification**
- **globular clusters**
- **Hertzsprung-Rusell diagram**
- **hierarchical structures**
- **Hubble parameter**
- **intergalactic medium**
- **interstellar medium**
- **Milky Way galaxy**
- **molecular clouds**
- **open clusters**
- **planetary nebulae**
- **planetary systems**
- **planetesimals**
- **primordial nucleosynthesis**
- **protogalaxies**
- **protoplanetary disks**
- **protostars**
- **star-forming regions**
- **stars life cycle**
- **stellar populations**
- **supernova**
- **supernova remnants**

CHAPTER 3

Cosmic Distribution of the Main Biogenic Elements

"The four most abundant elements in the universe, with the exception of the noble gases, are hydrogen, oxygen, carbon and nitrogen, which are also precisely the four major constituent elements of organic compounds and of living matter."

(John Oró, 1963)

3.1 BIOCHEMICAL ASPECTS OF THE PERIODIC TABLE

In addition to the main biogenic elements we introduced in Section 1.2, there are also other atoms, which are essential for many life forms, namely, Na, Mg, and Cl (belonging to the third row), K, Ca, Mn, Cr, Fe, Co, Cu, and Zn (in the fourth row) and Mo and I in the fifth row. These atoms are referred to as oligoelements, since they are required in small quantities only. Accordingly, a total of 19 elements, highlighted in Figure 3.1, are necessary for current living beings on earth.

Why are certain atoms used frequently in living systems while others are found rarely or never? Does it make any sense to characterize the atoms in a specifically biochemical sense? To properly address these two fundamental questions it is convenient to firstly consider the relative abundance of a given element in those places where life phenomena are supposed to start and develop, and secondly analyze the nature of chemical bonding (i.e., ionic, covalent, metallic) among the atoms constituting the different classes of biomolecules, as well as their resulting three-dimensional geometries.

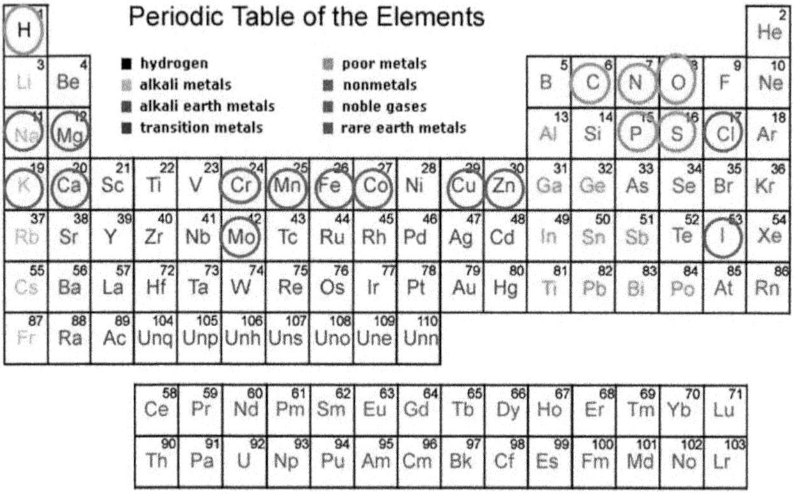

FIGURE 3.1 Location of the main biogenic elements (encircled in light gray) and the most important oligoelements (encircled in dark gray) in the periodic table. With the exception of H, Mo, and I, which stand alone, we observe these elements appear grouped in pairs or trios along a row.

As we mentioned in the two previous chapters, the relative abundances of the different elements present in the periodic table at a given place ultimately depend on two main factors, namely, the sequence of nucleosynthesis processes taking place in different stars at different stages of their evolution (which will be described in detail in Chapter 5), and the specific physicochemical processes occurring at different astrophysical environments in galaxies, within the overall galactic chemical evolution (see Figures 2.3–2.5), which determine how the chemical elements formed inside stars are distributed throughout the ISM. In fact, as we described in Section 2.2, chemical evolution in spiral galaxies is spurred by aged stars ejecting chemically enriched gas and dust grains into the ISM to form dense molecular clouds, which eventually gravitationally collapse to give rise to a new generation of young stars. The resulting elemental chemical inventory at a given time is then obtained by combining the nucleosynthetic products from the several populations of stars that existed over the age of the galaxy with those remaining from the primordial nucleosynthesis, to add up approximately to the currently observed abundance of the different elements of the periodic table (Matteucci, 2016).

Once the mechanism determining the elemental abundances of chemical elements has been disclosed, we should consider their ability to bind to each other, as this determines the resulting molecular inventory. Such ability depends on the electronic configuration and electronegativity of any given atom. In this sense, it is worth pointing out that the set formed by the main biogenic elements provides a minimal basis containing a representative (at least) of the most common valence states, i.e., 1 for H, 2 for O, and S, 3 for N, 2 and 4 for C, and 3 and 5 for P. According to their bonding ability the elements of biological interest can be classified attending to their atomic coordination and bond strength into skeletal, ligand-swapping, and ions. All of the main biogenic elements H, O, C, N, S, and P play structural skeletal functions in biomolecules, strongly linking these atoms among them via covalent bonds with bonding energies of about 3 eV. On the other hand, the metallic oligoelements Cr, Mn, Fe, Co, Cu, Zn, and Mo are among the commonest ligand-swapping atoms. The presence of partially filled *d* orbitals allow these atoms to form numerous hybrid bonds and thereby adjust to greatly different ligand configurations, generally involving lower bonding energies ~1 eV. Last but not least, we also find in living beings a significant number of ions, stemming from atoms prone to form ionic bonds, such as the alkaline and earth-alkaline representatives Na^+, K^+, Mg^{+2} and Ca^{+2}, the halogen Cl^-, and the acid-related molecular groups PO_4^{-3}, $SO_4^=$ and $CO_3^=$.

Along with the atoms which are usually present in living beings, it is also instructive to consider the case corresponding to those elements which are not. Since silicon is a relatively abundant element on a cosmic scale (see Figure 1.1), Si atoms, which like C also have a four valence state, have been sometimes proposed as a basis for an alternative biochemistry. In fact, its abundance in earth's crust (almost 30% by mass) and the fact that silicic acid is more abundant in Nature than phosphoric acid seem to provide some support to this proposal. Yet, no life form is known to have the ability to forge C-Si bonds, although it has recently been reported that certain proteins are able to catalyze the formation of organosilicon compounds under physiological conditions (Kan et al., 2016). Nevertheless, the chemical properties of C markedly contrast with those of Si. In the first place, unlike C, Si combines with O to form insoluble silicates or network polymers of SiO_2 instead of gaseous molecules like CO or CO_2. Accordingly, no reaction pathways similar to the synthetic processes of sugars in photosynthetic organisms are possible, and a metabolic scheme based on Si becomes impossible to conceive (Lambert et al., 2015). On the other hand, Si-polymers are unstable in the presence of water due to the larger size of Si atom as compared to C, making Si-Si bonds weaker than C-C bonds. Furthermore, under physiological conditions, the esters of silicic acid hydrolyze far too rapid, so it becomes an unsuitable substitute for phosphoric acid in nucleic acids backbone. Analogous arguments also apply to other possible alternatives, like sulfuric and arsenic acids (Wolfe-Simon et al., 2009, 2010, 2011; Huertas and Michán, 2012) or even the $O{=}P(OH)_2R$ acid derivative. In fact, as we will see in Section 4.6.1, phosphoric acid is specially adapted for its role in nucleic acids because it can simultaneously link two nucleotides and still remains doubly ionized. The resulting negative charge serves both to stabilize the diesters against hydrolysis and to retain the macromolecules within a lipid membrane (Westheimer, 1987).

3.2 THE ELEMENTAL ABUNDANCES OF MAIN BIOGENIC ELEMENTS

3.2.1 Solar System

Solar system elemental abundances for non-volatile elements are based on meteorites (mainly carbonaceous chondrites). However, the main biogenic elements H, C, N, O as well as noble gases are volatile elements, which are incompletely condensed in meteorites and thus solar and other astronomical data sources must be used. In fact, our Sun's chemical composition is a fundamental yardstick in astrophysics and geophysics. Solar abundances are based mainly on photospheric spectroscopic observations, augmented as needed by solar wind and the so-called solar energetic particles (SEP) observations (Anders and Grevesse, 1989; Asplund et al., 2009). In doing so, one must keep in mind that solar abundances cannot be determined straightforwardly from the solar spectrum by direct observation alone, but some degree of modeling is required as well (Scott et al., 2015). The elemental abundances of the main biogenic elements, as derived from meteorites, comet Halley dust particles, and the Sun (photosphere, corona, and solar wind contributions) are listed in Table 3.1.

TABLE 3.1 Elemental Abundances of the Main Biogenic Elements in the Solar System*

Element	Photosphere (1)	Corona (2)	SEP (2)	Meteorites (1)	Halley (4)
H	12.00	--	--	8.22 ± 0.04	9.47 ± 0.08
O	8.69 ± 0.05	8.30 ± 0.06	8.30 ± 0.03	8.40 ± 0.04	8.99 ± 0.05
C	8.43 ± 0.05	7.90 ± 0.06	7.92 ± 0.04	7.39 ± 0.04	8.64 ± 0.08
N	7.83 ± 0.05	7.40 ± 0.06	7.40 ± 0.03	6.26 ± 0.06	8.05 ± 0.12
S	7.12 ± 0.03	6.93 ± 0.05	6.93 ± 0.02	7.15 ± 0.02 (3)	7.44 ± 0.12
P	5.41 ± 0.03 (3)	5.24 ± 0.08	5.24 ± 0.06	5.43 ± 0.04 (3)	--

*Elemental abundances of the main biogenic elements relative to hydrogen in different solar system bodies expressed in the logarithmic scale $[X] = \log([X]/[H]) + 12$, usually referred to as dex units. SEP stands for solar energetic particles.
Source: (1) Asplund et al., 2009, (2) Anders and Grevesse, 1989, (3) Scott et al., 2015, (4) Jessberger et al., 1988.

By inspecting Table 3.1, we see that all biogenic elements are under abundant, as compared to their photospheric values, in the corona and SEP sources, which otherwise compare pretty well to each other. This expresses the presence of different physical processes affecting their relative abundances. In fact, neutral atoms are able to diffuse away from the photospheric gas and enter the corona, whereas ions are prevented from doing so by the Sun's magnetic field. Regarding the comparison between the comet Halley dust and solar photosphere compositions, we appreciate a large depletion of H in the former, which reflects the non-condensation of molecular hydrogen in the comet's nucleus. Quite remarkably C, N, O, and S atomic abundances, on the other hand, are significantly overabundant as compared to their solar values.

3.2.2 *Nearby Stars*

Recently obtained evidences indicate that the solar system elemental abundances are not the most representative of the galactic stellar disk population in general, but instead there exists an enhancement in the abundances of a number of species, including especially those thought to be the major components of interstellar dust. Thus, a differential elemental abundance analysis including ten solar analogues (dwarf stars belonging to the spectral classes G0 – G5) and eleven stars almost identical to the Sun (the so-called solar twins) disclosed that the Sun shows a characteristic chemical signature with about a 20% depletion of refractory elements relative to volatile elements in comparison with the solar twins mean value elemental abundances (Meléndez et al., 2009). Noteworthy enough, the most enhanced elements in the Sun are precisely the main biogenic elements N, S, P, C, and O, (in this order) as it can be seen by inspecting Figure 3.2.

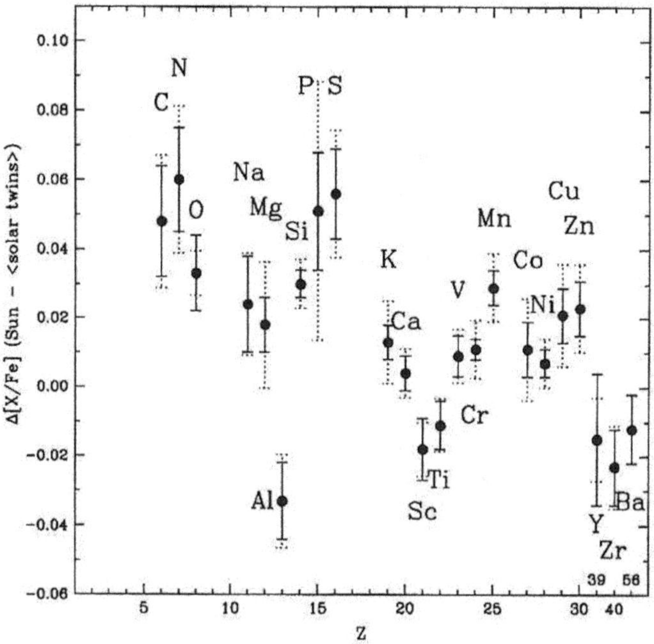

FIGURE 3.2 Differences between the elemental compositions of the Sun and the mean values obtained for eleven solar twins (normalized to iron) as a function of the atomic number Z. Solid (dotted) observational 1σ error bars refer to solar twins and both Sun and solar twins relative abundances, respectively.
Source: Meléndez, Asplund, Gustafsson, and Yong (2009). ©AAS. Reproduced with permission.

3.2.3 The Interstellar Medium (ISM)

As we have mentioned in previous chapters, the ISM is mainly enriched by heavy elements produced by stars during their forming and dying phases. Spectroscopic observations indicate that atomic and molecular hydrogen are the most abundant species by both number and mass in the regions between stars, which contain both gas and condensed phases (see Section 2.3.2). In the gas, the next most abundant element after hydrogen is helium, roughly 10% of hydrogen by atomic abundance. Oxygen, carbon, and nitrogen follow, with elemental abundances of 4.6×10^{-4}, 2.1×10^{-4}, and 6.6×10^{-5} relative to hydrogen in the interstellar abundance value adopted by Snow and Witt (1996). In the same reference, the sulfur abundance is 1.2×10^{-5} (Table 3.2).

Galactic cosmic-ray nuclei arriving near earth provide direct access to a sample of material from outside the solar system. These energetic particles pervade our Galaxy and penetrate interstellar clouds, regulating the chemical reaction networks inside them through the ionization degree of different chemical species. The composition and energy distribution of these nuclei reflects the nucleosynthetic processes at work in the original source, as well as the conditions experienced by these materials during acceleration and transport through the galaxy.

TABLE 3.2 Elemental Abundances of the Main Biogenic Elements Beyond Solar System*

	Solar (1)	**ISM** (2)	**GCR**(3)	**QSO1** (4)	**QSO2** (5)	**QSO3** (6)	**I Zw 18** (7)
O	8.69 ± 0.05	8.66	8.36 ± 0.02	8.1 ± 0.2	8.01 ± 0.02	7.0 ± 0.2	7.0 ± 0.1
C	8.43 ± 0.05	8.33	8.38 ± 0.02	7.13 ± 0.08	7.30 ± 0.03	--	6.0 ± 0.3
N	7.83 ± 0.05	7.82	7.74 ± 0.04	6.21 ± 0.02	6.26 ± 0.01	5.3 ± 0.1	5.2 ± 0.1
S	7.12 ± 0.03	7.09	6.70 ± 0.08	6.35 ± 0.03	--	5.3 ± 0.1	5.4 ± 0.1
P	5.41 ± 0.03	--	5.9 ± 0.2	4.8 ± 0.1	4.38 ± 0.04	3.2 ± 0.1	3.6 ± 0.3

*Elemental abundances of the main biogenic elements relative to hydrogen in the ISM, the galactic cosmic rays (GCR) for solar minimum, three quasars (QSO1: HE 2243–6031 with z = 2.33, QSO2: Q0347–3819 with z = 3.025, and QSO3: 0000–2620, with z = 3.3901) and the blue compact galaxy I Zw 18, expressed in the logarithmic scale [X] = log([X]/[H]) + 12, usually referred to as dex units. The elemental abundances in the Sun's photosphere are listed in the first column for the sake of comparison.

Source: (1) Asplund et al., 2009, (2) Snow and Witt, 1996, (3) George et al., 2009, (4) López et al., 2002, (5) Levshakov et al., 2002, (6) Molaro et al., 2001, (7) Leboutellier, 2013.

Nevertheless, the composition and energy spectra of light cosmic rays measured near the earth are not always representative of the distribution of cosmic rays at large distances from the Sun. This is so because the solar magnetic field, carried by the solar wind, regulates the electrodynamics inside a region named the heliosphere, which extends till about one hundred AU. Such an interplanetary magnetic field screens out the low energy light cosmic rays, and specially hydrogen, helium, and electrons, from the inner solar system region. In the third column of Table 3.2, we list the elemental abundances of the main biogenic elements measured in the GCR by the Cosmic Ray Isotope Spectrometer during the solar cycle 23 (George et al., 2009). By comparing them with the abundances measured for the energetic particles coming from the Sun in Table 3.1, we appreciate a remarkable coincidence in the abundances of all the elements but carbon, which is significantly enhanced in the galactic cosmic ray composition.

3.2.4 Galaxies and the Intergalactic Medium (IGM)

Phosphorus atoms (generally P^+ ions) have been detected in the ISM of several external galaxies, including the Large Magellanic Cloud (Friedman et al., 2000), NGC 1068 (Oliva et al., 2001), the giant HII region NGC 604 in the spiral galaxy M33 (Lebouteiller et al., 2006), the low-metallicity gas-rich star-forming dwarf galaxy I Zw 18 (Lebouteiller et al., 2013), as well as the high-redshift quasars QSO HE 2243–6031 with z = 2.33 (López et al., 2002), Q0347–3819 with z = 3.025 (Levshakov et al., 2002), and QSO 0000–2620 with z = 3.39, which exhibits the low value [P/H] = –2.31 ± 0.10 (Molaro et al., 2001). For more than three decades, the dwarf irregular galaxy I Zw 18 held the record as the most metal-deficient galaxy known, with an oxygen abundance of 7.17 ± 0.01 dex, but it has been recently displaced by the blue compact dwarf galaxies SBS0335–052W and DDO 68 with [O] = 7.12 ± 0.03, and 7.14 ± 0.03, respectively. It is thought that I Zw 18, located at 45 million light-years from the earth, is only about 500 million years old, so that Population III stars may be still forming there. Quite interestingly, by comparing the last two columns in Table 3.2, we realize that the elemental composition of this young galaxy is almost identical to that observed in high redshifted quasar QSO 0000–2620, displaying the elemental composition of a then younger universe.

Active galactic nucleus outflows play an important role in the chemical enrichment of the intergalactic medium (IGM). Thus, the study of the elemental abundances in galactic outflows, observed over a range of redshifts, provides us with a unique probe to investigate the chemical enrichment along with the Universe history and evolution over cosmological scales, which constrains star formation scenarios and evolution of the host galaxy. In several cases, huge enhancements of carbon, nitrogen, oxygen, and silicon by factors of tens to hundreds of times the solar values were reported in some objects, in contrast to the usual enhancement values (an order of magnitude or less) generally derived from the analysis of quasar emission lines. In the case of phosphorus, the enhancement is not so large, since measurements of abundances in Seyfert galaxies and quasars outflows yielded abundances of only a few times the solar abundance value. For instance, the detection of the highly ionized P^{+4} phosphorus ions in the quasar SDSS J1512+1119 outflows, with a column density indicating a phosphorus abundance within the range 0.5–4 times the solar value in this object (Borguet et al., 2012). Anyway, the detection of significantly high values of phosphorus in a high red-shifted source clearly indicates that this element was formed relatively soon, during the early chemical evolution stages of the universe.

3.3 DIFFERENTIAL ENHANCEMENT OF BIOGENIC ELEMENTS

In the previous section, we have seen that the elemental composition of nearest astrophysical objects does not coincide with those observed in farthest ones (see Table 3.2). Since these sources exhibit the features they had in the past (the farthest they are located the oldest they look) we conclude the cosmic abundances of elements are changing with time. To properly understand this trend, we should perform a detailed study of the elemental abundances of different heavenly bodies at different epochs. For instance, we know that the chemical composition of younger Population I main sequence stars largely coincides with the observed solar system abundances, whereas older Population II stars exhibit a significantly smaller content of heavy elements.

An important contribution to the heavier elements enhancement observed in Population I stars is due to the planetary nebula produced up to now by intermediate-mass ($0.8 \leq M_{\odot} \leq 8.4$) stars in their transition from red giants to white dwarfs. The wide range of masses covered implies that these planetary nebulae progenitor stars have also had a wide range of ages, since the time a star remains in the main sequence of the HR diagram significantly decreases as its initial mass increases (see Figure 2.6). For the sake of illustration, a typical planetary nebula contributes with about 1 $M_\odot$ to its surrounding medium. Some estimations indicate that about 15% of all matter expelled to the ISM in a year comes from planetary nebulae, of which about 40% is composed of the biogenic C, N, and O atoms (Kwok, 2000). Accordingly, the ubiquitous presence of these atoms in most molecules detected in the ISM (see Figures 1.4 and 1.5) can be properly justified as stemming from this continued enrichment over the years in our Galaxy (see Section 5.6).

This result is illustrated in Figure 3.3, where we observe the accumulation of the main biogenic elements in the Galaxy on a cosmic time scale measured in terms of the relative concentration of iron atoms as compared to hydrogen ones. It is generally assumed that these iron atoms come from successive supernova (SN) events occurred in the Galaxy since the first massive Population III stars burst, hence providing a natural chemical clock in order to measure the pace of chemical evolution in the Galaxy. In particular, the progressive enrichment of stellar atmospheres with P and S atoms with the passage of time then highlights the importance of massive stars as the main sources of these elements at early epochs in the universe, when hypernovae stars were important contributors to the phosphorus production (Jacobson et al., 2014).

Once the bioelements produced in the stars are liberated to the ISM the next stage of chemical evolution happens in the ISM, where the chemical synthesis of complex organic and inorganic compounds can take place over very short timescales (10^3 yr) in the low-density circumstellar environment (Kwok, 2004). This chemical trend correlates with a progressive enhancement of the relative abundance of bioelements in condensed matter due to the processing of chemical compounds during their transition from gas phase to condensed phase in both ISM and preplanetary stages. This point can be ascertained by comparing the elemental abundances of bioelements in the ISM gas phase with those measured in meteorites. Thus, phosphorus is found to be strongly depleted in dense star-forming molecular clouds (Turner et al., 1990).

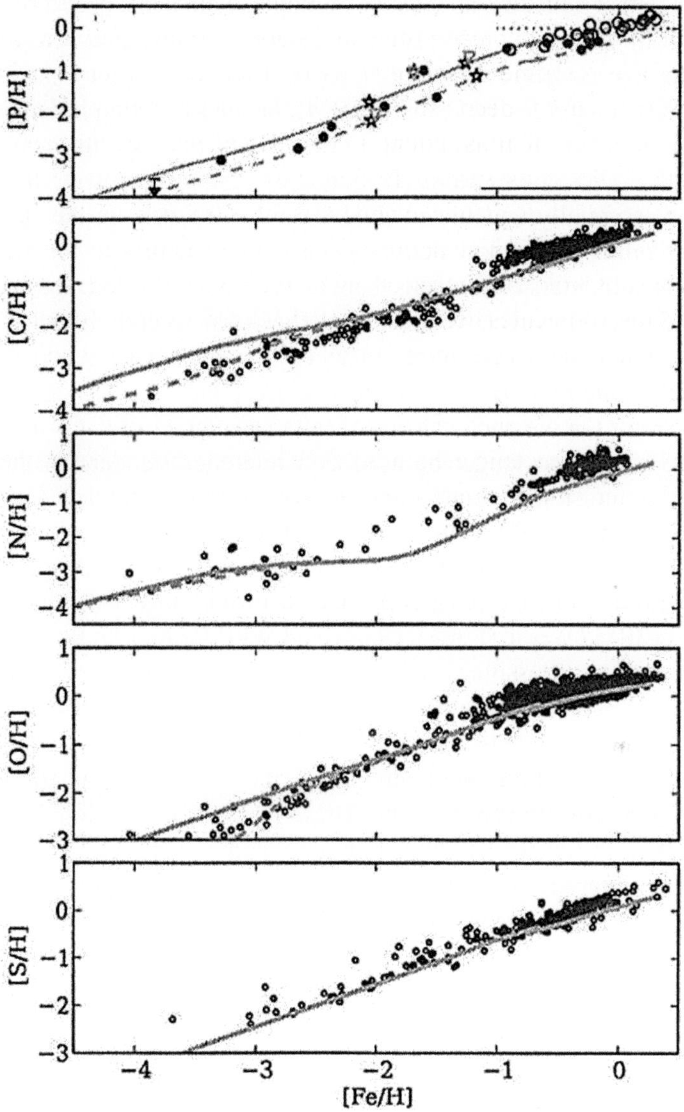

FIGURE 3.3 Elemental abundances of the main biogenic elements (relative to hydrogen) observed in the photospheres of a suitable stars sample as a function of the Galaxy age, which is measured in terms of their relative Fe/H ratio. The solid and dashed lines are predictions for the chemical evolution for each element according to different models.
Source: Jacobson, Thanathibodee, Frebel, Roederer, Cescutti, and Matteucci (2014). ©AAS. Reproduced with permission.

On the other hand, the abundance of biogenic elements observed toward the cold diffuse cloud ζ Oph relative to meteoritic abundances was reported to amount log(ISM/ʘ) $\simeq$ –0.1 for N, log(ISM/ʘ) $\simeq$ –0.4 for O and C, and log(ISM/ʘ) $\simeq$ –0.5 for P (Cardelli, 1994), hence indicating a depletion of these atoms in the observed cloud. In the case of phosphorus these results agree with earlier observations (Dufton et al., 1986) suggesting that phosphorus is essentially undepleted (P/H = 5.59 ± 0.5 dex) in the star sightlines containing warm, low-density neutral gas, whereas in the sightlines containing cold interstellar clouds there exists some depletion degree (P/H = 5.12 ± 0.5 dex). However, we must recall that it has recently been suggested that the solar system may have enhanced abundances of many elements (particularly, the main biogenic elements, see Figure 3.2), as compared to those measured in the photospheres of suitable samples of F, G, and K type stars, and should no longer be used as a reference standard in the interpretation of interstellar clouds elemental abundances or the development of interstellar dust models. Thus, the abundances of main biogenic elements toward the ζ Oph cloud, relative to the above mentioned stars sample read log(ISM/ʘ) $\simeq$ +0.1 for N, log(ISM/ʘ) $\simeq$ –0.1 for O, log(ISM/ʘ) $\simeq$ –0.2 for C, and log(ISM/ʘ) $\simeq$ +0.3 for S (Snow and Witt, 1996).

Accordingly, from these data we can conclude that most biogenic elements are depleted from the gas phase by different amounts in different ISM environments and they are concentrated in interstellar grains composing the solid phase, phosphorus being the most depleted biogenic element, and nitrogen the least one. These grains' composition varies with time: at their initial stages, they are mainly composed by a core of highly refractory materials, and have undergone little processing. As chemical evolution proceeds, due to their interaction with the gas phase, which has experienced substantial processing via ion chemistry molecular networks (see Figure 1.3), these grains are progressively covered by different ices (CO_2, NH_3, CH_4, H_2CO, H_2S).

Additional processing would occur in the nuclei of comets, formed by low-temperature aggregation of these interstellar grains during preplanetary disks formation. In fact, although most of the time comets remain far away from stars, they occasionally approach them due to small gravitational perturbations. During their approach to the star, their temperature increase promotes intense photochemical processing of their originally pristine materials. In the case of periodic orbit comets, these episodes assume a cyclic nature, undergoing successive condensation events due to the recurrent loss of most volatile materials forming their comae and tails. In this sense, periodic comets somehow mimic usual chemical handling in terrestrial laboratories, leading to the synthesis of complex organics by means of alternating sublimation/condensation processes in the course of their recurrent orbital motions.

In order to illustrate the relevance of this cometary chemistry in Figure 3.4, the elementary abundances of bioelements in the human body and the mean compositions of cometary ices are compared. As we can see, the biogenic elements content in comets compares well with that observed in living beings, particularly regarding H, O, C, and P atoms, and, to a lesser extent, for S and N as well. Conversely, the bioelements are overabundant in the living matter by several orders of magnitude as compared to their solar system abundances, especially in the case of phosphorus. Therefore, we can conclude that during their transition from the ISM to protoplanets the bioelements are efficiently incorporated to the raw materials out of which new planetary systems will be formed. In fact, among the molecules observed in the preplanetary disk surrounding the newborn star DM Tauri, the organic compounds HCN, H_2CO, CS, CN, and C_2H exhibit large depletion factors with respect to their gas-phase abundances, hence indicating these species are being significantly incorporated into the condensed phase (Ehrenfreund et al., 2002).

3.4 LIFE SUSTAINABILITY AND PHOSPHORUS AVAILABILITY

Since the elemental composition of current organisms is generally quite different from that of the habitat they live in, those organisms rely on their metabolic and genetic machinery to harvest, process, and store different biogenic elements from the environment to be able to live and reproduce in a variety of different ways (Sterner and Elser, 2002). As a consequence, despite the common use of the same biogenic elements in basic life processes among all organisms, specimens, and genotypes within species vary in their somatic elemental composition (Frost et al., 2006). For example, recent work indicates marked effects of elemental supply environments in disparate genomes that have generated biases in the usage of various amino acids (differing in the number of nitrogen atoms and their side chains) in proteomes of plants and animals (Elser, 2006). Accordingly, genome properties and environmental elemental supply are coupled.

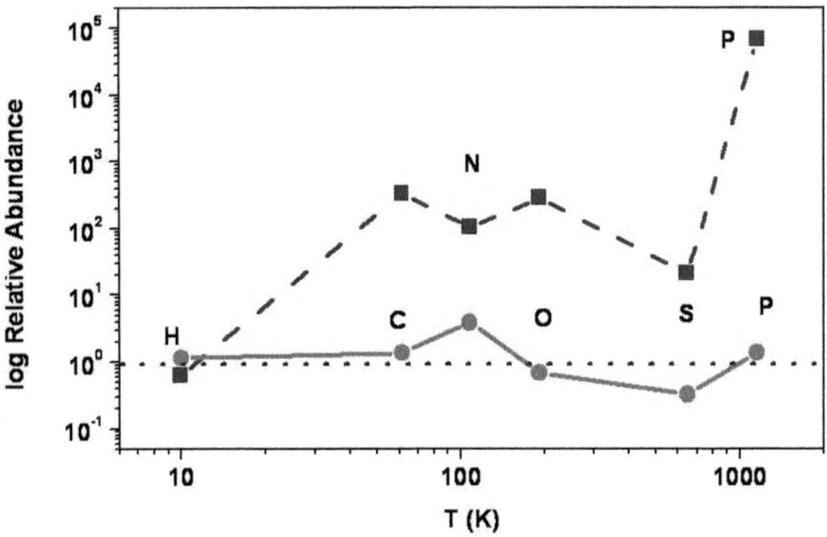

FIGURE 3.4 The logarithm of the abundances of biogenic elements in human body (dashed line; Emsley, 1998) and comets (solid line; Ehrenfreund et al., 2002) relative to meteoritic abundances (Grevesse and Sauval, 1998) plotted against condensation temperature. For a given element, X, its elemental abundance in human body is normalized to silicon, [X], and then divided by its corresponding solar system or cometary abundance, [X]′, respectively, to obtain log([X]′/[X]), where [X] are all expressed in atomic percent number. The cometary phosphorus abundance has been estimated from Rosetta's measurements suggesting a P/O ratio about two times the solar value (Altwegg et al., 2016).
Source: Maciá, (2005); with permission from the Royal Society of Chemistry.

The central roles of phosphorus in the structure and functioning of primary biological components listed in Table 1.1 suggest that the lack of this biogenic element would have important structural and functional consequences in these basic cellular components, which could eventually affect the organism fitness. Such a critical role of phosphorus in biological processes is striking because of its relative scarcity compared to other biogenic elements in the biosphere (Maciá, 2005). Accordingly, organisms have developed a variety of responses to phosphorus shortage, which can be observed at multiple levels of organization. Indeed, strong evidence for genetic variation in the sequestration and handling of this element progressively appear in the literature (Jeyasingh and Weider, 2007).

The first noticeable genetic response to differences in environmental phosphorus availability is via the so-called P-responsive genes, which are generally classified as "early" or "late" responding genes. Early genes respond rapidly and non-specifically to phosphorus deficiency, whereas late ones are those that alter the morphology, physiology, or metabolism under prolonged phosphorus stress. In addition to altered regulation of several genes, such as phosphorus transporters, environmental phosphorus availability also involves several alternative metabolic pathways. Generally, autotrophs alter their metabolic physiology to increase acquisition, retention, and processing efficiency of P atoms. These changes are achieved by increasing activity and/or transcription of proteins involved in alternative respiratory pathways. In this regard, it is interesting to mention the observation of a degradation of ribosomes upon request due to phosphorus lacking in the presence of extremely high arsenic concentrations in lake environments were certain extremophile bacteria can grow (Huertas and Michán, 2012).

3.5 SUMMARY AND REVIEW QUESTIONS

When comparing the relative abundances of the main biogenic elements H, O, C, N, S, and P, as they appear in the cosmic abundance curve, with their observed abundances in the living beings, a remarkable match is found, not only from a qualitative point of view, but from a quantitative viewpoint as well. The comparison between the elemental compositions of different astrophysical environments (ISM gas phase, ISM grains, and comets) with that of living beings suggests the probable existence of processes favoring a differential enhancement of the main biogenic elements during chemical evolution in the Galaxy. The relative enhancement of phosphorus is higher than that experienced by the remaining biogenic elements in about two orders of magnitude, implying an increase of about four orders of magnitude when referred to hydrogen content. Thus, phosphorus is remarkably concentrated in the living matter as compared to the other main bioelements.

Therefore, as chemical evolution proceeds a systematic enrichment of biogenic elements takes place as a result of a series of physicochemical processes which ultimately determine the basic stoichiometry of emerging biomolecules according to the following picture: most available P atoms condense in microsized grains in circumstellar envelopes (see Section 6.2.2) which interact with free molecules present in the ISM gas-phase exposed to the action of cosmic rays and UV radiation flux, thereby promoting an efficient chemical processing leading to the formation of icy mantles around these grains.

Noteworthy, albeit phosphorus is an important constituent of living matter, it is relatively scarce in the cosmic scale, with an elemental abundance of [P] = 5.41 relative to H. On the basis of this general perspective an estimation of the biomass capacity of the Galaxy can be derived by considering a rough model which assumes that bioavailability of P atoms is equated to the presence of this element in condensed matter forms, while gaseous states are considered unavailable. In doing so, a crude estimate would yield about five possible earth-sized biomasses per star in the Galaxy (Exercise 5).

Keeping this figure in mind let us conclude this chapter with a question to think about. Currently, the universe is mainly composed of dark energy (69.6%) and dark matter (25.8%), so that baryonic matter only amounts to 4.8% of its content (Staggs et al., 2018). Of this baryonic matter stuff, hydrogen and helium are by far the most abundant elements. With a concentration of 2.6×10^{-7} relative to hydrogen, phosphorus atoms only account for about 0.000001% of the universe content. Notwithstanding this marginal contribution, phosphorus atoms are essential for biological structures and metabolism. May this feature have something to do with the chemical properties of this element as compared to those exhibited by the other main biogenic elements?

KEYWORDS

- **biogenic elements**
- **chemical differential enhancement**
- **chemical evolution**
- **life-limiting bioelements**
- **oligoelements**

CHAPTER 4

Physical and Chemical Properties of Phosphorus Compounds

"Inter inventa nostra saeculi non minimum habendum est phosphorus igneus... [Not least amongst the discoveries of our time is phosphorus igneus...]."

(Gottfried Wilhelm Leibniz, 1710)

In 1669, German physician and alchemist Hennig Brand discovered a strange substance in human urine. By all indications, it seems that to obtain this substance Brand heated thick syrup from previously boiled-down urine on his furnace until a red oil distilled up from it, and draw that off, allowing the remainder to cool, where it consists of a black spongy upper part and a salty lower part. Then, he discarded the salt, mixed the red oil back into the black material, and heated that mixture strongly for 16 hours. First white fumes come off, then an oil. Brand poured the liquid in a jar and covers it, where it cooled and solidified while giving off a pale-green glow. Accordingly, the new substance was called phosphorus (light bearer in Greek), a name then common to various luminous substances. This discovery became public knowledge around the year 1677. About a century later, Lavoisier showed the elemental nature of the "Brand's phosphorus" in 1789, which was simply referred to as "phosphorus" since then (Krafft, 1969).

What was going on in the Brand's cauldron at a molecular scale? Without entering into the details, we can say that urine contains an ammonium-sodium phosphate salt, along with various carbon-based organic compounds. Upon heating this salt is converted into sodium metaphosphate ($NaPO_3$), and under stronger heat the oxygen atoms from the phosphate react with carbon ones to produce carbon monoxide (CO), leaving elemental phosphorus, which comes off as a gas consisting of very stable P_4 tetrahedral molecules (Figure 4.1a), which just start to decompose into P_2 molecules above 800°C.

By cooling down the gas phosphorus molecules condense to a liquid below about 280°C and then solidify (to the α-white phosphorus allotrope) below about 44°C (depending on purity), as is illustrated in Figures 4.1b-c. An analogous procedure, based on phosphate molecules decomposition is still used today (see Section 9.1.1), though now employing mined phosphate ores (instead of urine), coke for carbon, and much more powerful electric furnaces to provide the required thermal energy. Not surprisingly, modern industrial processes are much more efficient in producing elemental phosphorus as well (Exercise 6).

From this brief account on the discovery of phosphorus we readily see that, at variance with most of the elements in the periodic table, which were discovered from the study of inanimate minerals and rocks, phosphorus was first obtained by distilling compounds intimately related to metabolic processes taking place in living beings. In this way, the phosphorus enigma thrill appears again in the course of events, as we realize that solid elemental phosphorus does not naturally appear on earth, where this element is found in highly oxidized states, either as PO_4^{-3} phosphate ions within the regular atomic arrays of apatite crystals in the mineral kingdom, or among phosphorylated organic compounds in the vegetable and animal kingdoms. Is this phosphate chemistry prevalence a particular feature of our planet geological and biological evolution? Or is it a natural outcome of phosphorus chemistry all over the universe?

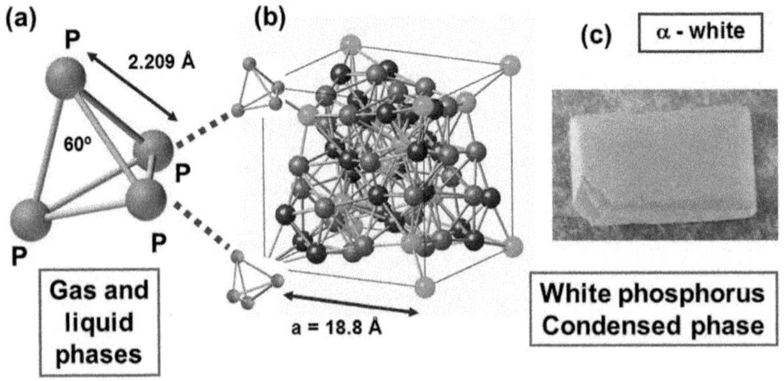

FIGURE 4.1 (See color insert.) The α-white phosphorus is a molecular crystal whose structure is based on stable P_4 tetrahedral molecules (a) with a bonding energy of 3.08 eV per atom, a P-P distance of d = 2.209 Å, and P-P-P bond angles of 60 ± 0.5°, which arrange in a complex body-centered-cubic unit cell shown in (b), with a large lattice constant of a = 18.8 Å, where each ball in the reference α – Mn structure (shown in green) is replaced by a P_4 molecule. In (c) we show a centimeter-sized crystal of α – white phosphorus.

4.1 PHOSPHORUS CHEMICAL BONDING

4.1.1 Molecular Structures of Phosphorus Compounds

Phosphorus is known to occur in at least eight oxidation states, including the phosphide (–3), diphosphide (–2), tetraphosphide (–1/2), elemental (0), hypophosphite (+1), phosphite (+3), hypophosphate (+4), and phosphate (+5) forms. The stereochemistry and bonding of phosphorus are very varied since the element is known in at least 14 structural geometries and its compounds can exhibit coordination numbers up to 9, although most of them have coordination numbers ranging from 3 to 6. Typical coordination geometries are illustrated in Table 4.1.

None of the compounds with a linear geometry and coordination number 1 is stable under normal conditions on earth, but some representatives, containing a carbon-phosphorus triple bond, can be synthesized under controlled conditions in laboratories. Thus, the phosphaethyne H-C≡P molecule was obtained by passing phosphine through an arc between graphite electrodes (Gier, 1961). Later workers showed that essentially pure HCP could be generated by pyrolysis of CH_3PCl_2 and low-temperature distillation of the products. HCP is quite stable at room temperature at pressures below a few Torr. By using microwave spectroscopy, the second member of the phosphaalkyne family (CH_3CP) was subsequently identified, along with a broad collection of phosphorus analogs of nitriles given by the general formula RC≡P, with R={NC, HCC, H_2CCH, C_6H_5} (Kroto, 1997). Clearly, C≡P is a viable functional group, and a whole family of such molecules exists. Quite interestingly, both HCP and NCCP, along with their related radicals CP and CCP, have been found under the unusual physical conditions prevailing in the circumstellar shell around IRC+ 10216 (see Table 1.2 and Section 6.2.2).

The simplest phosphorus bearing molecule with coordination number 2, containing a carbon phosphorus double bond, namely, H_2C=PH, was also synthesized by Kroto's group (1980), and it adopts a bent geometry at the P atom position (Figure 4.2).

The most representative phosphorus-bearing molecule with coordination number 3 is phosphine PH_3. It has a trigonal pyramidal structure with P-H bond length of 1.419 Å, dissociation energy of 320 $kJmol^{-1}$ (3.31 eV) and the H-P-H angle 93.6°. The distortion with respect to a planar triangular configuration is due to the presence of a lone electron pair in the P atom, which is indicated by the letter A in Table 4.1.

TABLE 4.1 Stereochemistry of Illustrative Phosphorus Compounds*

Coordination	**1**	**2**	**3**	**4**	**5**	**6**		**7**	**8**	**9**
Example compound	H–C≡P	$H_2C{=}PH$ H-P=Si	PH_3	P_4O_{10} PO_4^{3-}	PF_5 PCl_5	PF_6^-	Rh_4P_3	Ta_2P	Rh_2P	Fe_3P
Spatial structure	linear	bent	trigonal pyramid	tetrahedral	trigonal bipyramid	octahedral	trigonal prism	capped trigonal prism	cube	Triply capped trigonal prism
Schematic diagram			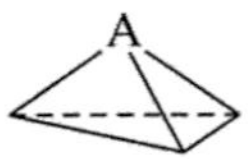	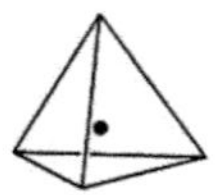	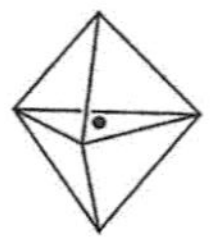	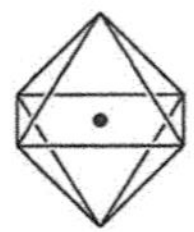	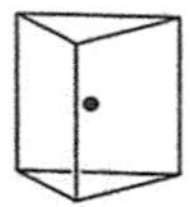	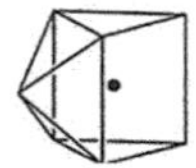	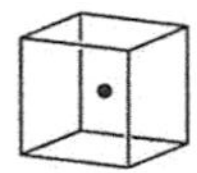	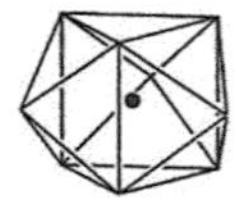

*Geometries of representative phosphorus compounds with coordination numbers ranging from 1 up to 9.

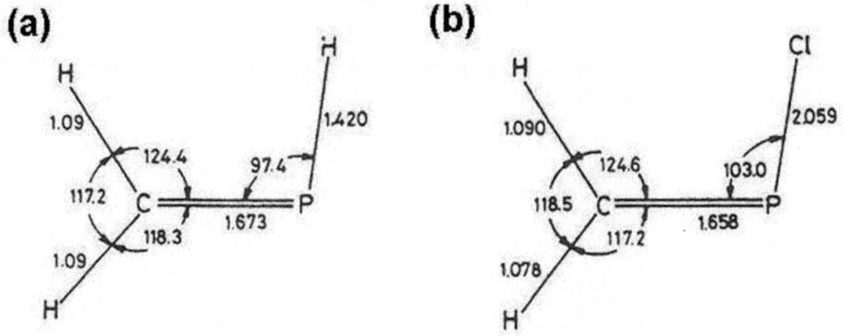

FIGURE 4.2 The structures of the first phosphaalkynes synthesized: (a) $H_2C = PH$ and (b) $H_2C = PCl$, as determined by microwave spectroscopy. Bond lengths are measured in Å. Reprinted with permission from Kroto, (1997), Copyright (1997) by the American Physical Society.

There are two characteristic phosphorus bearing molecules with coordination number 4, namely, the orthophosphate anion PO_4^{3-} and the phosphorus oxide P_4O_{10} (see Figure 4.10), which also share the phosphoryl PO_4 group structural unit, where the four oxygen atoms are arranged about the central phosphorus atom with tetrahedral geometry. The tetrahedral configuration of phosphorus atom in this unit is indicative of sp^3 hybridization, although there are a total of *five* bonds to phosphorus (four σ bonds and one delocalized π bond) in the phosphoryl group (see Figure 4.3b), which cannot be readily accounted for by simply attending to the P atom (Z = 15) electronic configuration $1s^22s^2p^63s^23p^3$, which suggests a trivalent atom due to the presence of three available *p* orbitals for bonding. The reason why phosphorus can break the 'octet rule' is that it is on the third row of the periodic table, and thus also has *d* orbitals available for bonding.

In the hybrid orbital picture for phosphate ion (PO_4^{3-}), a *single* 3*s* and three 3*p* orbitals combine to form four sp^3 hybrid orbitals with tetrahedral geometry (Figure 4.3a). In this way, four of the five valence electrons on phosphorus occupy sp^3 orbitals, and the fifth occupies an unhybridized *d* orbital. The phosphorus is thus able to form five bonding interactions in phosphorus oxides and phosphates, rather than three as it does in phosphine. The four sp^3 orbitals on phosphorus are each able to overlap with one sp^2 orbital on an oxygen atom (forming a tetrahedral structure), while a *delocalized* fifth bond (a π bond) is formed by side-by-side overlap of the *d* orbital on phosphorus with $2p$ orbitals on the oxygens, so that each P – O bond in the phosphoryl group can be considered as exhibiting some double bond resonant character, as is illustrated in Figure 4.3b (Soderberg, 2016).

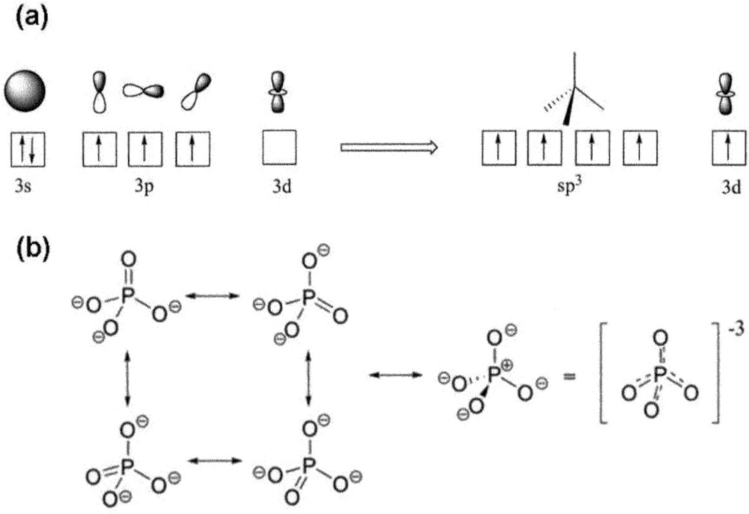

FIGURE 4.3 (a) Hybridization of 3*s* and 3*p* orbitals in the phosphoryl group. (b) Resonance of the P-O bond in the phosphoryl group. The –3 charge on a fully deprotonated phosphate ion is spread evenly over the four oxygens, and each phosphorus-oxygen bond can be considered to have some double-bond character.
Adapted from Soderberg, (2016); http://chem.libretexts.org/Textbook_Maps/Organic_Chemistry_Textbook_Maps/Map%3A_Organic_Chemistry_With_a_Biological_Emphasis_(Soderberg)/10%3A_Phosphoryl_transfer_reactions/10.1%3A_Overview_of_phosphates_and_phosphoryl_transfer_reactions.

Illustrative examples of phosphorus bearing molecules with coordination numbers 5 and 6 are respectively given by the phosphorus halides PF_5 and PF_6^-, respectively. Some time ago it was thought that higher than 5 valence values of phosphorus could be attained by invoking an explicit hybridization of high energy *d* orbitals with lower energy *s* and *p* ones. However, an intrinsic problem with sp^3d hybridization is that it would involve the energetically unfavorable $sp^3d \rightarrow sp^2d^1$ excitation, which requires promotion energy. In addition, *d* orbitals in non-metals are heavily shielded by the more penetrating *s* and *p* electrons. As a result, they have poor overlap with the orbitals of the neighboring atoms. In fact, an sp^3d hybridization, which would lend to the formation of five *equivalent* molecular orbitals, does not provide the best explanation to the fact that in most of the PB_5-type main group compounds the P-B axial bond distance is constantly significantly greater than the distance of the P-B equatorial bond. For example, in phosphorus pentahalide PF_5, the axial and equatorial lengths are respectively given by 1.58 Å and 1.52 Å. Similarly, for PCl_5, we have 2.14 Å and 2.02 Å, respectively (Sun, 2002a). Accordingly, the trigonal bipyramidal structure of the PF_5 and PCl_5 molecules are better accounted for by considering the central phosphorus atom adopts the energetically favorable sp^2 rather than sp^3d hybridization, with the unhybridized p_z orbital adopting the axial orientation, perpendicular to the sp^2 equatorial plane. Each of the three equatorial bonds is then formed by sp^2 $-p$ overlap and the linear axial bonds are formed by p-p_z-p overlap. Thus, in these molecules, the oxidation state (+5) in phosphorus is achieved by partial electron transfer from the phosphorus $3s3p$ subshells to the ligand p orbitals. This partial electron transfer occurs especially via the p_z orbital delocalization in the axial orientation, which is less electronegative than the equatorial sp^2 orbitals (Sun, 2002a).

The structure of the hexacoordinated PF_6^- molecular ion can be explained by invoking a completely analogous electron transfer, now involving unhybridized p orbitals of P atoms. In this way, the six fluorine ligands approach to the central phosphorus atom along the x, y, and z Cartesian axes adopting an octahedral geometry (see Table 4.1). Thus, the ligand p orbitals have effective overlap with all the orthogonal phosphorus $3p_x$, $3p_y$, and $3p_z$ orbitals in six orientations (Sun, 2002b).

PCl_5 and PF_6^- compounds nicely illustrate that many of the similarities and differences between the chemistry of carbon and its compounds and that of phosphorus and its compounds can be explained on the basis of the orbital theory of covalent bonding along with the valence-shell electron-pair repulsion theory (Müller, 2007). In fact, since carbon is a second-row element, it is restricted to use s and p atomic orbitals, according to the electronic configurations $1s^22s^2p^2$ (fundamental state, valence two) and $1s^22s^1p^3$ (excited state, valence four). Thus, carbon can have no more than four σ bonds, corresponding to the tetrahedral sp^3 hybridization of the atomic orbitals. In order to exhibit π bonding, the carbon atom must go to lower coordination numbers, such as the planar triangular sp^2-hybrid and the linear sp-hybrid. In comparison phosphorus, because of its lower tendency to form hybrid orbitals, favors arrangements with a lone pair over those with a triple bond (Lattanzi et al., 2010), as it occurs in the case of the HP = Si molecule.

Examples of phosphorus compounds exhibiting coordination numbers larger than six are found in metal phosphides representatives (see Table 4.1). Among them, we can highlight the case of schreibersite (see Section 4.3.2), a tetragonal crystal of composition $(Fe,Ni)_3P$, named after the Austrian scientist Carl von Schreibers (1775–1852), who was one of the first to describe this compound from iron meteorites.

4.1.2 *Chemical Reactivity*

The bond dissociation energy of P ≡ P molecule (5.03 eV) is about half that of N ≡ N molecule (9.79 eV). This energy difference accounts for the preference of elemental phosphorus for condensed forms rather than for the gaseous molecular form P_2, which is the most obvious distinction from N_2 elemental nitrogen. In fact, for nitrogen, the triple bond is preferred since it has more than 3 times the energy of a single N –N bond (3.06 eV), whereas for phosphorus the triple bond energy is significantly less than 3 times the single-bond P – P one (3.08 eV), and so compounds having three single bonds per P atom are more stable than those with a triple bond (Greenwood and Earnshaw, 1997). For the sake of information in Table 4.2, we list the dissociation energies and bond lengths of P-bearing diatomic molecules containing main bioelements.

TABLE 4.2 Properties of P-Bearing Diatomic Molecules*

Molecule	D (kJ mol^{-1})	D (eV)	d (Å)
P ≡ N	757.7	7.84	1.491
P = C	665.6	6.89	1.562
P = O	603.2	6.24	1.447
P ≡ P	490.0	5.07	1.895
P – P	297.6	3.08	2.209
P – S	438 ± 10 [(a)]	4.5 ± 0.1[(a)]	1.908 [(b)]
P – H	343.0	3.55	1.421

*Dissociation energies, D, and bond lengths, d, of P-bearing diatomic molecules as determined from experimental data. The P-H bond length value corresponds to phosphine and that of P – P bond to tetraphosphorus molecule P_4. Conversion factors: 1cal = 4.186 J; x kJ mol^{-1} = 10.35x meV. Energy values taken from Greenwood and Earnshaw, 1997; (a) Lodders, 2004. Bond lengths sources: Tables of Interatomic Distances and Configuration in Molecules and Ions, L. E. Sutton, ed., London: The Chemical Society, 1958; (b) Viana and Pimentel, 2007.

The multiple bonded molecules, P≡P, P≡N, and P=O, are stable only at elevated temperatures, above 1273 K. Nevertheless, diatomic phosphorus molecules which are unstable under ordinary terrestrial conditions, have been detected in different astrophysical objects, including circumstellar envelopes (CP, PN, PO), cold molecular clouds (PO), star-forming regions (PN, PO), and protoplanetary nebulae (PN).

Generally speaking, the compounds of phosphorus are considerably more reactive than similar compounds of carbon. The compounds based on triply connected phosphorus are quite reactive because the unshared pairs of electrons favor the formation of an activated complex. Silane and phosphine are generally more reactive than methane, and diphosphine $H_2P\text{-}PH_2$ is more reactive than ethane. The compounds in which the phosphorus is quadruply connected are less reactive in comparison with triply connected ones, but even so those phosphorus compounds exhibit reactivities of the same order of magnitude as those of the majority of biologically active organic compounds. The greater reactivity is readily explained in terms of the electronic structure. By converting π bonds into σ bonds, a number of activated complexes are available involving the *d* orbitals. Since these bonds are, at the same time, both energetic and labile, the resulting complexes lie in an intermediate position between the extremes of being highly stable or highly labile. Accordingly, in spite of the fact that the reactions can go essentially to completion, they proceed relatively slowly (Soderberg, 2016). These properties are very convenient for the energy storage and transfer processes in metabolic processes, hence explaining the predominant role of phosphates in biology.

Recently, computational studies revealed a new type of bonding known as phosphorus bonding (Joshi et al., 2017), which resembles hydrogen bonding interaction. In these studies, the H–P…N interaction was found to be twice more stable than the generally expected P–H…N interaction. Furthermore, the abilities of different donor atoms on the strength of the phosphorus bonding have been compared and revealed the following trend: P…N > P …O > P…S > P…π.

4.1.3 *Aromatic Phosphorus Compounds*

The concept of aromaticity was developed to express some aspects related to reactivity and chemical behavior of a certain class of molecules. These chemical properties are associated with certain physical properties of these molecules, like resonance energy and magnetic susceptibility. Moreover, as the connection between π electrons and aromatic properties has been progressively understood on theoretical bases, it has been recognized that aromaticity can be expected wherever conditions of stereochemistry, availability of orbitals, and a number of electrons allow for electron delocalization to occur in a cyclic system. From this point of view, the kind of atom participating in the delocalized electronic system is not important; the kind of orbital is.

The essential features for aromaticity, in a first approximation, are that the molecules can be formulated so that contiguous atoms each provide a *pπ* orbital, and that the total number of electrons occupying the orbitals in closed cycles is 4n + 2. The next stage of generalization is to admit not only different atoms, say P instead of C, but also different orbital types, for example, *dπ* rather than *pπ*. According to their symmetry properties, $3d_{xz}$ orbitals in P are the analogs to 2*pπ* carbon orbitals, and they lead to benzenoid aromaticity.

For the sake of illustration, on the left column of Figure 4.4 the bonding energy of the aromatic cyclohexaphosphabenzene is compared to those of other moieties containing six P atoms each (taking the P_4 molecule energy as a reference). Although this cyclic molecule is 0.2 eV more stable than three separate P_2 molecules, this benzene-like planar hexagon is the least stable P_6 isomer in the plot and distorts spontaneously to nonplanar, less symmetrical 6-membered rings. Its aromaticity is also less pronounced than in benzene. In fact, in the 2*pπ* aromatic compounds, the orientation of the 2*pπ* orbital is always normal to the plane of the ring. On the contrary, a 3*d* orbital of phosphorus is orientated only rather weakly by the tetrahedral field of the σ bonds. Consequently, the 3*dπ* orbital can continue to overlap its neighboring orbitals substantially in a puckered ring system only (Craig, 1959).

Other examples of aromatic P-bearing compounds include structures with alternating *p* and *d* orbitals, like phosphonitrilic halides of the general form $(PNX_2)_n$, where X stands for the halide atom, which can adopt a cyclic hexagonal geometry analogous to that of benzene (Craig, 1958). Their high stability, low reactivity (as compared to other acid chlorides) and the maintenance of the ring structure in a variety of derivatives are all compatible with aromatic character. Moreover, the P–N bond lengths are all equal (d = 1.61 ± 0.04 Å). These compounds may even present fused ring systems, completely analogous to the widespread polycyclic aromatic hydrocarbon (PAH) molecules. Consequently, aromatic phosphorus compounds may represent a promising reservoir of phosphorus, exhibiting stability comparable to that of organic PAHs, which will allow them to survive the hostile conditions imposed by intense UV ambient radiation in the diffuse ISM.

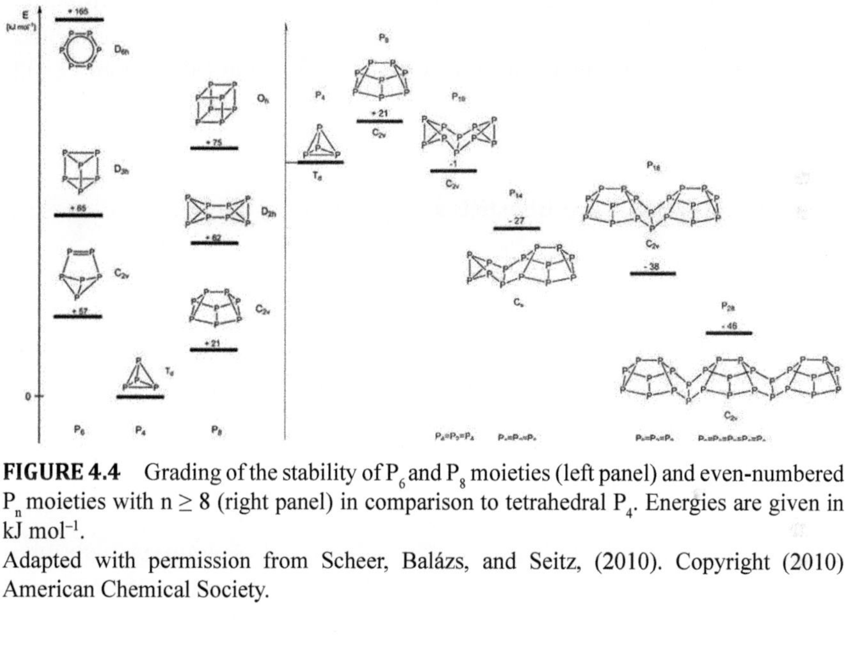

FIGURE 4.4 Grading of the stability of P_6 and P_8 moieties (left panel) and even-numbered P_n moieties with $n \geq 8$ (right panel) in comparison to tetrahedral P_4. Energies are given in kJ mol^{-1}.

Adapted with permission from Scheer, Balázs, and Seitz, (2010). Copyright (2010) American Chemical Society.

4.1.4 Polyphosphorus Species

In Figure 4.4, the bonding energies of several polyphosphorus compounds containing from 6 up to 28 phosphorus atoms are compared to the P_4 tetrahedral molecule bonding energy. Tetrahedral P_4 molecule is 2.08 eV more stable than P_2 molecule, which is the most unstable moiety among free P_n units. As we see, the cubic octahedral P_8, which was longly thought to be an accessible and interesting species, is more unstable than P_4 by 0.78 eV, albeit once formed its dissociation into P_4 is forbidden. Calculations for other stable P_8 molecules reveal that the cuneane structure (pictured at the bottom of the second left column in Figure 4.4) is the most stable species (Scheer et al., 2010). From P_{10} up to P_{28} cages (Figure 4.4 right panel), there is a change in the stability as compared to the tetrahedral P_4's. The predominantly favored P_n aggregates are structurally more related to motives present in red -V phosphorus, which will be discussed in the next section. So, for those P_8 cuneane-like motifs which are connected by P_2 molecules, the formed cages are energetically more favored in comparison to tetrahedral P_4. On the other hand, icosahedral P_{80}, P_{180}, P_{320}, P_{500}, and P_{720} ring-shaped polyphosphorus compounds, made of 5- and 6-membered rings have been theoretically shown to be more stable than P_4 molecule, and only the pentagonal icosahedron P_{20} is less stable (Karttunen et al., 2007).

4.2 ELEMENTAL PHOSPHORUS

In the solid phase, elemental phosphorus exists in a wide variety of structural arrangements (allotropes), which are typically divided into three main classes, namely, white, red, and black phosphorus, since they considerably differ in their physical and chemical properties.

4.2.1 *White Phosphorus*

The reference state of solid phosphorus is the so-called white phosphorus. We currently know three modifications of white phosphorus, respectively labeled as α-, β-, and $\gamma - P_4$, since they consist in tetrahedral P_4 units arranged in different ways (Figure 4.5). The α-white form is usually formed by condensation from the gaseous or liquid states. Earlier X-ray diffraction studies by Natta and Passerini (1930) suggested a body-centered-cubic (bcc) structure with a lattice parameter a = 7.17 Å, containing four molecules of P_4 per unit cell. Subsequent studies indicated that it actually consisted of P_4 tetrahedra arranged in the body-centered cubic α-Mn structure (A12), which contains 58 Mn atoms in its cubic unit cell, with a lattice parameter of 18.8 Å (Corbridge and Lowe, 1952). Thus, the room temperature crystal structure of white phosphorus consists of a loosely aggregation of P_4 molecules dynamically rotating around their centers of gravity (Figure 4.5a), so that they are randomly orientated (Simon et al., 1987, 1997).

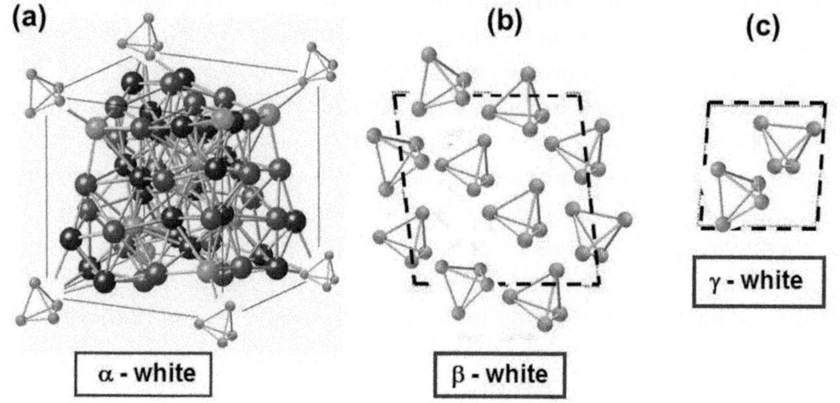

FIGURE 4.5 The white phosphorus molecular crystal structures are based on P_4 tetrahedral molecules arranged in (a) a cubic structure (α – white form), (b) a triclinic structure (β – white form), and (c) a monoclinic structure (γ – white form). The corresponding crystallographic data are listed in Table 4.3. The α – white form is obtained by replacing the Mn atoms in the α-Mn structure by P_4 tetrahedral molecules, as it is schematically shown in (a).

White phosphorus is sensitive to visible light and oxidizes spontaneously in air. Due to its easy oxidation α-white phosphorus is actually cream color (Figure 4.1c), so it is also known as yellow phosphorus. It exhibits a waxy appearance, has a very high vapor pressure (0.181 mm Hg at room temperature) and burns upon contact with air (its spontaneous ignition temperature in air is ~35°C, so that it is highly flammable). It is highly toxic as well, and ingestion (the fatal dose is about 50 mg), inhalation, or even contact with the skin must be avoided.

By decreasing the temperature below T = –76.9°C at ambient pressure (or under higher than 1.0 GPa pressure at room temperature), white phosphorus adopts a triclinic crystal structure called $\beta - P_4$ form with 24 atoms in the unit cell (Figure 4.5b). The $\beta - P_4$ form can be maintained as a solid up to 64.4°C under high pressure (11,600 atm), whereas the α form melts at 44.1°C to give a liquid which consists of symmetrical P_4 molecules. The liquid boils at 280.5°C, and the same P_4 molecule is found in the gas phase, which is the predominant form below 1246°C. Above this temperature, the presence of the diatomic molecule P ≡ P becomes noticeable and increasingly dominant above 1806°C. In fact, at atmospheric pressure, dissociation of P_4 into 2 P_2 reaches 50% at ~1800°C, and dissociation of P_2 into two P atoms reaches 50% at ~2800°C. Triatomic P_3 can be also present, although its thermodynamic stability relative to those of P_2 and P_4 moieties is negligible (Schlesinger, 2002).

Finally, the existence of yet another low-temperature white phosphorus modification was reported in 1974 and referred to as the $\gamma - P_4$ form. This modification adopts a monoclinic structure, which forms when α-P_4 is quenched to –165°C and kept at that temperature. $\gamma - P_4$ transforms into the β form when warmed to –115°C, without a chance to recover $\gamma - P_4$ by cooling the sample again (Simon et al., 1997). The asymmetric unit of $\gamma - P_4$ contains eight P atoms in the primitive cell (Figure 4.5c) arranged in such a way that two P atoms in a molecule are related by a mirror plane which bisects the cell. The centers of gravity of these P_4 molecules show a distorted body-centered cubic arrangement and the four apices of the P_4 tetrahedron point to the largest possible voids formed by neighbor molecules (Okudera et al., 2005).

4.2.2 *Red Phosphorus*

Red phosphorus is the most common available variant of solid elemental phosphorus. It is stable in air and much less poisonous than white phosphorus, and it can be obtained by different procedures (Figure 4.6), including heating white phosphorus, exposing it to visible light, by crystallization from the liquid phase, high-temperature condensation of the vapor phase, or precipitation from supersaturated solution in molten lead (Schlesinger, 2002). It is also possible to induce the photochemical conversion of the white to the red forms by irradiation with UV photons. The diatomic molecule P_2 is a key intermediary in this process, which is initially generated and subsequently polymerizes (Cummins, 2014). When prepared by the first three methods, it initially adopts an amorphous structure designated red – I, which is thermodynamically and chemically more stable than white phosphorus (even in air).

Chronologically, this amorphous form was first obtained in 1848 by heating white phosphorus above 250°C out of contact with air for several days. Its structure is considered to be a cross-linked polymeric network of different building units based on tubular structures of eight- and nine-membered P rings (respectively labeled as P_8 and P_9 in Figures 4.6 and 4.7a, b) connected to each other by P_2 dimers. All P atoms in the network are three-fold coordinated with an average bond length of 2.219 Å and an average bond angle of 100.9°. While this network exhibits only short or intermediate range order, further heating or irradiation leads to various crystalline structures belonging to the red phosphorus family (Figure 4.6).

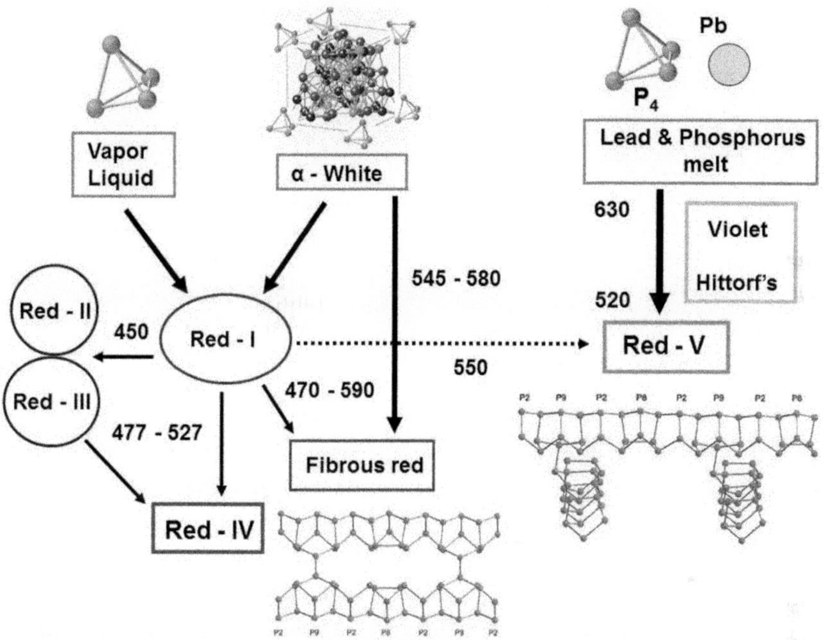

FIGURE 4.6 Possible routes to obtain the different members of the red phosphorus family. The numbers indicate temperature values in Celsius degrees.

Thus, heating red – I phosphorus to 450°C gives rise to two crystalline forms of red phosphorus designated red – II and III, respectively. Their crystalline structures are still unknown, since their pure forms have not yet been obtained. Both forms are highly polymeric and contain three-dimensional networks formed by breaking one P-P bond in each P_4 molecule and then linking the remaining P_4 units into chains or rings. Holding red – I, II or III phosphorus at higher temperatures (~477–527°C) generates the red – IV one (Figure 4.7c). The structure of this phase seems to be monoclinic, although different investigators have claimed it to be hexagonal, tetragonal, monoclinic, or even triclinic (see Table 4.3). On the other hand, red – I phosphorus held at still higher temperatures (~550°C) renders the so-called red – V form. The structure of this crystalline type of phosphorus, determined in 1969 by Thurn and Krebs, was identified with a previously known form: the so-called violet or Hittorf's phosphorus (Figure 4.7d).

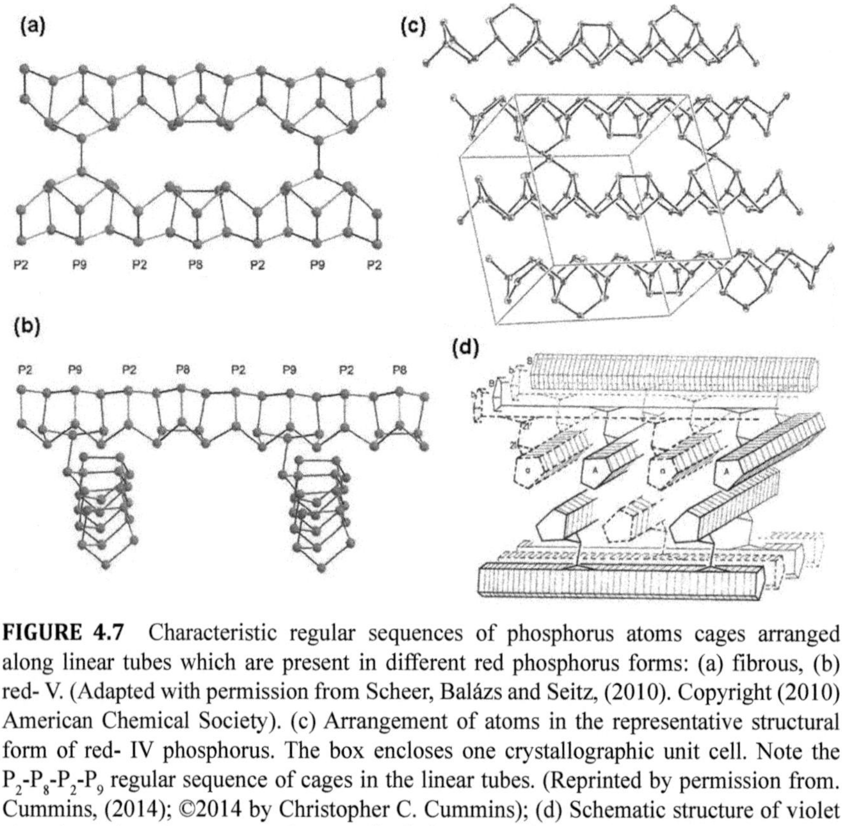

FIGURE 4.7 Characteristic regular sequences of phosphorus atoms cages arranged along linear tubes which are present in different red phosphorus forms: (a) fibrous, (b) red- V. (Adapted with permission from Scheer, Balázs and Seitz, (2010). Copyright (2010) American Chemical Society). (c) Arrangement of atoms in the representative structural form of red- IV phosphorus. The box encloses one crystallographic unit cell. Note the P_2-P_8-P_2-P_9 regular sequence of cages in the linear tubes. (Reprinted by permission from. Cummins, (2014); ©2014 by Christopher C. Cummins); (d) Schematic structure of violet phosphorus based on arrays of perpendicularly arranged pentagonal cross-section tubes. (Reprinted with permission of the International Union of Crystallography from Thurn and Krebs, (1969)).

In fact, early in 1865 Hittorf had found that phosphorus could be crystallized by slowly cooling (from 630 to 520°C) of a lead-phosphorus melt, giving red-to-violet colored platelets. This form crystallizes in a monoclinic space group with 84 atoms per unit cell. Its structure is relatively complex and consists of P_8 and P_9 groups linked alternately by pairs of P atoms to form tubes of the pentagonal cross-section with a repeating unit of P_2-P_8-P_2-P_9 sequence of cages (Figure 4.7b). These tubes are stacked (without direct covalent bonding) to form sheets and are linked by P-P bonds to similar chains which lie at right angles (Figure 4.7d).

Finally, by heating amorphous red phosphorus at 470–590°C one can obtain the so-called fibrous red phosphorus, whose atomic structure exactly coincides with that of the red – IV form, as shown in Figures 4.7a-c (Ruck et al., 2005). Thus, both violet (red – V) and fibrous (red – IV) allotropes consist of tubes with pentagonal cross-sections, arranged either perpendicular (violet) or parallel (fibrous red) to each other. Due to its polymeric structure, red phosphorus forms have a much higher melting point (600°C in the case of amorphous one) than the white forms (44°C and 65°C for α-P_4 and $\beta - P_4$ forms, respectively), and they are much less reactive as well. Regardless of the adopted crystalline structure red phosphorus always sublimes to P_4 vapor when properly heated, the sublimation temperature accordingly depending on the actual structural type (Schlesinger, 2002).

4.2.3 Black Phosphorus

Black phosphorus was originally made by P. W. Bridgman in 1914 by heating the white form to 200°C under a pressure of 12,000 atm. The black phosphorus crystal structure thus obtained is orthorhombic (with eight atoms per unit cell) at atmospheric pressure (Figure 4.8). By further increasing pressure one first obtains a rhombohedral phase (stable in the 5.5 to 10 GPa pressure range) and then a simple cubic form (stable up to 32 GPa, Kikegawa, and Iwasaki, 1983). While the rhombohedral structure is common to the group 15 elements, the simple cubic one is the less common among elements, and only polonium is known to crystallize in this structure at low temperatures. The orthorhombic form has a layered structure which is based on a puckered hexagonal net of triply coordinated P atoms with two interatomic angles of 103.69° and one of 98.15°. The atoms are arranged in double layers as indicated in Figure 4.8a with three nearest neighbors at distances of 2.164 Å and 2.207 Å, respectively, within the layers and four next-nearest neighbors in adjacent layers 3.592 Å and 3.801 Å away.

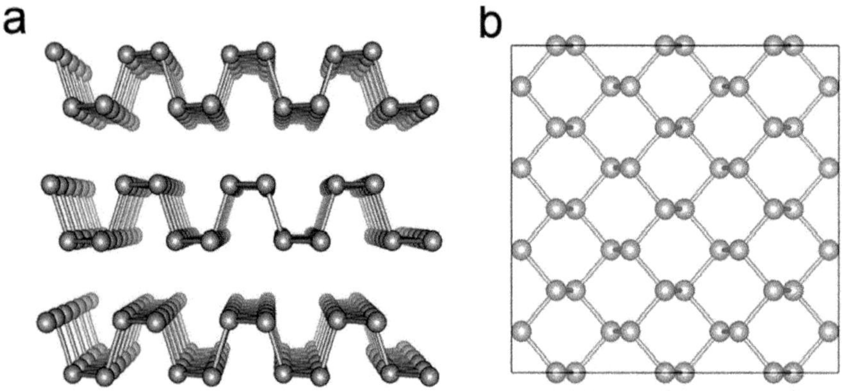

FIGURE 4.8 Black phosphorus displays a layered structure of sheets with the phosphorus atoms ordered in a puckered honeycomb lattice. Adjacent layers interact by weak van der Waals forces and follow an ABA stacking order. (a) Lateral view. (b) Top view.
Source: http://pubs.rsc.org/services/images/RSCpubs.ePlatform.Service.FreeContent.ImageService.svc/ImageService/Articleimage/2015/CP/c4cp03890h/c4cp03890h-f1_hi-res.gif.

Although black phosphorus only forms under special conditions, it is regarded to be the most thermodynamically stable form of the element under normal conditions and, at about 550°C, it transforms to the violet (red V) phosphorus form. A closely related allotrope of black phosphorus is phosphorene, which can be considered as a single layer of black phosphorus, just as graphene is a single layer of graphite (see Section 9.2.1). In fact, phosphorene was first isolated in 2014 by mechanical exfoliation of a black phosphorus crystal (Liu et al., 2014).

In summary, despite significant structural differences, all phosphorus allotropes consist of covalently bond molecular building blocks, based upon P-P single bonds, three for every phosphorus atom, which are held together by weak van der Waals interactions. Indeed, the propensity of phosphorus atoms to catenate into chains, rings, and clusters is thus a quite remarkable feature of most of its compounds. Keeping into account the different structural modifications reported within the three main allotropes families, at least nine crystalline polymorphs of elemental phosphorus, belonging to all possible crystalline symmetry classes, are known, in addition to several amorphous forms as well. The main structural data of the most important ones are listed in Table 4.3, which properly illustrates the rich structural diversity of the elemental phosphorus condensed phases.

TABLE 4.3 Allotropic Forms of Phosphorus*

Name	Crystal Symmetry	a, b, c (Å)	α, β, γ (°)	T_m (°C)	ρ (g cm^{-3})
$\alpha - P_4$	bcc cubic (I43m)	18.8	90	44.1	1.828 (2)
$\beta - P_4$	Triclinic (P1) (1)	5.4788	94.285		1.88
		10.7862	92.695	64.4	
		10.9616	100.680		
$\gamma - P_4$	Monoclinic (C2/m) (4)	$a = 9.1709$	90		
		$b = 8.3385$	$\beta = 90.311$	—	1.981(4)
		$c = 5.4336$	90		
Red (I)	Amorphous	—	—		2.16 (2)
Red (IV)	cubic		90		
	tetragonal		90		2.34
	monoclinic (5)	$a = 9.21$	90		
		$b = 9.15$	$\beta = 106.1$	600	
		$c = 22.60$	90		
Red (V)	Monoclinic (3)	9.21	90		2.35(2)
Violet	(P2/c)	9.15	106.1		
(Hittorf)		22.60	90		
Black	Orthorrombic (4)	3.31	90	610	2.69 (2)
	(Cmca)	10.23	90		
		4.38	90		
	Rhombohedral				3.56 (2)
	simple cubic				3.88 (2)

*Crystal system, unit cell parameters, melting temperature and density of different elemental phosphorus modifications.

Source: (1) Simon et al., 1997; (2) Goodman et al., 1983; (3) Thurn and Krebs, 1969; (4) Okudera et al., 2005; (5) Oxtoby et al., 2015.

4.3 REDUCED PHOSPHORUS COMPOUNDS

4.3.1 Phosphine

Phosphine, PH_3, has been known since the birth of modern chemistry, and its discovery was credited to Philippe Gengembre (1764–1838), a student of Lavoisier, who first obtained phosphine in 1783 by heating phosphorus in an aqueous solution of potassium carbonate. It is a highly reactive, colorless gas with a faint garlic odor at concentrations above about 2 ppm by volume. Phosphine is extremely poisonous, approximately two times more toxic to humans than HCN, and has a short-term exposure limit of 1 mg m^{-3} (WHO, 1988). Phosphine is the most stable hydride of phosphorus, intermediate in thermal stability between NH_3 and AsH_3. It is the first representative of a homologous open-chain series P_nH_{n+2} whose members rapidly diminish in thermal stability, though pure P_2H_4 (diphosphine) and P_3H_5 (triphosphane) have been isolated. Phosphine melting and boiling points are –133.5° and –97.7°C, respectively. It is rather insoluble in water at atmospheric pressure, but is more soluble in organic liquids. Aqueous solutions are neutral, and there is little tendency for PH_3 molecule to protonate or deprotonate. Phosphine is instead a strong reducing agent: many metal salts are reduced to the metal form. It is currently known that phosphides are introduced to the environment from degradation of corroding metals, such as iron, from the anaerobic processes in biosphere, from the combustion of coal in power plants, from burning of gas, and from their use as fumigants (Morton and Edwards, 2005).

4.3.2 *Schreibersite*

Schreibersite has the general composition $(Fe,Ni,Co)_3P$, with $Fe > Ni > Co$. There is a reasonably wide range of melt compositions which give a pure phosphide. This is not surprising, since both Fe_3P and Ni_3P decompose on melting. Accordingly, the solid solution is generally known by its mineral name schreibersite, and it is given the general chemical equation $(Fe,Ni)_3P$. The oxidation state of phosphorus is approximately –1, based on binding energies obtained from x-ray photoelectron spectroscopy studies (Pirim et al., 2014). This is effectively equivalent to phosphorus forming a metallic alloy with Fe and Ni. The iron-nickel phosphide solid solution Fe_3P-Ni_3P is a common component of many iron-rich meteorites and is also found as an accessory mineral in many silicate-rich meteorites (see Section 7.3.3). Schreibersite occurs naturally in one location on earth: Disko Island off the coast of Greenland, inside a basaltic intrusion into a coal seam, which creates a highly enough reduced environment.

In molten iron containing carbon and phosphorus, one can observe metastable mixtures of iron plus iron carbide and iron phosphide or as stable mixtures of iron plus graphite and iron phosphide. On slow cooling, some of the C of the iron carbide diffuses out and forms a fringe of pearlite around the region of the mixture, but in diffusing into the surrounding solid solution of P in Fe, the C precipitates some Fe_3P, which appears as laminae in juxtaposition with the pearlite. Thus, fringes consisting of Fe saturated with P and C, iron phosphide and iron carbide are formed.

The crystal structure of schreibersite is identical to those of both synthetic Fe_3P and Ni_3P alloys, all of them belonging to the tetragonal (I-4) space group, which is non-centrosymmetric and chiral, so that there exist two different atomic arrangements dubbed normal (or regular) and inverse, respectively (Figure 4.9).

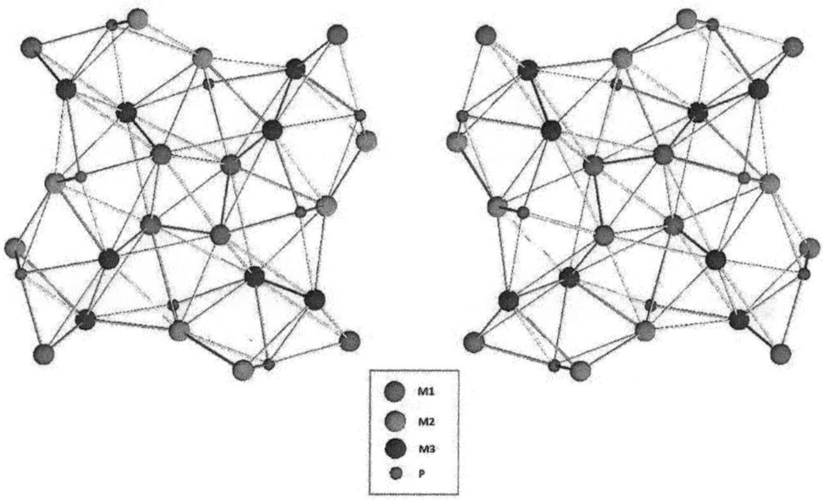

FIGURE 4.9 (See color insert.) The regular (on the left) and inverse (on the right) enanthiomorphic crystal structures of schreibersite.
By courtesy of Cameron Pritekel, (2015).

4.4 PHOSPHORUS OXIDES

Elemental phosphorus is especially prone to oxidation and species formed in this process have attracted scientific interest since the discovery by Boyle in 1680 of a green-white luminescence which occurs when white phosphorus is oxidized. In fact, oxidation is the most important reaction of elemental phosphorus, mainly producing the cage-like molecules phosphorus pentoxide P_4O_{10}, plus some amount of phosphorus trioxide, P_4O_6, according to the reaction sequence shown in Figure 4.10.

$$P_4 + O_2 \rightarrow P_4O + O,$$

$$P_4O + O_2 \rightarrow P_4O_2 + O, \ldots$$

$$P_4O_5 + O_2 \rightarrow P_4O_6 + O,$$

$$P_4O_6 + O_2 \rightarrow \ldots \rightarrow P_4O_{10}$$

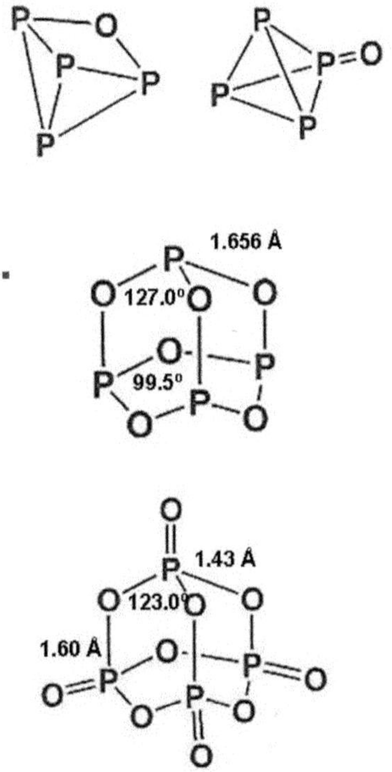

FIGURE 4.10 Sequence of phosphorus oxides. The structures of P_4O_6 and P_4O_{10} molecules exhibit the tetrahedral coordination characteristic of the most stable phosphorus compounds.

The oxidation of P_4 molecules provides a nice example of the link between chemistry and geometry: starting from the tetrahedral P_4 scaffold the subsequent oxides are obtained by either inserting an oxygen atom in the P-P edges or at the P vertices, as it is illustrated in the molecular structures depicted in the upper right corner of Figure 4.10. Accordingly, the tetrahedral P_4O_6 structure is determined by inserting six oxygen atoms in the available P-P edges present in the original P_4 tetrahedral molecule, whereas the P_4O_{10} structure is obtained from the previous P_4O_6 one by attaching an oxygen atom at every P atom occupied vertex.

Although P_4O_{10} does not occur naturally on earth, the oxidation cycle described in Figure 4.10 may take place in the expanded atmospheres of oxygen-rich evolved stars, and P_4O_6 is predicted to occur in both protostars and brown dwarf stars (see Figure 6.5). Once formed, P_4O_{10} molecules can cluster among them to develop extended networks, which may ultimately grow to form large grains, as sketched in Figure 4.11. Indeed, the structure of the thermodynamically most stable condensed form of phosphorus pentoxide (O'- P_2O_5) consists of infinite layers built from six-membered rings of three corner linked PO_4 tetrahedra (Stachel, 1995, see Figure 4.12a). These sheets stack together defining an orthorhombic unit cell (Pnma) with cell parameters a = 9.193 Å, b = 4.890 Å, and c = 7.162 Å (Figure 4.12b). On the other hand, the structure of vitreous P_2O_5 is thought to consist of a three-dimensional network of corner-sharing PO_4 tetrahedra, each one decorated with a non-bridging P=O bond (Suzuya, 1998).

Since phosphorus pentoxide is extremely hygroscopic, the absorption of water molecules may easily lead to the formation of phosphoric acid in small grains of phosphorus oxide. This opens an interesting route to the possible formation of phosphates in the frozen mantles covering dust particles in the ISM (see Sections 6.3.1 and 10.2.3).

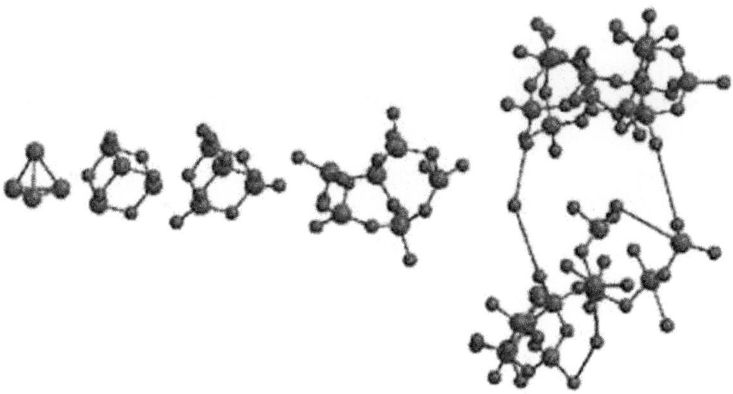

FIGURE 4.11 Sketch illustrating the oxidation sequence of phosphorus which might occur in the extended atmospheres of evolved, oxygen-rich stars. The sequence starts with the oxides P_4O_6 and P_4O_{10}, whose structures exhibit a characteristic tetrahedral symmetry. The formation of P_4O_{10} clusters paves the way to the formation of condensed P_2O_5 sheets, based on linked PO_4 tetrahedra. Depending on the environmental physical conditions, these cage-like structures will ultimately crystallize or will acquire a glassy structure instead. Updated from Maciá, (2005); with permission from the Royal Society of Chemistry.

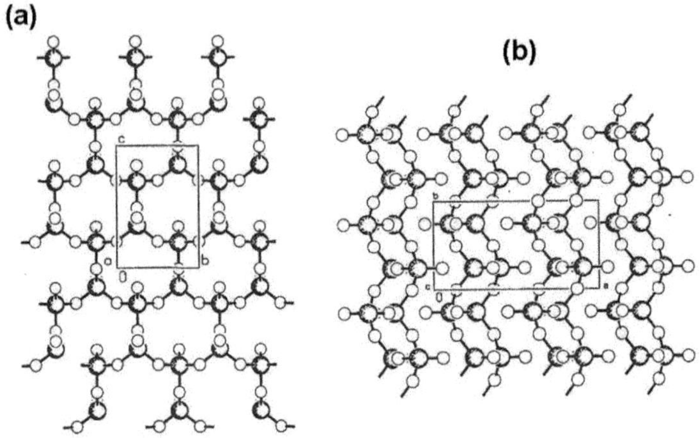

FIGURE 4.12 (a) The structure of a sheet of phosphorus pentoxide. (b) The stacking of the sheets in phosphorus pentoxide.
Reprinted with permission of the International Union of Crystallography from Stachel, (1995).

The phosphorus analogs of nitrogen oxides, namely, the lower oxides of phosphorus P_2O, PO, and PO_2 are observed only under carefully controlled experimental conditions such as in matrices, in vacuum or molecular beams and cannot be considered as chemical reagents in the normal sense of the term. Of all these lower oxides, PO deserves a special attention, since it has been detected in several astrophysical objects. Furthermore, under the ISM conditions, lower phosphorus oxides could combine with relatively abundant transition metal atoms to form condensed phases. The bonding of PO molecules to transition metal atoms is interesting for several reasons: (i) PO is the heavy congener of NO and analogies between the bonding of these two diatomic molecules to transition metals would be expected, (ii) differences in metal-PO and metal-NO bonding might arise as a result of the availability of 3*d* orbitals in PO which are absent in NO and the possibility of higher coordination numbers at phosphorus in complexes of PO (Sterenberg, 2002).

4.5 PHOSPHORIC ACIDS AND PHOSPHATES

Phosphorus oxoacids are more numerous than any other element's, and probably only silicon exceeds its number of oxoanions and oxo-salts.

Many of them are technologically important (Chapter 9), and some of their derivatives are crucial in many biological processes as well (Chapter 8). Fortunately, this extensive array of compounds share relatively simple structural principles that can be summarized as follows: 1) all P atoms in the oxoacids and oxoanions are 4-coordinated and contain at least one P = O bond; 2) all P atoms in the oxoacids have at least one P-OH bond, which is ionizable as proton donor; 3) catenation is either by P-O-P links or via direct P-P bonds. In the former only corner-sharing of tetrahedra occurs. These structural principles are illustrated in Figures 4.13–4.15. Although each P atom is penta-covalent, the oxidation state of P is +5 only when it is directly bound to four oxygen atoms; the oxidation state is reduced by 1 whenever a P-OH bond is replaced by a P-P one, and by 2 each time a P = O bond is replaced by a P-H one (Muller and Greenwood, 1997).

4.5.1 *Orthophosphoric Acid*

The chemical structure of the orthophosphoric acid molecule, H_3PO_4, is shown in Figure 4.13a. As it was illustrated in Figure 4.3b, H_3PO_4 should be properly regarded in terms of a resonating system in which the double bond is delocalized. This acid can be obtained pure in the crystalline state (Figure 4.13b), with each $PO(OH)_3$ tetrahedral molecule linked to six others by hydrogen bonds of two lengths, 2.53 and 2.84 Å, respectively (Figure 4.13a). The shorter bonds link OH and P = O groups whereas the longer hydrogen bonds involve two OH groups on adjacent molecules. Addition of an appropriate amount of water to anhydrous H_3PO_4 yields the so-called hemihydrate compound. The unit cell parameters of both the anhydrous and the hemihydrate orthophosphoric acid forms are listed in Table 4.4.

Due to extensive H bonding persistence on fusion, orthophosphoric acid behaves as a viscous syrupy liquid that readily supercools. At 45°C the viscosity is 76.5 cP, and it increases to 177.7 cP at 25°C (for the sake of comparison the viscosities of water and glycerine at 20°C are 1 cP and 1499 cP, respectively). Albeit this high viscosity, fused orthophosphoric acid conducts electricity extremely well because of the efficient proton transfer via rapid proton-switch among phosphoryl groups. In dilute aqueous solutions, H_3PO_4 behaves as a strong acid although only one of the hydrogens is readily ionizable, the second and third ionization constants decreasing successively by factors of $\sim 10^5$.

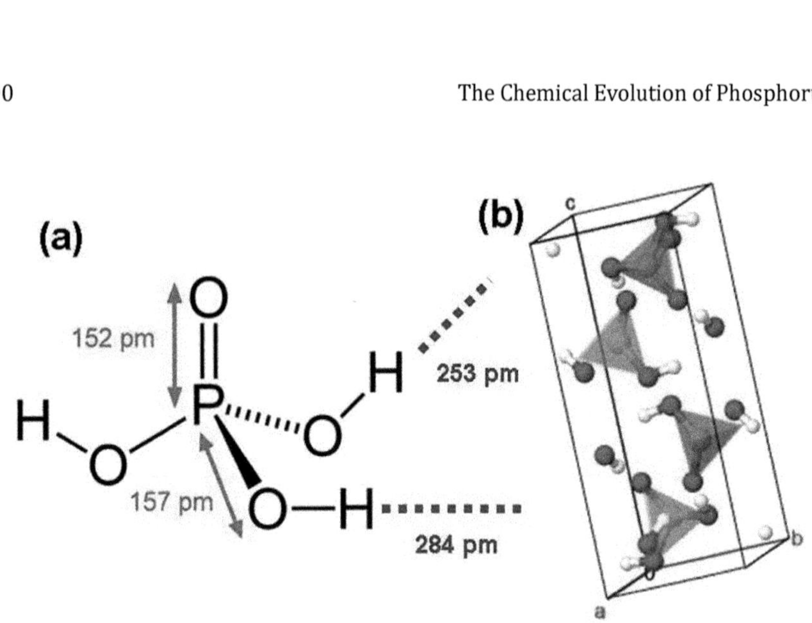

FIGURE 4.13 **(See color insert.)** (a) Molecular structure of the orthophosphoric acid molecule H_3PO_4 exhibiting tetrahedral symmetry; (b) crystal structure of the anhydrous orthophosphoric acid crystal containing four tetrahedral H_3PO_4 molecules in the unit cell (Souhassou et al., 1995). The P, O, and H atoms are shown in orange, red, and white, respectively. The 3D unit cell view has been generated employing the *Jmol* online service provided by IUCr journals.

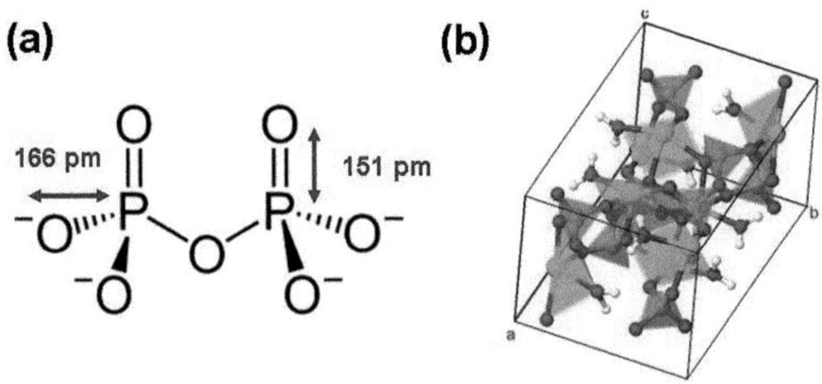

FIGURE 4.14 **(See color insert.)** (a) Molecular structure of the pyrophosphoric acid molecular ion $P_2O_7{}^{4-}$ (angle and bonds obtained from Jmol); (b) crystal structure of the calcium pyrophosphate dihydrate ($Ca_2P_2O_7{\cdot}2H_2O$) containing four pyrophosphate molecules (in orange) in the unit cell (Gras et al., 2016). The Ca, P, O, and H atoms are shown in green, orange, red, and white, respectively. The 3D unit cell view has been generated employing the *Jmol* online service provided by IUCr journals.

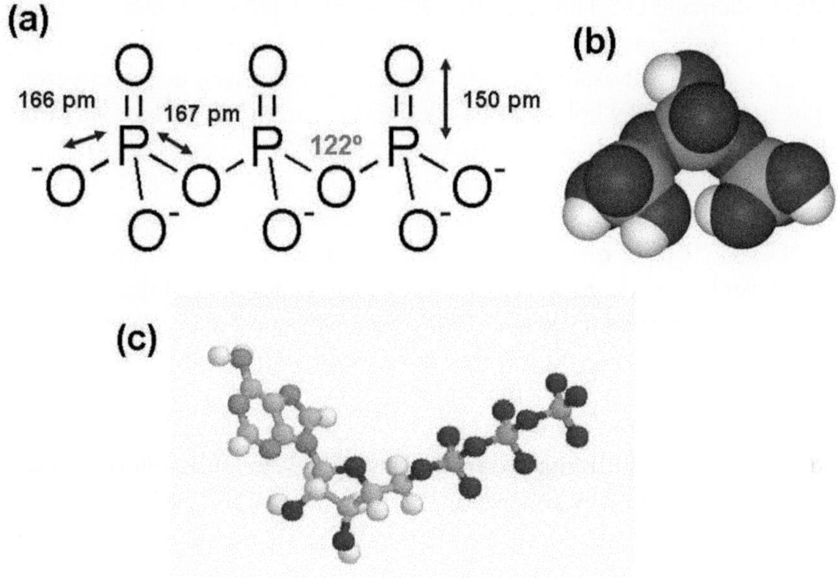

FIGURE 4.15 Molecular structures of (a) the tripolyphosphate ion $P_3O_{10}^{5-}$; (b) the tripolyphosphoric acid; (c) the adenosine triphosphate nucleotide (ATP).

TABLE 4.4 Crystal Structure of Orthophosphoric Acid Crystals*

Name	Crystal Symmetry	a, b, c (Å)	α, β, γ (°)	T_m (°C)	ρ (g cm^{-3})
Anhydrous	Monoclinic ($P2_1/c$) (1)	5.7572 (13)	90	42.35°	2.030 (1)
H_3PO_4		4.8310 (17)	95.274(12)		
		11.5743 (21)	90		
Hemihydrate	Monoclinic ($P2_1/a$) (2)	7.922	90	29.3°	1.99 (2)
$H_3PO_4{\cdot}\frac{1}{2}H_2O$		12.987	109.9		
		7.470	90		

*Crystal system, unit cell parameters, melting temperature and density of different orthophosphoric acid crystals.

Source: (1) Souhassou et al., 1995; (2) Mighell et al., 1969.

4.5.2 *Pyrophosphoric Acid and Polyphosphates*

Pyrophosphoric acid, also known as diphosphoric acid (Figure 4.14a), is colorless, odorless, and extremely hygroscopic. It is soluble in water, diethyl ether, and ethyl alcohol. In 1827, a "Mr. Clarke of Glasgow" gave the name pyrophosphoric acid and was credited with its discovery by heating a sodium phosphate salt to red heat, which was readily converted to phosphoric acid by hot water (Beck, 1834). Currently, it is best prepared by ion exchange from sodium pyrophosphate or by reacting hydrogen sulfide with lead pyrophosphate.

The anhydrous acid has two different crystalline forms, which melt at 54.3°C and 71.5°C, respectively. An equilibrium mixture of phosphoric acid, pyrophosphoric acid, and polyphosphoric acids is rapidly formed by melting pyrophosphoric acid. The pyrophosphoric acid percentage by weight is around 40%, and it is difficult to recrystallize from the melt.

Pyrophosphoric acid is a medium-strong inorganic acid, which is corrosive, but otherwise nontoxic. It hydrolyzes in aqueous solution, and eventually, equilibrium is established between phosphoric, pyrophosphoric, and polyphosphoric acids. When highly diluted an aqueous solution of pyrophosphoric acid contains only phosphoric acid, according to the reaction $H_4P_2O_7 + H_2O \rightarrow 2H_3PO_4$.

Pyrophosphates are anions, salts, and esters of pyrophosphoric acid (Khaoulaf et al., 2019). A pyrophosphate mineral, named pyrophosphorite $Ca_9Mg(PO_4)_6(HPO_4)$ (analogous to whitlockite, Palache et al., 1951), was earlier reported by Shepard (1878). Canaphite ($CaNa_2P_2O_7.4H_2O$) was the first identified polyphosphate mineral (Peacor et al., 1985) and wooldridgeite ($CaNa_2Cu_2\ (P_2O_7)_2.10H_2O$) was discovered next (Hawthorne et al., 1999). Wooldridgeite has an unknown origin and has been found as a weathering crust on primary sulfide minerals. In addition to these four minerals, three other have been proposed, namely, arnhemite $(K,Na)_4Mg_2(P_2O_7)_2.5H_2O$, pyrocoproite $(K,Na)_2MgP_2O_7$, and the confusedly named pyrophosphate, $K_2CaP_2O_7$ (Forti et al., 2016).

To date, two different types of calcium pyrophosphate crystals have been identified in joint tissues of arthritis patients (see Section 9.3): monoclinic and triclinic calcium pyrophosphate dihydrate ($Ca_2P_2O_7{\cdot}2H_2O$) crystals (see Figure 4.14b and Table 4.5).

The tripolyphosphate sodium salt $Na_5P_3O_{10}$ was introduced in the mid-1940s as a building agent for synthetic detergents. The molecular structures of the tripolyphosphate ion $P_3O_{10}^{5-}$ and the tripolyphosphoric acid are shown in Figures 4.15a and 4.15b, respectively. Thus far, only four polyphosphate minerals have been recognized by the International Mineralogical Association. Kanonerovite ($MnNa_3P_3O_{10}.12H_2O$) was the first triphosphate mineral (Popova et al., 2002), followed by hylbrownite ($MgNa_3P_3O_{10}.12H_2O$), which is structurally similar (Elliot et al., 2013). Kanonerovite is associated with coarse-grained igneous rocks characterized by high concentrations of rare elements and the occurrence of unusual minerals. Altogether, these polyphosphate minerals likely comprise about 10 kg of material, suggesting few geologic sources of polyphosphates were present on the early earth in mineral form (Pasek and Kee, 2011).

TABLE 4.5 Crystal Structure of Pyrophosphate Crystals*

Chemical Formula	Crystal Symmetry	a, b, c (Å)	α, β, γ (°)	T_m (°C)	ρ (g cm^{-3})
$Ca_2P_2O_7{\cdot}2H_2O$	Monoclinic ($P2_1/n$) [(1)]	12.60842(4)	90	1353	3.69
		9.24278(4)	104.9916(3)		
		6.74885(2)	90		
	Triclinic (P1) [(2)]	7.365(4)	102.96(1)	1353	3.69
		8.287(4)	72.73(1)		
		6.691(4)	95.01(1)		

*Crystal system, unit cell parameters, melting temperature and density of different pyrophosphate crystals.

Source: (1) Gras et al., 2016; (2) Mandel, 1975.

Conversely, purine and pyrimidine nucleotide diphosphates and triphosphates (Figure 4.15c) are universally occurring in all known living forms on earth, and it has been suggested that polyphosphates may have played an important role in the prebiotic synthesis earlier stages (Baltscheffsky et al., 1999). In fact, linear polyphosphate chains obeying the general formula $(P_nO_{3n+1})^{(n+2)-}$, with n from less than 10 to over 5000 orthophosphate units, have been found in all organisms ranging from bacteria to mammals at 25–120 μM concentration levels. This ubiquitous occurrence is related to a number of important physiological roles including 1) source of energy, 2) phosphate reservoir, 3) chelator for divalent cations, 4) buffer against alkaline stress, 5) regulator of development, and 6) a possible structural element in competence for DNA (Gabel and Thomas, 1971; Kumble and Kornberg, 1995). More recently, it has been shown that polyphosphates can be produced by mammalian mitochondria, where they play an important role in the mechanisms underlying Ca^{2+}-dependent cell death.

Polyphosphates may also act as neurotransmitters, mediating communication between astrocytes in the mammalian brain (Holmström et al., 2013).

4.5.3 Phosphonic Acids

Phosphonic acids and their derivatives are chemically and structurally related to phosphorus acid, by replacing the P-O bond by a P-R one, where R stands for a suitable organic molecule (Figure 4.16a). Accordingly, phosphonic acids and phosphonates belong to the enormous family of organophosphorus compounds, characterized by the presence of simple C–P bonds. Phosphonates can be extremely persistent in the environment as the C–P bond (with a lower polarity) does not hydrolyze as readily as the P–O or P–N bonds.

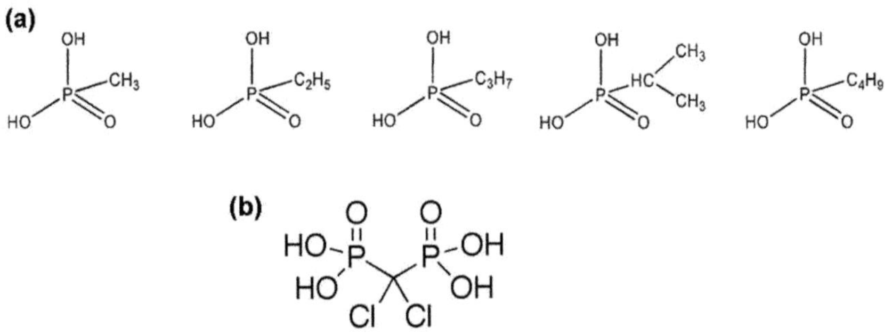

FIGURE 4.16 Molecular structure of (a) five alkyl phosphonic acids with methyl (CH_3), ethyl (C_2H_5), n-propyl (C_3H_7), isopropyl ($CH(CH_3)_2$), and n-butyl (C_4H_9) groups detected in the Murchison meteorite; (b) clodronic acid, a bisphosphonate used as a drug in osteoporosis treatment.

Phosphonic acids and phosphonate salts are typically white, nonvolatile solids that are poorly soluble in organic solvents, but soluble in water and common alcohols. Many commercial compounds are phosphonates, including glyphosate ($C_3H_8NO_5P$), the herbicide "Roundup" (see Section 9.1.2), and ethephon, a widely used plant growth regulator. Biphosphonates (Figure 4.16b) are popular drugs for the treatment of osteoporosis (Svara et al., 2008).

Bi- or polyphosphonates have not been found in minerals, but phosphonic acids are naturally synthesized by various organisms and are relatively common in living organisms (about 3% of total phosphorus content in plankton is present as phosphonates), from prokaryotes to eubacteria and fungi, mollusks, insects or marine invertebrates (Hildebrand, 1983). The predominant form is the aminoalkyl phosphonic acid, first identified in 1959, which is present in both lipids and proteins of the membranes. The biological role of the natural phosphonates, so far found in over 80 animal species including humans, is still poorly understood, although it is interesting to note that certain strains of bacteria, as well as some yeasts and fungi, can enzymatically cleave the C-P bond of phosphonates (Schowanek and Verstraete, 1990). Mammals do not synthesize phosphonates by themselves, but they obtain them through ingestion.

Homologous series of alkyl phosphonic acids (including methyl, ethyl, propyl, and butyl representatives) along with orthophosphate were identified in water extracts of the Murchison meteorite (Cooper et al., 1992). These moieties could be synthesized by UV irradiation of sodium phosphite, Na_2HPO_3, in the presence of formaldehyde, alcohols, or acetone (Graaf et al., 1995). Another possibility may be that interstellar HPC condensed onto grains had undergone thermally-induced, anaerobic hydrolysis to alkyl phosphinic acids (R-H_2PO_2) and that phosphonic acids could result from subsequent photo-induced redox processes in aqueous solution (Gorrell et al., 2006). Alternatively, the required phosphorus acid may be produced by hydrolysis of schreibersite (Pasek and Lauretta, 2005; Bryant and Kee, 2006), a common mineral in iron meteorites (see Section 4.3.2). This finding suggests the possibility of delivery of these water-soluble, phosphorus-containing molecules by meteorites or comets to the early earth (Maciá et al., 1997; Pasek, 2008; Pasek et al., 2013).

4.5.4 Hydroxyapatite and Related Mineral Phosphates

Almost all naturally phosphorus compounds occurring on earth are orthophosphates, based on the phosphoryl PO_4^{3-} group, in which the P atom exhibits *sp*^{3} tetrahedral symmetry. Due to the large variety of possible substitutions in their crystal structure, it is very difficult to write a general chemical formula for the apatites, although the following one is generally used $M_5X(PO_4)_3$, where M = Ca, Pb, Na, K, Sr, Mn, Zn, Cd, Mg, Fe, Al, and Ce; X= F, OH, Cl, and Br. Their general structure consists of a pseudo-hexagonal network of PO_4 tetrahedra with Ca^{2+} ions in the interstitial sites and columns of anions oriented along the c axis. Among them calcium apatites are the commonest and their general formula reads $Ca_5(PO_4)_3X$, where X is a negative ion, generally F^- (fluroapatite), Cl^- (chlorapatite) or OH^- (hydroxyapatite). Fluorapatite, $Ca_5(PO_4)_3F$, is hexagonal at room temperature (see Table 4.6) and its crystalline structure is usually taken as a model for the apatite materials (Figure 4.17a). A possible phase transition has been reported at –143.18°C, the lower temperature phase being a monoclinic arrangement. Chlorapatite, $Ca_5(PO_4)_3Cl$, has a monoclinic structure at room temperature. Differential thermal analyses have shown that it also undergoes a phase transition at 320°C, with the high-temperature phase remaining monoclinic, albeit with different unit cell parameters. The hydroxyapatites are of prime interest to biologists because they make up the mineral portion of teeth and bones in all animals.

TABLE 4.6 Room Temperature Crystal Structure of the Commonest Apatites*

Name	Crystal Symmetry	a, b, c (Å)	α, β, γ (°)	T_m (°C)	ρ (g cm^{-3})
Fluorapatite	Hexagonal ($P6_3/m$) (1)	$a = 9.418$	90	1.660	3.16 (1)
		$c = 6.875$	120		
Chlorapatite	Monoclinic ($P2_1/b$) (2)	$a = 9.421$	90	1.530	3.15(2)
		$b = 18.842$	90		
		$c = 6.881$	120		
Hydroxyapatite	Hexagonal ($P6_3$) (3)	$a = 9.440$	90	1.670	3.13 (3)
		$c = 6.856$	120		
	Monoclinic ($P2_1/b$) (3)	$a = 9.451$	90		3.15 (3)
		$b = 18.816$	90		
		$c = 6.852$	119.97		
	Monoclinic ($P2_1$) (3)	$a = 9.450$	90		3.16 (3)
		$b = 18.86$	90		
		$c = 6.841$	119.86		

*Crystal system, unit cell parameters, melting temperature and density of different apatite minerals.

Source: (1) Hughes et al., 1989; (2) Elliot, 1994; (3) Haverty et al., 2005.

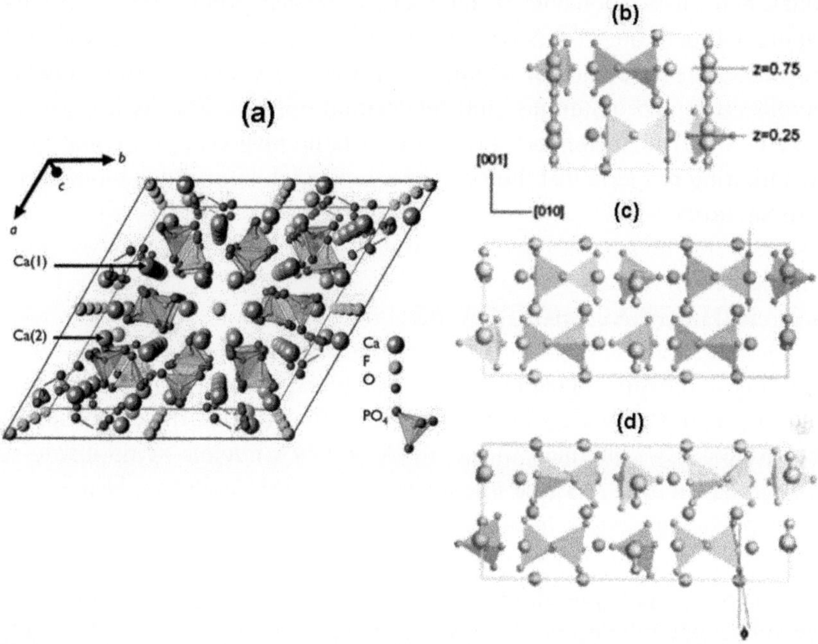

FIGURE 4.17 (See color insert.) (a) Structural model for the fluorapatite, with the two calcium sites indicated. Ca, F, and O atoms are in blue, green, and red, respectively, and phosphoryl groups are shown explicitly. (Source: Reproduced from Jay, Rushton, and Grimes, (2012); with permission of The Royal Society of Chemistry). Different structural models of hydroxyapatite: (b) disordered $P6_3/m$ structure, having the superposition of the reflected OH^- and the $O_3 - O_3$ edge parallel to the c axis, (c) superstructure with the $P6_3/m$ arrangement of the Ca and PO_4 framework, but with an ordered alternate arrangement of the OH^- ions, (d) the monoclinic $P2_1/b$ structure having a relaxed Ca-PO_4 arrangement in addition to the alternate orientation of the OH^- ion. (Reprinted figure with permission from Haverty, Tofail, Stanton, and McMonagle, (2005). Copyright (2005) by the American Physical Society.)

According to diffraction data (see Table 4.6), two phases have been suggested (Elliot, 1994): a hexagonal structure (Figure 4.17b), containing some degree of disorder, and a monoclinic structure. The disorder in the hexagonal phase stems from the presence of F^- and Cl^- ions interspersed at random among the OH^- ions. The central problem associated with this model is the reflection of the hydroxyl ions in the mirror planes z = 0.25 and z = 0.75, which leads to the superposition of oxygen atoms. In order to circumvent this problem a lower symmetry hexagonal structure without mirror planes (space group $P6_3$) was proposed, though recent density functional calculations favor monoclinic (space groups $P2_1/b$ and $P2_1$, Figure 4.17d, Table 4.6) models instead (Haverty et al., 2005). In that case, the absence of a center of symmetry in the $P2_1$ structure points towards pyroelectric and potential piezoelectric properties for hydroxyapatite which should shed light on the origin of bone piezoelectricity and the *in vivo* binding properties of the hydroxyapatite platelets used in biomaterials (see Section 9.3.3).

4.6 BIOLOGICAL PHOSPHATES

In living organisms, phosphates, and polyphosphates are profusely found, and they reside primarily in five reservoirs in cells (in viruses phosphates are exclusively contained in the RNA or DNA nucleic acids inside the protein capside): 1) as free phosphate ions within cellular plasma; 2) as sugar-phosphate esters in nucleic acids (Figure 4.18a); 3) as condensed phosphates in ATP (Figures 4.18b and 4.15c) or triphosphate (Figure 4.15a); 4) as phospholipids in membranes (Figure 4.18c); and 5) as phosphorylated metabolic compounds (Figure 4.18d–f). In modern cells, dissolved free phosphate has a concentration of about 10^{-4} M, and is highly useful as a buffer, thus ensuring a static chemical system with respect to acidity. However, most of the phosphate in cells is actually bond through covalent bonds, where it plays several important biological roles. These phosphate biomolecules can be classified according to the following energetic hierarchy (Pasek and Kee, 2011): I) unreactive, stable phosphorylated biomolecules in which phosphorus atoms provide a structural or binding handle, archetypical examples being the sugar-phosphate DNA backbone and phospholipids in membranes, II) energetic condensed phosphates which store metabolic energy, a prototypical representative being ATP, and III) reactive phosphorylated molecules generated during metabolism, which transfer phosphates and energy to condensed phosphates forms for energy storage. In reactive phosphorylated molecules, the phosphate group is bonded to carbon atoms with sp^2 hybridization, whereas in stable biomolecules it bounds with sp^3 hybridization instead. The addition of a charged phosphate group increases the solubility of the required metabolic molecules, allowing them to be properly shuffled about within the cell. Furthermore, phosphate bearing molecules are strongly attracted to divalent cations such as Mg^{+2} or Ca^{+2}, allowing certain molecules to be directed to an enzyme-site by attracting the molecule's phosphoryl group with a bound divalent cation.

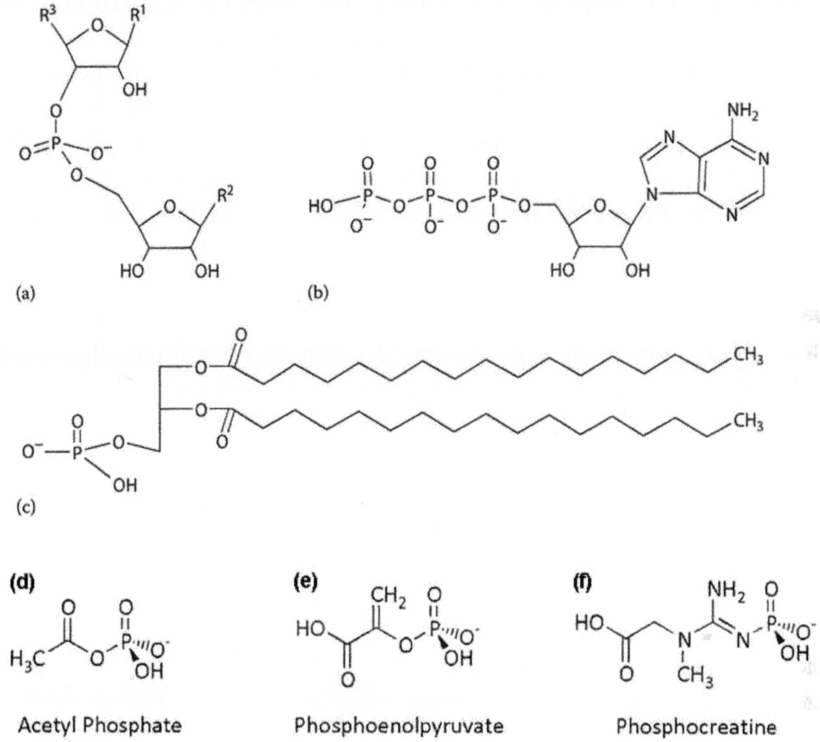

FIGURE 4.18 Phosphate bearing biomolecules acting in (a) replication (RNA) where R^1 and R^2 are nucleobases such as adenine or guanine, and R^3 indicates the continuous repetition of the shown basic structure, (b) energy transfer and storage (ATP), (c) cellular membranes (phospholipid), (d-f) metabolism. The indicated ionizations are appropriate for pH 6.

4.6.1 The Sugar-Phosphate Backbone

An extremely important role for phosphates in biology is the formation of the sugar-phosphate backbone of the nucleic acids RNA and DNA. Both nucleic acids consist of three main structural components (Figure 4.18a): 1) a nucleobase to be chosen among the purines adenine (A) and guanine (G), or the pyrimidines cytosine (C), thymine (T), or uracil (U), that serve as the letters of the genetic code, 2) a sugar pentose, either ribose (in RNA) or deoxyribose (in DNA), and 3) a phosphate group that binds the sugars together in succession to build the skeletal sugar-phosphate backbone.

In doing so, the phosphate group is able to form the required bridges between the sugars through a C–O–P ester bond, while maintaining a negative charge in the phosphate group at physiological conditions. To become a long, charged polymer ensures that this macromolecule cannot easily leak out through the cellular membrane (see Section 4.6.3), which is quite permeable to hydrophobic substances, but not to charged ones. The presence of negative charges also contributes to stabilizing the diester bonds against hydrolysis (Westheimer, 1987), which guarantees the required genetic materials endurance (Schrödinger, 1944). In this regard, we note that the charge must be negative to this end, since this sign sharply diminishes the rate of nucleophilic attack on the ester bond stemming from the presence of both OH^- radicals and polarized water molecules usually present around the nucleic acids at physiological conditions.

It is worthy to highlight that all these features can also be displayed by either arsenic or (to a lesser extent) silicic acids, hence apparently providing alternative compounds to be used in the nucleic acids backbone. However, both arsenic and silicic acids esters are unsuitable since they hydrolyze remarkably fast in neutral water at room temperature, and furthermore, the silicic acid is too weak to become fully ionized, as required at these conditions (Westheimer, 1987).

4.6.2 The Energy Carrier Nucleotides

A phosphate group is able to bond to another phosphoryl group to make a polyphosphate chain (Figure 4.15a in Section 4.5.2), entailing a loss of water and a gain of energy. Thus, polyphosphates such as ATP or ADP molecules, are essential compounds to keep the cellular metabolism running: when a cell is at rest, it burns sugars to make ATP and CO_2, a process called glycolysis. When the cell needs to do work, it relies on its stored ATP to liberate the necessary chemical energy via hydrolysis of the ester polyphosphate bonds (Westheimer, 1987). ATP is not stored in great quantities in the cells, but it is rapidly turned-over catalytically (Exercise 7). The energy produced by hydrolysis of ATP is about –30 $kJmol^{-1}$ (0.31 eV), thereby occupying an intermediate position between highly energy-releasing compounds, such as acetic anhydride (–91 $kJmol^{-1}$, 0.94 eV) or phosphoenolpyruvate (–61.9 $kJmol^{-1}$, 0.64 eV, Figure 4.18e), and lesser exothermic compounds, such as sugar phosphates (ranging from about –21 to –13 $kJmol^{-1}$, 0.22–0.13 eV). Thus, ATP is particularly well-suited to its role as the energy currency in metabolic processes conducting the metabolic breakdown of organic molecules through several chemical steps, each one involving a moderate energy exchange. If the energy transfer molecule was a highly-energy compound like phosphorus pyruvate, then very few of these breakdown steps would provide too much energy to preserve the chemical integrity of more fragile moieties. Conversely, if the metabolic currency was a low energy compound like a sugar-phosphate, the amount of energy extracted from each step would be too low, and catabolism would be a wasteful process. Therefore, optimizing the energy extracted from each step of metabolic networks, ATP is able to act as a rechargeable molecular battery (Pasek and Kee, 2015).

4.6.3 Phospholipid Membranes

The phosphate group is strongly hydrophilic, whereas the fats are strongly hydrophobic. As a result, molecules containing both hydrocarbon chains and phosphate groups, referred to as phospholipids, spontaneously rearrange themselves in the presence of water to create lipid bilayers. Indeed, a significant amount of phosphate in cells is in the phospholipids that make up the membranes that separate the cell from the outside world in prokaryotes, or the nucleus and the organelles from the outside cytoplasm in eukaryotes. Phospholipid biomolecules can broadly be grouped into two classes, namely, the isoprenoid ether lipids and the fatty acid acyl lipids, which are found in diverse types of organisms. The fatty acid ester type is widely distributed across the eubacteria, and eukaryotic domains of life and consists of a phosphate group bonded to a glycerol molecule, which is in turn bound to two saturated or unsaturated fatty acids long chains (Figure 4.18c). These chains typically contain 16–18 carbon atoms. The isoprenoid ether type is widely distributed among the archeobacteria, particularly extremophilic species, probably as a consequence of their greater resistance to chemical degradation at extreme salinity, pH, and temperature. These consist of one or more polyisoprenoid chains, attached by ether linkages to a phosphate group.

4.7 SUMMARY AND REVIEW QUESTIONS

Phosphorus exhibits a very rich and diverse chemistry naturally leading to a broad palette of different kinds of diverse compounds. Indeed, for one thing, phosphorus is known to occur in at least eight oxidation states, including the phosphide (–3), diphosphide (–2), tetraphosphide (–1/2), elemental (0), hypophosphite (+1), phosphite (+3), hypophosphate (+4), and phosphate (+5) forms. On the other hand, the stereochemistry and bonding geometry of phosphorus compounds are very varied since the element is known in at least 14 structural geometries and the resulting molecules can exhibit coordination numbers up to 9, although most of them have coordination numbers ranging from 3 to 6 (see Table 4.1). Examples of phosphorus compounds exhibiting coordination numbers larger than 6 are found in metal phosphides representatives. In particular, we can highlight the schreibersite crystal of composition $(Fe,Ni)_3P$, which is a common accessory mineral in most iron meteorites.

Elemental phosphorus can adopt a great number of crystalline structures, ranging from the highly symmetrical cubic to the low symmetry monoclinic and triclinic systems, in addition to several amorphous forms as well (see Table 4.3). Their related physical and chemical properties significantly differ as one goes from white (yellow), red, violet, or black forms. Notwithstanding this, despite significant structural differences, all phosphorus allotropes consist of covalently bond molecular building blocks, based upon P – P single bonds, three for every phosphorus atom, which are held together by weak van der Waals interactions. Indeed, the propensity of phosphorus atoms to catenate into chains, rings, and clusters is thus a quite remarkable feature of most of its compounds.

The most important reaction of elemental phosphorus is oxidation, whose main final products are the cage-like molecules phosphorus pentoxide P_4O_{10}, plus some amount of phosphorus trioxide, P_4O_6, following the reaction sequence shown in Figure 4.10. The oxidation of P_4 molecules shows the link between chemistry and geometry: starting from the tetrahedral P_4 scaffold the subsequent oxides are obtained by inserting an oxygen atom either in the P-P edges or at the P vertices, as it is illustrated in the molecular structures depicted in the upper right corner of Figure 4.10. Thus, the tetrahedral P_4O_6 structure is obtained by inserting six oxygen atoms in the original P_4 molecule P-P edges, while the P_4O_{10} structure is formed by attaching an oxygen atom at every P vertex of the previously obtained P_4O_6.

The oxoacids, oxoanions, and oxo-salts of phosphorus are more numerous than those of almost any other element. Many of them are of great importance either technologically and/or for biological processes. Anions, salts, and esters of pyrophosphoric acid are called pyrophosphates. A few pyrophosphate minerals have been reported to date, and only four polyphosphate minerals have been recognized by the International Mineralogical Association. Altogether, these polyphosphate minerals likely comprise about 10 kg of material, suggesting few geologic sources of mineral polyphosphates were present on the early earth (Pasek and Kee, 2011).

Conversely, purine and pyrimidine nucleotide diphosphates and triphosphates (Figure 4.15c) are always present in all known living forms on earth, and it has been suggested that polyphosphates may have played a crucial role in the prebiotic synthesis earlier stages. In fact, polyphosphates such as ATP or ADP, are essential compounds for the cellular metabolism. ATP is not stored in great quantities in the cells, but it is rapidly turned-over catalytically. On the other hand, there are linear polyphosphate chains with the general formula $(P_nO_{3n+1})^{(n+2)-}$, where n varies from less than 10 to over 5000 orthophosphate units, in all organisms from bacteria to mammals at 25–120 μM concentration levels. It has also been shown that mammalian mitochondria can produce polyphosphates, which play an important role in Ca^{2+} – dependent cell death. Polyphosphates may also act as neurotransmitters, mediating communication among mammalian brain astrocytes (Holmström et al., 2013).

Phosphonic acids and phosphonates are chemically and structurally related to phosphorus acid, by substituting the P-O bond by a P-R one, where R is an organic molecule (Figure 4.16a). Accordingly, they are included in the wide family of organophosphorus compounds, containing C–P bonds. Phosphonates are environmentally persistent since the C–P bonds lower polarity prevents them from hydrolyzing as readily as the P–O or P–N bonds. Mineral bi- or polyphosphonates have not been found, but phosphonic acids are relatively common in living organisms, from prokaryotes to eubacteria and fungi, mollusks, insects or marine invertebrates (Hildebrand, 1983), predominantly as aminoalkyl phosphonic acid, present in both lipids and proteins of the membranes. The biological role of the natural phosphonates is still poorly understood, although certain strains of bacteria, and some yeasts and fungi, can enzymatically cleave the C-P bond of phosphonates (Schowanek and Verstraete, 1990). Mammals obtain phosphonates through ingestion.

Water extracts of the Murchison meteorite (Cooper et al., 1992) showed homologous series of alkyl phosphonic acids (including methyl, ethyl, propyl, and butyl representatives) along with orthophosphate, compounds which could be synthesized by UV irradiation of sodium phosphate in the presence of formaldehyde (Graaf et al., 1995). Another possible route may be that interstellar HPC condensed onto grains underwent thermally-induced, anaerobic hydrolysis to alkyl phosphinic acids (R-H_2PO_2) and subsequent photo-induced redox processes in aqueous solution resulting in phosphonic acids (Gorrell et al., 2006). Alternatively, hydrolysis of schreibersite (common in iron meteorites, see Section 4.3.2) may produce the required phosphorus acid (Pasek and Lauretta, 2005). This finding suggests the possibility of delivery of these water-soluble, phosphorus-containing molecules by meteorites or comets to the early earth (Maciá et al., 1997; Pasek, 2008; Pasek et al., 2013).

Almost all naturally phosphorus compounds occurring on earth are orthophosphates, based on the phosphoryl PO_4^{3-} group. Among them, calcium apatites are the commonest, and they have the general formula $Ca_5(PO_4)_3X$ where X is a negative ion, generally F^- (fluroapatite), Cl^- (chlorapatite) or OH^- (hydroxyapatite).

In living organisms, phosphates, and polyphosphates are profusely found, and they reside primarily in five reservoirs in cell, namely, as free phosphate within the cellular plasma, as sugar-phosphate esters in nucleic acids, as condensed phosphates in ATP, as phospholipids in membranes, and as phosphorylated metabolic compounds (see Figure 4.18). An extremely important role for phosphates in biology is the formation of the sugar-phosphate backbone of the nucleic acids RNA and DNA, where phosphate group binds the sugars together in succession to make the skeletal sugar-phosphate backbone. In doing so, the phosphate group is able to form the required bridges between the sugars through a C–O–P ester bond, while maintaining a negative charge in the phosphate group at physiological conditions. To become a long, charged polymer ensures that this macromolecule cannot easily leak out through the cellular membrane.

An interesting analogy between organic chemistry and phosphorus chemistry refers to the possible extension of the very concept of aromaticity, which was originally introduced to account for the chemical properties of benzene molecules, to compounds containing P atoms instead of C ones. Thus, aromatic P-bearing compounds include structures with alternating *p* and *d* orbitals, like phosphonitrilic halides of the general form $(PNX_2)_n$, where X stands for the halide atom, which adopts a cyclic hexagonal geometry analogous to that of benzene. These compounds may even present fused ring systems, completely analog to the widespread PAHs. Consequently, aromatic phosphorus compounds may represent a promising reservoir of phosphorus, which will allow them to survive the hostile conditions imposed by intense UV ambient radiation in the diffuse ISM.

KEYWORDS

- **adenosine triphosphate (ATP)**
- **aromatic phosphorus compounds**
- **black phosphorus**
- **hybrid orbitals**
- **hydroxyapatite**
- **phosphates**
- **phosphides**
- **phosphine**
- **phospholipids**
- **phosphonic acids**
- **phosphorene**
- **phosphoric acids**
- **phosphorus oxides**
- **phosphoryl group**
- **polyphosphates**
- **pyrophosphoric acid**
- **red phosphorus allotropes**
- **schreibersite**
- **violet phosphorus**
- **white phosphorus allotropes**

CHAPTER 5

The Nucleosynthesis of Phosphorus

"The problems of determining the abundances of the chemical elements and their isotopes and accounting for them in terms of a theory of nucleosynthesis and the evolution of our Galaxy involves virtually every branch of astronomy, and many related sciences."

(Virginia Trimble, 1975)

By cosmic standards chlorine ([Cl] = 5.50 ± 0.30) and phosphorus ([P] = 5.41 ± 0.03, see Table 3.1) are the less abundant species among the third-row elements of the periodic table (Asplund et al., 2009). Furthermore, within the group of the main biogenic elements phosphorus is the least abundant representative (see Figures 1.1 and 1.2). What is the reason for the relatively low cosmic abundance of this chemical element? In order to answer this question in the following sections we will review the possible astrophysical scenarios where the formation of ^{31}P nuclei could possibly take place, namely, (a) during the primordial nucleosynthesis episode, (b) in nuclear reactions occurring in the first outburst of stars in the universe (Population III), (c) in thermonuclear reactions taking place in the core shells of massive stars belonging to later generations of stars (Populations II and I), and (d) explosive processes undergoing in types I and II supernovae (SN) events and classical nova stars. Let us then start from the very beginning.

5.1 PRIMORDIAL NUCLEOSYNTHESIS

In Section 2.1-we mentioned that the emergence of the main constituents of atomic nucleus, namely, protons (p) and neutrons (n), took place during the first seconds elapsed after the universe's initial state, by means of the two coupled reactions $p + e^- \leftrightarrow n + \upsilon$, $p + \overline{\upsilon} \leftrightarrow n + e^+$, involving electrons ($e^-$), positrons ($e^+$), neutrinos ($\upsilon$) and antineutrinos ($\overline{\upsilon}$). These reactions lead to the simultaneous creation and annihilation of protons and neutrons, the latter particle decaying through the reaction $n \rightarrow p + e^- + \overline{\upsilon}$, whose rate controls the relative abundance of neutrons, thereby determining the equilibrium n/p ratio at a given temperature. Now, since the temperature of the universe is continuously decreasing in a marked non-linear way due to space-time expansion, the temperature-dependent weak interaction governing the neutron decay rate, $\Gamma_W(T) \propto G_F^2 T^5$, where $G_F = 4.54 \times 10^{14}$ J^{-2}, is the so-called Fermi constant which determines the strength of the weak interaction, froze-out below the Hubble parameter expansion rate $H(T) \propto \sqrt{G}T^2$, where G is the gravitation constant, so that the mean-time for neutron decaying became lower than the age of the universe. The freeze-out condition is then given by $H(T) \equiv \Gamma_W(T)$, leading to the threshold temperature value $T^* = 9.5 \times 10^9$ K we mentioned in Section 2.1 (Cyburt et al., 2016). As a consequence, the relative number of protons and neutrons was frozen to the ratio p/n ~ 5, when the expanding universe reached $T = T^*$ (Exercise 2). Thereafter, free neutrons decay dropped the ratio to p/n ~ 7 before primordial nucleosynthesis began by the fusion of protons and neutrons to form deuterons (^{2}H), followed by further capture of deuterons, protons, and remaining neutrons leading to the production of tritium (^{3}H), ^{3}He, and ultimately ^{4}He nuclei. The amount of helium produced during the primordial nucleosynthesis stage strongly depends on the neutron mean life (about 880 s), as the available free neutrons must be captured and processed to helium on a time scale short enough compared to their own decay time. Assuming, to a good approximation, that almost all of the neutrons originally present end up in ^{4}He nuclei, the helium mass fraction is given by $\dfrac{2n}{p+n} = \dfrac{2(n/p)}{1+n/p} = \dfrac{2/7}{8/7} = \dfrac{1}{4}$, thereby leading to a primordial helium abundance ~ 25% of the current cosmic value. Accordingly, the main element produced during the primordial nucleosynthesis is helium (Cyburt et al., 2016).

As we will see in the following sections, ^{4}He nuclei (also referred to as α particle) play a very important role in the subsequent nuclear reactions taking place in the stellar cores, substantially contributing to the chemical enrichment progressively occurring in the galaxies. Nevertheless, helium atoms formed when these nuclei transform into neutral atoms by getting two electrons each, become chemically inert (helium belongs to the noble gas group in the periodic table), so that this element does not appreciably contribute to the resulting molecular inventory during the next chemical evolution stages in the ISM and planetary systems. Indeed, no known molecule or mineral form contains He on the earth, and we should remind that the element itself was spectroscopically discovered in the Sun's photosphere. Thus, helium is the main character in the nucleosynthesis arena which disappears from the scene when chemistry takes over the evolution game in the unfolding universe drama.

At the same time, the high stability of ^{4}He nuclei prevented further nucleosynthesis to occur via successive neutron or α particles fusion, because no stable nuclei exist at atomic mass 5 and 8. Indeed, ^{5}He and ^{8}Be nuclei disintegrate almost as fast as they are made through the reactions $^4He + n \leftrightarrow {}^5He$ and $\alpha + \alpha \leftrightarrow {}^8Be$, respectively. Although Be nuclei are extremely unstable at the high temperatures prevailing at the early universe, and they are readily destroyed by the thermonuclear reactions that occur in stellar interiors, once they are formed (mainly from spallation reactions involving heavier nuclei in stellar atmospheres and SN shells, Wallerstein et al., 1997) these nuclei are not radioactive, and they form stable atoms which are incorporated in molecules and minerals, such as emerald $(SiO_3)_6Al_2Be_3$.

Therefore, the only elements produced during primordial nucleosynthesis were hydrogen, helium, and some traces of lithium via the nuclear reaction $^2H + \alpha \rightarrow {}^6Li$, involving deuterium and helium nuclides. Accordingly, primordial nucleosynthesis must be disregarded as a suitable place to explain the origin of ^{31}P nuclei in the universe.

5.2 THE FIRST GENERATION OF STARS

Recent evidences indicate that stars existed by 180 million years after the origin of the universe, permeating the space around them with high energy ultraviolet (UV) light (Bowman et al., 2018). The first stars in the universe (referred to as Population III) are predicted to have been much more massive than the Sun, and they generally outweigh the more massive stars currently known. Population III stars emerged about 13 billion years ago through gravitational contraction of primordial gas clumps containing between 50–300 $M_{\odot}$, mainly composed of atomic hydrogen and helium as well as some molecular hydrogen, along with their respective ions. Inside these gas clouds, H_2 molecules were formed through the three-body reaction $H + H +$

$H \rightarrow H_2 + H$, where one of the hydrogen atoms serves to stabilize the bond formation during the collision process by releasing kinetic energy. Once formed, the hydrogen molecules significantly contribute to cooling down the gas temperature through microwave radiation emission, hence favoring further condensation of the gas by decreasing its pressure. As the dense, central parts of the gas clouds with masses ~100 $M_\odot$ reached temperatures about $T \approx 100$–200 K and a hydrogen density of $n \sim 10^4$ cm^{-3}, they underwent rapid contraction, ultimately leading to the formation of stars with a probable minimum mass value close to ~30 $M_\odot$, although a lower number of ~1 $M_\odot$ stars cannot be excluded (Abel et al., 2002). Since most Population III stars were very massive, they evolved very quickly (stars larger than 100 $M_\odot$ will explode within ~2 million years), presumably ejecting to the space their external envelopes in extremely energetic SN explosions.

Indeed, the question as to whether first-generation stars ejected any metals (in astrophysical jargon by "metal" one refers to any atom heavier than helium) into the ISM at the end of their lives is of utmost importance for the early cosmic chemical enrichment, as well as for how subsequent Population II stars formation proceeded. In principle, stars below roughly 40 $M_\odot$ will be able to eject freshly synthesized material, through type-II SN explosions or intense stellar winds. The same goes for stars in the mass range $140 \leq M/M_\odot \leq 260$, which are completely disrupted as they explode as pair-instability SN explosions. Calculations suggest that when these stars exploded, they did not leave a black hole behind. Instead, all of their mass was ejected into the space to be incorporated into the next generation of stars. On the contrary, Population III stars in the range $100 \leq M/M_\odot \leq 140$ are expected to undergo direct collapse without an explosion (Heger and Woosley, 2002), thereby sequestering all their nucleosynthesized elements from any further chemical evolution (cf. Exercise 5).

In this way, while the universe got progressively colder due to the space-time expansion, the inner cores of the first generation of stars offered a unique site with sufficiently high temperature and density values to allow for nuclear reactions able to form elements heavier than primordial lithium to proceed. In the stellar interiors, nucleosynthesis was driven by nuclear reactions on the hydrogen and helium fuels resulting from the primordial nucleosynthesis period. Due to the absence of C, N, and O atoms, the CNO cycle (see Section 5.4.1) cannot operate during the early stage of hydrogen burning in Population III stars, and the star's core contracts until the central temperature rises sufficiently high for the 3α reaction (see Section 5.4.2) to produce ^{12}C with mass fraction ~10^{-10}, along with some O and N nuclei. Afterwards, Population III stars can undergo CNO cycle at a much higher central temperature ($T \cong 1.5 \times 10^{8}$ K) than it is usually attained in later generations of stars (Populations II and I). Thus, shortly after the first massive stars were formed the primordial gas was enriched with heavier than helium atoms nucleosynthesized in their cores and ejected when they become SN to yield metallic abundances as small as ~0.01–0.1% the Sun's. In this gas, emission cooling by singly ionized carbon and neutral oxygen atoms can then lead to the formation of low-mass stars (~1 $M_{\odot}$) by allowing cloud fragmentation of the condensing molecular cloud to smaller clumps (Brumm and Loeb, 2003).

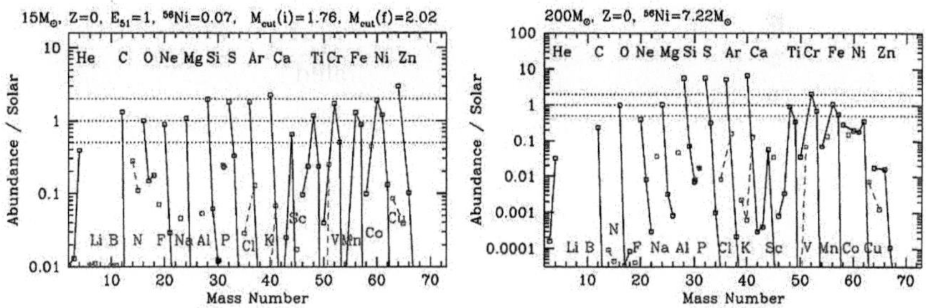

FIGURE 5.1 The abundance pattern in the ejecta of a 15 $M_{\odot}$ (left) and a 200 $M_{\odot}$ (right) model Population III star with initial zero metallicity, normalized to the solar ^{16}O. Isotopes of even-Z and odd-Z elements are connected by solid and dashed lines, respectively. *Source:* Reprinted figure with permission from Karlsson, Bromm, and Bland-Hawthorn, (2013). Copyright 2013 by the American Physical Society.

Albeit the light originally emitted by this early-generation stars has vanished from our space-time sector a long time ago, their SN explosion flares, stretched into infrared wavelengths by the expansion of the universe, can still be spotted by high-sensitive infrared detectors today. Thus, while waiting for an observational spectroscopic confirmation of the presence of phosphorus coming from Population III stars, we must rely on the numerical models results to conclude that the first generation of super-massive stars contributed a significant but minor fraction of the current phosphorus element inventory in the universe. In fact, detailed numerical studies on nucleosynthesis of elements in Population III type-II SN indicate that ^{31}P nuclei are formed in these stars, especially in massive stars within the range 15–30 $M_{\odot}$. Furthermore, the ratio of the pre- SN phosphorus yield divided by the post-explosion yield in a 20 $M_{\odot}$ model star is slightly less than unity, hence indicating that ^{31}P nuclei are only moderately consumed during the explosion itself (Umeda and Nomoto, 2002). Notwithstanding this, the overall yield of ^{31}P nuclei due to Population III stars is expected to be relatively low, within the interval $0.02 \leq [P]/[P_{\odot}] \leq 0.25$, as compared to current solar abundances (Karlsson et al., 2013), as can be appreciated from the data shown in Figure 5.1.

5.3 THE FORCES THAT GOVERN STELLAR EVOLUTION

"We cannot get away a deep understanding of the laws that govern the physical world without entering the world of mathematics"

(Roger Penrose, 2006)

"A book without equations would mean using words to try to convey the beauty of the theoretical (mathematical) basis for the physics (...)"

(Baker and Blackburn, 2005)

5.3.1 The Role of Mass and Pressure

A typical star keeps along most part of its life-cycle a state of hydrostatic equilibrium, maintaining a local balance between the inward gravitational force and the outward pressure gradient. By assuming the star is spherically symmetric, this balance can be mathematically expressed in the form (force per volume units):

$$\frac{dP}{dr} = -\frac{GM(r)}{r^2}\rho(r), \tag{1}$$

where, G is the constant of gravitation, $P(r)$ denotes the total pressure, and $\rho(r)$ and $M(r)$ are the density and the mass inside a sphere of radius r measured from the center of the star, respectively. There are two contributions to the total pressure: one is due to the thermal motion of the enclosed matter, P_m, and the other is due to the escaping photons constituting the electromagnetic radiation field, P_υ. If one assumes that stellar matter behaves as a perfect gas, its mechanical pressure depends on the local density and absolute temperature, $T(r)$, according to the expression:

$$P_m(r) = \frac{k_B}{\mu_m m_H} \rho(r) T(r), \tag{2}$$

where, k_B is the Boltzmann constant, m_H is the mass of the hydrogen atom, and μ_m is the mean molecular weight, which under normal stellar conditions is close to unity. On the other hand, the pressure due to radiation is given by:

$$P_\upsilon(r) = \frac{\sigma}{3} T^4(r), \tag{3}$$

where, σ is the Stefan-Boltzmann constant. A central result concerning current theories of stellar structure and evolution of normal stars refers to the role played by radiation pressure in their hydrostatic equilibrium state, a feature first emphasized by Eddington (1926). To see this, let us assume that radiation contributes a fraction β to the total pressure, that is, $\beta \equiv \frac{P_\upsilon}{P}$, so that we can write $P = P_\upsilon + P_m = \frac{P_\upsilon}{\beta}$, or alternatively, $P = \frac{P_m}{1-\beta}$. By equating both expressions for P we have $\frac{P_m}{1-\beta} = \frac{P_\upsilon}{\beta}$, and taking into account Eq. (2) and (3), we find:

$$T(r) = \left(\frac{\beta}{1-\beta} \frac{3k_B}{\mu_m \sigma m_H} \right)^{1/3} \rho^{1/3}(r). \tag{4}$$

Making use of Eq. (3) and (4) we can now eliminate the temperature from $P = \frac{P_\upsilon}{\beta}$, to obtain:

$$P(r) = \left(\frac{3}{\sigma} \right)^{1/3} \left(\frac{\beta^{1/4}}{1-\beta} \frac{k_B}{\mu_m m_H} \right)^{4/3} \rho^{4/3}(r). \tag{5}$$

Now, for a star of mass M, we have the upper bond for the pressure at the center of a star, P_c (Exercise 8).

$$P_c \leq \frac{G}{2} \left(\frac{4\pi}{3} \right)^{1/3} \rho_c^{4/3} M^{2/3}, \tag{6}$$

Thus, evaluating Eq. (5) at the center, equating it to Eq. (6) and solving for the mass, we obtain:

$$M \geq \frac{3\sqrt{15}}{2\pi^3 \mu_m^2 m_H^2} \left(\frac{hc}{G} \right)^{3/2} \left(\frac{\beta_c^{1/4}}{1-\beta_c} \right)^2, \tag{7}$$

where, β_c indicates the radiation pressure fraction contribution at the star's center, and we have used the Stefan-Boltzmann constant explicit value $\sigma = \frac{8\pi^5 k_B^4}{15h^3c^3}$, where h is the Planck's constant and c the speed of light. In this way, we realize that the stable existence of stars requires a precise relation between their mass and their central radiation pressure value to be fulfilled, namely:

$$\left(\frac{\beta_c^{1/4}}{1-\beta_c} \right)^2 \leq \frac{40\sqrt{15}\pi^3}{26235} \frac{M}{M_\odot} \cong 0.183 \frac{M}{M_\odot}, \tag{8}$$

where, we have assumed a mean molecular weight equal to 1, and made use of the following relationship (Chandrasekhar, 1984) among fundamental constants (see the Appendix).

$$\frac{1}{m_H^2} \left(\frac{hc}{G} \right)^{3/2} = \frac{583}{20} M_\odot \cong 29.15 M_\odot \tag{9}$$

Eqn. (8) provides an upper limit to the central radiation pressure for a star of a given mass to be structurally stable. For instance, for a star of the solar mass, the radiation pressure at the center cannot exceed about 3% of the total pressure (Exercise 9).

5.3.2 Energy Balance Equation

The first law of thermodynamics, or the principle of conservation of energy, states that the internal energy of a system can be changed by two forms of energy transfer: heat and work. Consider a small element of mass, *dm*, within a star, over which the temperature, density, and composition may be taken as approximately constant. Let u be the internal energy per unit mass and P the pressure. The sources of the heat of the mass element are: a) the release of nuclear energy inside its volume, if any, and b) the balance of the heat fluxes streaming into the element and out of it, F. The rate of nuclear release per unit mass is denoted by ε. The energy balance in a given mass element can be written in the form (Prialnik, 2011).

$$\frac{du}{dt} = \varepsilon - \frac{dF}{dm} - P\frac{d\rho^{-1}}{dt}. \tag{10}$$

In thermal equilibrium, the temporal derivatives vanish, and we get $\varepsilon = \frac{dF}{dm}$, so that integrating over the mass one obtains:

$$\int_0^M \varepsilon(m)dm = \int_0^M dF = F(M) - F(0), \tag{11}$$

where $F(0) \equiv 0$ and $F(M) = L$ denotes the star's luminosity, and thus thermal equilibrium implies that energy is radiated away by the star at the same rate as it is produced in its interior. Therefore, energy in stars is generated in their cores through nuclear reaction networks and transported by radiation and convection outwards up to their surfaces. Radiative transport dominates in the core region where thermonuclear reactions take place. For instance, just about 2.6% of the Sun by mass (the solar envelope) is convective, whereas radiative transport dominates in the interior ($r \leq 0.72R_{\odot}$).

5.4 MAIN NUCLEAR REACTIONS IN STARS

The nature of the thermonuclear reactions occurring in the cores of stars depends on their mass, average chemical composition, and central temperature. Stars in the main sequence generally generate their energy through hydrogen burning, which can occur in two main ways. The so-called *p-p* chain, which fuses four protons to give a helium nuclide, is important for small mass stars located in the lower temperature region of the Hertzsprung-Russell (HR) diagram (see Figure 2.6). For more massive stars (say, $M \geq M_{\odot}$) the *p-p* chain does not produce enough energy to stabilize the star against gravitational contraction, and the so-called *CNO* cycle, which is based on a sequence of proton capture process on carbon, nitrogen, and oxygen nuclei, becomes dominant. It must be kept in mind, however, that whereas the *CNO* cycle is almost not relevant for the energy generation of older, low-metallicity Populations III and II stars, both processes are simultaneously running in younger Population I stars possessing significant enough amounts of the required C, N, and O nuclei. In this case, the relative role played by each process significantly depends on the star's core temperature. For instance, the relatively cool temperature of the solar core ($T \cong 1.5 \times 10^6$ K) favors the *p-p* chain, which produces about 99% of the Sun's energy (Adelberger et al., 2011), whereas the *CNO* cycle dominates the energy production in more massive main-sequence stars such as Sirius, Vega, and Spica.

When hydrogen fuel is almost exhausted the stellar core, mainly containing helium nuclei produced during the hydrogen-burning phase, contracts under its own weight, while a residual hydrogen burning zone expands outward, and the star enters the so-called red giant phase (see Section 2.3.2). The contraction of the core continues until the temperature, and pressure values in the inner zone are high enough to trigger the helium-burning phase, in which three helium nuclides fuse to yield a carbon nucleus. In low- and intermediate-mass stars (say, below ~ 8 $M_\odot$) nucleosynthesis does not longer proceed beyond He-burning, and the star's core collapses to form a white dwarf, while its outer layers are ejected giving rise to the emergence of a planetary nebula. This explains the relatively high cosmic abundance of the main nuclear products of the He-burning stage, namely, O, C, and N.

In this way, we understand why bioelements P and S are relatively scarce, since their nucleosynthesis requires very high core temperatures, which can only be attained in the minor group of very massive stars, able to ignite atomic fuels heavier than He. In fact, in these more massive stars, further energy production stems from several burning stages which follow in succession in their progressively hotter inner cores, and are named after their principal fuel as carbon, neon, oxygen, and silicon burning (Wiescher and Langanke, 2015). As it is illustrated in Figure 5.2, each fuel ignites at a progressively increasing temperature, which is indicated by a step in the graph.

These steps are a direct consequence of the strong dependence of nuclear reaction rates on the relative velocity of the colliding nuclei. Indeed, the rate of a nuclear reaction is essentially the product of a cross-sectional area (related to the target nucleus) and the relative velocity of the interacting particles, *v*, which determines their kinetic energy. Now, in order to induce a nuclear reaction, the colliding nuclei have to come within a distance comparable to the range of strong nuclear force ($d \approx 10^{-15}$ m, see Figure 2.1). To this end, since both nuclei are positively charged, they must overcome the electrostatic Coulomb force which tends to separate them. This force thus imposes an effective barrier, and introduces a threshold distance d_*, that in the most favorable case of a linear collision is given by:

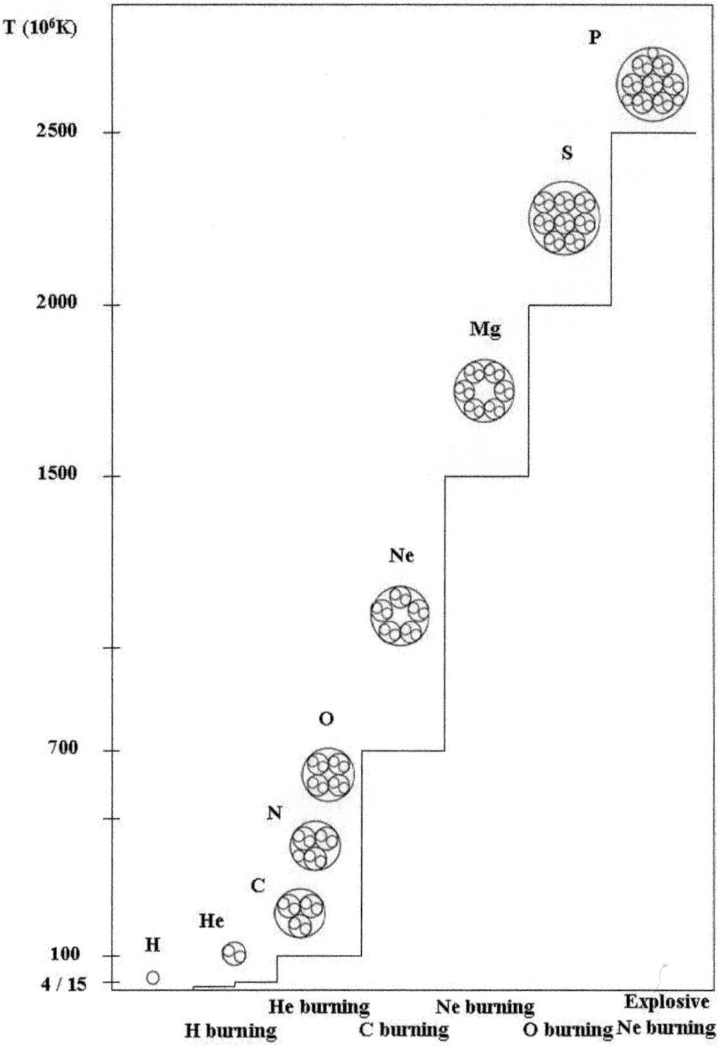

FIGURE 5.2 Nucleosynthesis of most elements heavier than lithium up to iron takes places in stars' cores. In this figure, we show in the abscissas the successive burning stages giving rise to the main biogenic elements, C, O, N, S, and P, along with their approximate ignition temperature threshold (in ordinates, not to scale). The inner structure of the different nuclei is illustrated in terms of protons (small light gray encircled balls) and α particles (dark gray circles containing two protons each) in order to highlight the relevant role of both ^{1}H and ^{4}He nuclei in the nuclear reactions giving rise to the synthesis of the chemical elements (Norman, 1993).

Source: Adapted from Maciá, (2005); with permission from the Royal Society of Chemistry.

$$d_* = \frac{1}{2\pi\varepsilon_0}\frac{Z_i Z_j e^2}{\mu v^2}, \tag{12}$$

So that the kinetic energy of the nuclei equals the electric potential energy, where ε_0 is the vacuum dielectric constant, $\mu = \frac{mM}{m+M}$, is the reduced mass of the colliding particles, *m*, and *M*, respectively, and $Z_{i,j}$ are the respective atomic numbers of the nuclei. The required ignition temperature for the nuclear reaction can then be estimated making use of the energy equipartition expression:

$$\frac{1}{2}\mu v^2 = \frac{3}{2}k_B T. \tag{13}$$

However, if we plug into the resulting expression, the average stellar core temperatures indicated in Figure 5.2 one gets a threshold distance d_* value which generally exceeds the typical range of strong nuclear force by about three orders of magnitude (Exercise 10). Indeed, a proper evaluation of nuclear reaction rates requires the explicit consideration of quantum tunneling effects, a phenomenon discovered by Gamow in 1928. In doing so, the resulting reaction rates read (Prialnik, 2011; Fowler, 1984).

$$r \propto (k_B T)^{-2/3} \exp\left(-\frac{3}{2}\left(\frac{\pi Z_i Z_j e^2}{h\varepsilon_0}\right)^{2/3}\left(\frac{\mu}{k_B T}\right)^{1/3}\right). \tag{14}$$

As we see, the nuclear reaction rate increases with increasing temperature and decreases with increasing both the charges and masses of the interacting nuclei. Accordingly, a fusion of progressively heavier nuclei would require higher and higher temperatures in a marked non-linear way. Eq. (14) holds for non-resonant nuclear reactions. The reaction rate is substantially increased in the presence of quantum resonant effects involving suitable nuclear energy levels between the colliding atoms, as occurs in the case of the 3α process proceeding via the ground state of ^{8}Be (see Section 5.4.2), whose reaction rate is given by (Rolfs and Rodney, 1988; Oberhummer, 2000).

$$r_{3\alpha} = 3^{3/2} N_\alpha^3 \left(\frac{h^2}{2\pi M_\alpha k_B T}\right)^3 \frac{2\pi\Gamma}{h} \exp\left(-\frac{\varepsilon}{k_B T}\right), \tag{15}$$

where, N_α and M_α are the number density and the mass of α particles, Γ = 3.7 ± 0.5 meV is the so-called radiative width, and ε = 379.47 ± 0.18 keV is the resonance energy value in the center-of-mass frame relative to the 3α threshold (Shirley, 1996).

For the sake of illustration, let us now consider Eq. (12) and (14) above in the context of the phosphorus nucleosynthesis. There are six reported isotopes of phosphorus nucleus, the only stable of which is the ^{31}P one, consisting of 15 protons and 16 neutrons. The binding energy per nucleon of phosphorus nucleus ^{31}P is 8.4 MeV, a figure close to the maximum binding energy per nucleon (8.7 MeV) exhibited by iron nuclide. This indicates that the nuclear structure of phosphorus is very stable. On the other hand, the electrostatic repulsion barrier associated to the nuclear charges of the colliding nuclei leading to the formation of ^{31}P nuclide (see Figure 5.3) implies the existence of quite large activation energies for the nuclear reactions (Exercise 11). Consequently, extremely high kinetic energies, corresponding to temperatures above 10^9 K, are required for phosphorus nucleosynthesis in the absence of resonance energy levels in the colliding nuclei.

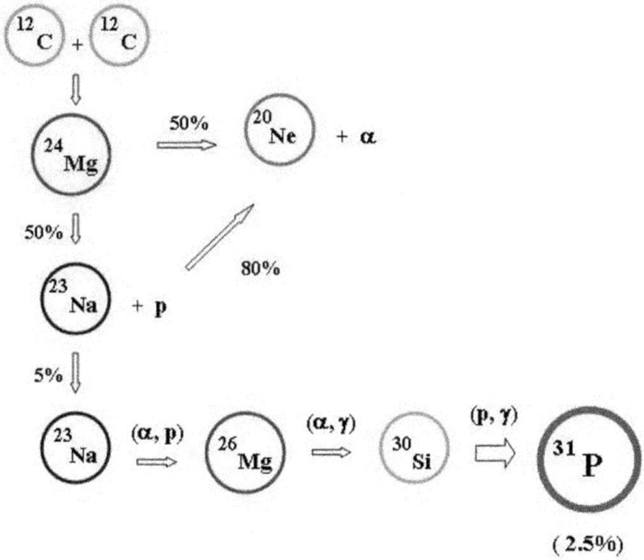

FIGURE 5.3 Nuclear reactions network leading to the synthesis of phosphorus nuclei in the core of massive stars (Clayton, 1988). The indicated branch ratios are those corresponding to the explosive neon burning phase described in Sec. 5.4.7.
Source: Updated from Maciá, (2005); with permission from the Royal Society of Chemistry.

In the following subsections we will briefly describe the main thermonuclear reactions taking place in the cores of stars, paying a special attention to those aspects directly related to the formation (or not) of phosphorus, in order to ascertain the most likely primary sources of this main biogenic element in chemical evolution (Maciá et al., 1997; Maciá, 2005).

5.4.1 *Hydrogen Burning*

The so-called proton-proton (*p-p*) chain consists of a sequence of nuclear reactions fusing four protons into helium, according to the overall relation $4p \rightarrow {}^4He + 2e^+ + 2\nu_e$, under the release of energy (26.7 MeV) in the form of photons, neutrinos, and the kinetic energy of other reaction products. Due to the electrostatic Coulomb repulsion, the probability of a simultaneous encounter of four protons is vanishingly small, even at the high densities of main-sequence stellar cores ($\rho \sim 10^2$ g cm^{-3}). Thus the hydrogen fusion process happens not at once but gradually, through a chain of reactions involving the interaction of only two particles each time. The first reaction in the full sequence is the conversion from two protons to a deuteron $p + p \rightarrow d + e^+ + \nu_e$. As this reaction is based on the weak nuclear interaction, which converts one proton into a neutron, it is extremely slow ($\tau \cong 7\times10^9$ yr) and ultimately causes the long life of all the small mass main sequence stars, including our Sun (Wiescher and Langanke, 2015). On the contrary, the second step in the *p-p* chain sequence, $p + d \rightarrow {}^3He + \gamma$, where γ indicates the emission of a high energy photon, takes place in just about $\tau \cong 4$ s. In comparison the aging scale length of more massive main-sequence stars (see Figure 2.6), whose energy production is dominated by the CNO cycle, is determined by the reaction ${}^{14}N + p \rightarrow {}^{15}O + \gamma$, with $\tau \cong 3\times10^7$ yr, a figure about two orders of magnitude quicker than that corresponding to the overall *p-p* chain.

5.4.2 *He-Burning*

Stars spend most of their lifetime burning ^1H and producing ^4He. Near the end of the main sequence stage, the hydrogen of the stellar core becomes nearly exhausted, and stars undergo a gravitational collapse, further compressing the core, which contracts and heats up through the conversion of gravitational potential energy as it gets denser. At the same time, the outer layers of the star expand and cool down to temperatures typical of the redder spectral classes K and M, although their luminosity ($L = 4\pi R^2 \sigma T_S^4$) substantially increases owing to the resulting larger surface area of the expanded atmosphere. Consequently, when arriving at this evolution stage, stars depart from the main sequence strip and move along the upper right region of the HR diagram becoming red giants (see Figure 2.6).

With the contraction of the stellar core up to typical densities around $\rho \approx 10^5$ g cm^{-3}, and temperatures of about $T \approx 10^8$ K, helium-burning by fusion among three helium nuclei, the so-called triple-alpha (3α) process: $\alpha + \alpha \leftrightarrow {}^8Be + \alpha \leftrightarrow {}^{12}\mathrm{C} + \gamma$, becomes possible. In the first step the very unstable ^{8}Be nuclide, with a lifetime of only 2.6×10^{-16} s, is formed in reaction equilibrium with two α particles. In the second step, an additional α particle is captured and an energetic γ photon emitted. The short lifetime of the ^{8}Be nucleus acts as a bottleneck for the whole process, so that the relatively large observed abundance of carbon in the universe cannot be explained by considering this two-step process alone. Indeed, the probability of a given α particle to be captured by a beryllium nucleus is greatly enhanced by the existence of a resonant energy level in the ^{12}C nuclide (Fowler, 1984), without which the 3α reaction rate would be much too low to account for the observed ^{12}C abundance in our universe (Exercise 12). A reaction requiring the (almost) simultaneous collision of three ^{4}He nuclei is highly improbable in most environments, so red giant stars such as Antares and Betelgeuse, where the resonant 3α reaction takes place, provide a unique place to yield substantial amounts of carbon, hence accounting for the predominance of organic chemistry in the universe (Oró, 1963; Kwok, 2004).

Some ^{12}C nuclei are further processed during the He-burning phase by subsequent α-capture to produce a significant amount of ^{16}O nuclei. The rate of the reaction ${}^{12}\mathrm{C} + \alpha \rightarrow {}^{16}\mathrm{O} + \gamma$, relative to that of 3α, determines how much of the carbon is eventually converted to oxygen, leading to an abundance ratio in the universe of ^{12}C: ^{16}O ≈ 1:2 (Rolfs and Rodney, 1988, see Table 3.1), a ratio which is mainly determined by the ejecta of low- to intermediate-mass stars to the ISM during the subsequent planetary nebula phase. In this way, He-burning accounts for the existence of two main biogenic elements essential for the development of water-based organic life (Oberhummer et al., 2000).

On the other hand, in high-mass stars, the relative amounts of C and O at the end of the He-burning phase set the initial conditions for the next series of thermonuclear reactions inside their cores. Indeed, in low- to intermediate-mass red giants stars, He-burning terminates with ^{16}O synthesis because further α particle captures occur too slowly to be significant at the prevailing temperatures and densities attained in their cores. This is due to the absence of a suitable excited state of ^{20}Ne nuclei near the $\alpha + {}^{16}O$ reaction threshold able to serve as a resonant assistance for α capture on ^{16}O (Wallerstein et al., 1997). Thus, when all He nuclei in the center of the star are consumed the 3α reaction proceeds in shells around this core, so that two burning concentric regions are present, respectively burning H (outer envelope) and He (inner shell around the core), and are separated by a region of radiative energy transport. In stars with masses below about 8 $M_\odot$ this double shell-burning is thermally unstable, creating thermal pulses leading to high rates of mass loss, so that these stars eventually lose the layers above the H-burning shell, creating planetary nebulae with a white dwarf in the center.

Therefore, the formation of nuclei heavier than ^{16}O will take place through successive carbon, neon, oxygen, and silicon burnings occurring at the cores of massive enough stars. An important new feature of the carbon and ensuing burning stages is that the star's dominant luminosity (hence, its energy loss) is by neutrino emission directly from the core rather than by electromagnetic radiation from star's surface. Consequently, in order to compensate for this more efficient energy loss mechanism, the star's core temperature is increased, nuclear reaction rates are greatly sped up, and C-burning and the following thermonuclear reactions will have progressively shorter lifetimes (Wallerstein et al., 1997).

5.4.3 C-Burning

When the helium supply is near exhaustion, the high-mass stars core consists largely of carbon and oxygen. Gravitational contraction increases the core density again, and temperatures reach values close to $T \approx 10^9$ K. Under these conditions, carbon nuclei combine with an energy release comparable to that of the previous He-burning phase, thereby preserving the star stability. Accordingly, the principal nuclear reaction during C-burning, starting at about $T = 7\times10^8$ K, is the fusion of two ^{12}C nuclides to produce a highly excited magnesium nucleus, that is, $^{12}C + ^{12}C \rightarrow ^{24}Mg^{*}$, which can then decay through two possible channels, namely, $^{24}Mg^{*} \rightarrow ^{23}Na + p$, and $^{24}Mg^{*} \rightarrow ^{20}Ne + \alpha$, each one having about the same probability to occur (Figure 5.3). When keeping into account all possible reaction networks among different atomic species one obtains that the principal nuclei resulting from the C-burning phase are ^{16}O (a survivor from He-burning), $^{20,21,22}Ne$, ^{23}Na, $^{24,25,26}Mg$, and $^{26,27}Al$, with small amounts of ^{30}Si and ^{31}P nuclei, which are obtained through the reaction $^{26}Mg + \alpha \rightarrow ^{30}Si + \gamma$, followed by $^{30}Si + p \rightarrow ^{31}P + \gamma$ (Woosley et al., 2002). Sequences of this sort can be conveniently denoted in the more compact form $^{26}Mg(\alpha,\gamma)^{30}Si(p,\gamma)^{31}P$, as it is indicated in Figure 5.3. However, once the C-burning has been completed and Ne-burning starts, most of the previously formed ^{31}P nuclei are destroyed through the reaction $^{31}P + p \rightarrow ^{28}Si + \alpha$, leading to the synthesis of the highly stable silicon nuclide. Consequently, the hydrostatic C-burning stage cannot be considered as the main source of ^{31}P nuclide.

5.4.4 Ne-Burning

Following C-burning the composition of the star's core consists chiefly of ^{16}O (70%), ^{20}Ne (23%), ^{24}Mg (2%) and ^{23}Na (2%) nuclei. Among them oxygen has the smallest Coulomb barrier, corresponding to an ignition temperature of about $T = 1.8 \times 10^9$ K, but before this temperature is reached the ambient electromagnetic radiation, mainly composed of energetic γ photons, is able to photodisintegrate the nuclei with lowest α-particle threshold energy, that is ^{20}Ne (with $\Delta E = -4.73$ MeV), via the process $^{20}Ne + \gamma \rightarrow ^{16}O + \alpha$, starting at about $T = 1.2\text{–}1.5 \times 10^9$ K and $\rho \sim 2 \times 10^6$ g cm^{-3}. The α particles liberated in this way can react with the remaining ^{20}Ne nuclei in a coupled process which is called Ne-burning:

$$^{20}Ne + \gamma \rightarrow ^{16}O + \alpha,$$

$$^{20}Ne + \alpha \rightarrow ^{24}Mg + \gamma$$

The net result is that for each two ^{20}Ne nuclei that disappear, one ^{16}O nucleus and one ^{24}Mg nucleus appear, i.e., $2\ ^{20}Ne \rightarrow ^{16}O + ^{24}Mg$. Although the photodisintegration is endothermic, hence requiring energy to proceed, the overall combination of subsequent fission and fusion reactions is exothermic (+4.59 MeV), as required to maintain the star stability (Woosley et al., 2002). Energy balance conditions give a lifetime of a few months for Ne-burning and an energy yield of about 25% of that resulting from C-burning. Albeit this small energy efficiency, Ne-burning nucleosynthesis is very important for chemical evolution. In fact, along with the coupled process involving ^{20}Ne, ^{16}O, and ^{24}Mg nuclei, there are also other secondary reactions preferentially leading to the formation of ^{28}Si, as well as phosphorus nuclei according to the nuclear reactions path indicated in the bottom row of Figure 5.3. Thus, these secondary reactions set is of interest to phosphorus nucleosynthesis, although it is not so relevant to the energy generation budget (Woosley et al., 2002).

In summary, at the end of the Ne-burning the star's core main composition is ^{16}O (83%), ^{28}Si (6%), ^{24}Mg (5%), ^{32}S (1.4%), ^{29}Si (1.4%) and ^{30}Si (1%), with traces of ^{25,26}Mg, ^{26,27}Al, and ^{31}P nuclei (Chieffi et al., 1998). However, neon burning cannot be considered as the main source of ^{31}P nuclide either, for many of the phosphorus nuclei formed during this stage will be consumed during the forthcoming oxygen burning phase.

From this stage onward, the location of the star in the H-R diagram will not change anymore, since the evolutionary timescales are much shorter than the reaction timescales of the star's outer envelope layers.

5.4.5 O-Burning

Following Ne-burning, the core further contracts and heats to reach $T = 1.8–2\times10^9$ K and $\rho \sim 4 \times 10^6$ g cm^{-3}, conditions under which oxygen nuclei are able to burn. The oxygen fusion produces excited nuclear states of ^{32}S that may decay through the following possible channels (Woosley et al., 2002):

$$^{16}\mathrm{O} + {}^{16}\mathrm{O} \rightarrow {}^{32}\mathrm{S}^* \rightarrow {}^{28}\mathrm{Si} + \alpha + 9.59\ \mathrm{MeV}$$

$$\rightarrow {}^{31}\mathrm{P} + \mathrm{p} + 7.68\ \mathrm{MeV}$$

$$\rightarrow {}^{31}\mathrm{S} + \mathrm{n} + 1.45\ \mathrm{MeV}$$

$$\rightarrow {}^{30}\mathrm{P} + \mathrm{d} - 2.41\ \mathrm{MeV}$$

Additional phosphorus forming reactions during this stage include ^{28}Si(α,p)^{31}P and ^{27}Al(α,γ)^{31}P (Emir et al., 2013). Thus, the principal immediate products are ^{28}Si, ^{31}P, and ^{31}S. The channel that produces ^{31}P requires the core temperature to be high enough and therefore only in massive stars would this reaction proceed. Thus, stars with masses lower than 10 $M_\odot$ do not seem to produce any significant amount of phosphorus through this reaction (Emir et al., 2013). Unfortunately, these promising initiating reactions are followed by a host of secondary reactions leading to the synthesis of nuclei up to A~ 40 at the expense of ^{31}P and ^{31}S nuclei. In fact, when the temperature rises above 2.3 × 10^9 K, the ^{32}S and ^{34}S nuclei begin to burn themselves, leading to a further increase of ^{28}Si nuclei and to the production of isotopes up to Ca (Figure 5.4). The most relevant reactions in regard to phosphorus nucleosynthesis during this episode are $^{31}P(p,\gamma)^{32}S$, $^{31}P(\gamma,p)^{30}Si$, and $^{31}P(p,\alpha)^{28}Si(\alpha,\gamma)^{32}S$ (Chieffi et al., 1998). When all reactions are considered, the chief products of O-burning are ^{28}Si, ^{32,33,34}S, ^{35,37}Cl, ^{36,37,38}Ar, ^{39,41}K, and ^{40,41,42}Ca. Of these, ^{28}Si and ^{32}S constitute the bulk (~90%) of the final composition (Woosley et al., 2002). Consequently, O-burning is not an adequate stage for obtaining phosphorus, not only due to the consumption of the phosphorus nuclei formed in the previous C- and Ne-burning phases, but also since the products of central hydrostatic O-burning are probably never ejected into the ISM, but remain trapped inside the degenerate core which after the SN explosion event will appear as a neutron star or a black hole (cf. Exercise 5).

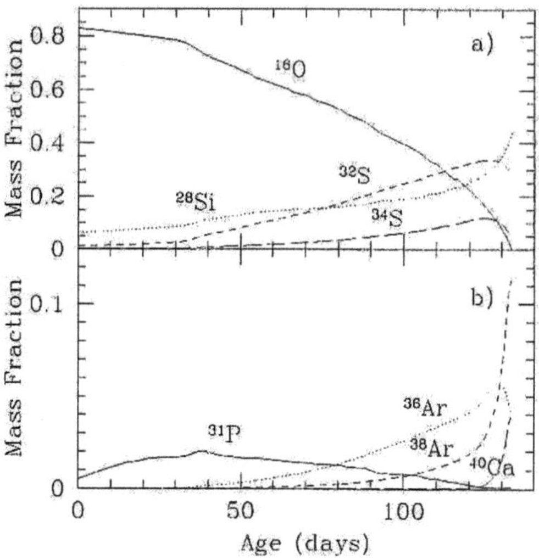

FIGURE 5.4 Core abundances of selected nuclear species as a function of time during the oxygen burning phase.
Source: Chieffi, Limongi, and Strainiero (1998); © AAS. Reproduced with permission.

5.4.6 Si-Burning

The stellar core's temperature progressively increases through O-burning stage and eventually, at $T = 3–3.5\times10^9$ K, O-burning merges into silicon burning, which can be described as the photodisintegration of nuclei to produce neutrons, protons, and specially α particles, resembling, in some ways, the previous Ne-burning process. Thus, a portion of the ^{28}Si melts by a sequence of photodisintegration reactions into the chain $^{28}Si(\gamma,\alpha)^{24}Mg(\gamma,\alpha)^{20}Ne(\gamma,\alpha)^{16}O(\gamma,\alpha)^{12}C(\gamma,2\alpha)\alpha$. The resulting free α particles, protons, and neutrons then build nuclei up to the most stable nucleus ^{56}Fe. An equilibrium is maintained between the α particles and the free nucleons by the existence of chains such as $^{28}Si(\alpha,\gamma)^{32}S(\gamma,p)^{31}P(\gamma,p)^{30}Si(\gamma,n)^{29}Si(\gamma,n)^{28}Si$, each reaction being in equilibrium with its inverse (Woosley et al., 2002). Although the ^{31}P nuclei play an active role in these network reactions, when properly combining the entire involved reactions, one realizes that no net production of ^{31}P nuclei is attained during the Si-burning stage. Instead, most of the resulting nuclei cluster around the iron group elements, while the silicon abundance substantially decreases.

In summary, we conclude that none of the above considered thermonuclear processes, namely, C-, Ne-, O-, and Si-burnings, are able to produce phosphorus nuclei in the proportions observed in the solar system abundance curve. Whence then phosphorus nuclei came from?

5.4.7 Explosive Nucleosynthesis

It has been known for some time that, albeit the relative proportions of ^{12}C, ^{16}O, ^{20}Ne, ^{24}Mg, and ^{28}Si nuclei in a massive star at the completion of the hydrostatic burnings described in the previous pages are in good accord with their solar system abundance values, the detailed distribution of the products of hydrostatic C- and O-burnings in the range $A = 20–40$ (which includes phosphorus, $A = 31$) does not agree with those abundances. Thus, the stellar materials must have undergone additional nuclear reactions under explosive rather than hydrostatic conditions during the events that released them to the ISM.

Massive stars at their late evolution stages exhibit an onion-like internal structure caused by their successive burning stages. Nearest to the center of the star are the shells undergoing the hottest and heavier nuclear fusion, while regions farther away, near the surface, host the coolest and lighter nuclear fusion reactions, and the shells sustaining C- and Ne-burning are comprised in between. In fact, the central regions of a massive star are capable of undergoing static nuclear reactions all the way up to the synthesis of the iron-peak elements. Beyond that, no further nuclear reactions can be exothermic. Thus, a dense, inner iron core gradually builds up and when its mass reaches roughly the so-called Chandrasekhar limit ($\sim$1.4 $M_{\odot}$), it becomes unstable to a variety of disasters which have a general tendency to cause a rapid collapse of the core and the liberation of large amounts of gravitational potential energy in the form of a shock wave, a process which is referred to as a Type-II SN event (Trimble, 1982, 1983).

What will be the fate of the ^{31}P nuclei produced in the previous hydrostatic burning stages during the final SN explosion event? Is it possible that additional ^{31}P nuclei will be nucleosynthesized in the inner core? Indeed, at the onset of the explosive nucleosynthesis phase, the central core is mainly composed of the heavy nuclei ^{16}O, ^{24}Mg, and ^{28}Si, it is denser and hotter than the C- and Ne-rich outer shells and will be still more heated by the SN event itself. Thus, it makes sense to consider nuclear processes in stellar matter consisting of ^{16}O, ^{24}Mg, and ^{28}Si in various proportions, when it is suddenly heated and then expands and cools in a short timescale. The effects on the chemical composition of the outer layers of an evolved massive star, which might be caused by a sudden heating of the layers, followed by expansion and rapid cooling, constitutes the general scheme named explosive burning. Depending on the fuels present and on the maximum temperature reached, these processes can conveniently be called explosive O-, Ne-, and Si-burning processes. The explosive conditions, taking place between roughly T = 2–3×10^9 K, convert common isotopes produced in the red giant phase, into more rare ones quite efficiently, and therefore only a small portion of the ejected stellar material has to be processed in the explosion to bring the less common isotopes up to their solar system abundances. In fact, relatively little matter is processed because the material cools too quickly.

Therefore, the ratios of the more abundant nuclei will not be disturbed by the explosive nucleosynthesis, which just puts the finishing touches on the abundance distribution and makes it match the details of the observed distribution (Trimble, 1975). This is properly illustrated in Figure 5.5, in which the relative abundances of the products of explosive nucleosynthesis as compared to the observed solar system abundances in the range A = *1–72* are shown (Woosley et al., 2002). In particular, it is noteworthy that the phosphorus yield can be nicely accounted for in terms of explosive nucleosynthesis processes, its relative abundance being even a little bit enhanced over the solar measured value. The most relevant reactions involving ^{31}P nuclide during these processes, occurring after ^{28}Si photodisintegration, are listed in Table 5.1.

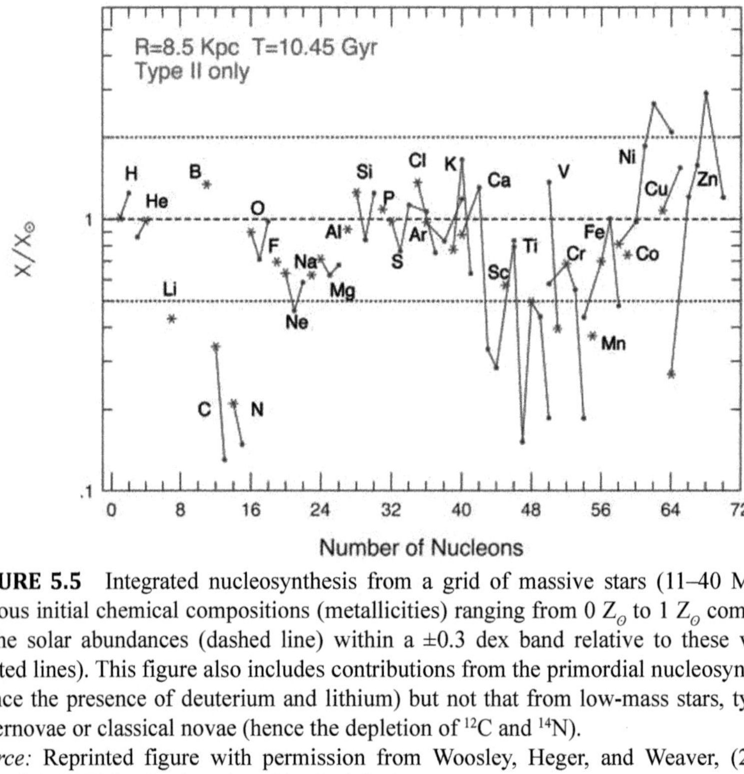

FIGURE 5.5 Integrated nucleosynthesis from a grid of massive stars (11–40 $M_\odot$) of various initial chemical compositions (metallicities) ranging from 0 $Z_\odot$ to 1 $Z_\odot$ compared to the solar abundances (dashed line) within a ±0.3 dex band relative to these values (dotted lines). This figure also includes contributions from the primordial nucleosynthesis (hence the presence of deuterium and lithium) but not that from low-mass stars, type-Ia supernovae or classical novae (hence the depletion of ^{12}C and ^{14}N).
Source: Reprinted figure with permission from Woosley, Heger, and Weaver, (2002). Copyright 2002 by the American Physical Society.

TABLE 5.1 Relevant Channels for the Formation and Consumption of ^{31}P Nuclei During a Supernova Explosion

^{31}P creation	**^{34}S (p, α) ^{31}P**	**Main**
	^{28}Si (α, γ) ^{32}S (γ, p) ^{31}P	Major
	^{35}Cl (γ, α) ^{31}P	Subsidiary
	^{28}Si (n, γ) ^{29}Si (n, γ) ^{30}Si (p, γ) ^{31}P	Subsidiary
^{31}P destruction	^{31}P (p, α) ^{28}Si	Major

Modified from Arnett et al., (1965).

In some works, it was proposed that ^{31}P, along with other odd atomic number nuclei such as ^{23}Na and ^{27}Al, may be produced through neutron capture on ^{30}Si, so that its abundance should be sensitive to the neutron flux intensity in massive stars, in turn depending on the star metallicity. Accordingly, a decreasing ratio of ^{31}P to elements non-dependent on neutron flux, such as the α-related elements O, Ne, and S, is expected on this basis. On the contrary, measured abundances indicate a nearly constant phosphorus-to-sulfur ratio of [P/S] = 0.10 ± 0.10 (Caffau et al., 2011), hence suggesting that ^{31}P production is insensitive to the neutron excess, and thus other processes, such as α-particle capture on ^{27}Al (Emir et al., 2014) or proton capture on ^{30}Si (as shown in Figure 5.3), should be considered instead. In a recent study, the local nucleosynthetic yield of ^{31}P for a spherically symmetric SN model for a progenitor star of mass 15 $M_\odot$ was numerically obtained, and it was observed that ^{31}P production is enhanced in the oxygen-rich layer (ranging from about 0.45 to 0.6 *M(r)/M(core)*). This makes the [P/Fe] ratio in this layer about two orders of magnitude higher than that of the outer helium-rich layer, a ratio which is essentially equal to the cosmic abundance value (Lee et al., 2013). A close inspection of the [P/Fe] versus *M(r)/M(core)* profile within the enhanced ^{31}P region reveals the presence of two significant broad peaks: a peak extending from 0.55 to 0.73 *M(r)/M(core)* is the result of hydrostatic Ne-burning in the pre- SN stage, whereas that located at *M(r)/M(core)* ~ 0.5 is due to explosive Ne-burning during the explosion phase (Lee et al., 2013).

In the light of these results we can conclude that the primary origin of natural ^{31}P nuclei is the inner zone of stars massive enough (say, above 15 $M_\odot$) to undergo hydrostatic C- and Ne-burnings followed by explosive Ne-burning between roughly $T = 2 - 3 \times 10^9$ K, according to the overall network of nuclear reactions depicted in Figure 5.3, ultimately relying on the proton capture on ^{30}Si nuclei (Clayton, 1988). From the given branching ratios, we realize that the final ^{31}P yield is below 2.5%. Although this figure is certainly small, it is enough to account for the observed solar abundances when one properly adds up all the nucleosynthesis products for a representative set of massive stars, as it is shown in Figure 5.5.

Accordingly, some freshly synthesized phosphorus should be expected to be found in young SN remnants resulting from the explosion of massive enough stars (Rauscher et al., 2002). Indeed, the first reported detections of atomic P^+ line at 32.8 μm toward the RCW 103 remnant and at 60.6 μm toward the 3C 391 remnant (Oliva et al., 1999; Gerardy and Fesen, 2001) showed line brightness consistent with a solar abundance of phosphorus, although the question regarding whether some or all of the detected P atoms stemmed from preexisting ISM grains vaporized by the SN event strong shock wave, rather than being direct nucleosynthetic products, remained open at that time (Reach and Rho, 2000). Quite remarkably, an abundance ratio of phosphorus to the major nucleosynthetic product iron of $[P/Fe] = 8.1\times10^{-3}$ (about 100 times the average cosmic ratio), has subsequently been reported in Cassiopeia A SN material, thereby confirming that anew phosphorus atoms have been produced in this SN explosion (Koo et al., 2013). Furthermore, recent detailed spectroscopic studies have revealed that the P/Fe abundance ratio varies almost two orders of magnitude among different locations of this remnant (Lee et al., 2017), hence providing additional evidence that ^{31}P nuclides are efficiently produced in SN explosions.

5.5 CLASSICAL NOVAE AND TYPE I SUPERNOVA (SN) OUTBURSTS

So far, we have considered nuclear processes taking place in single stars. Nonetheless, a significant number of the stars in the galaxy are members of gravitationally bounded double or even multiple stellar systems. Then, some words are in order concerning the possible influence of a companion star in the course of stellar evolution. In particular, we will consider close binary systems consisting of a white dwarf and a large mass giant star. In this kind of systems, the giant companion overfills its so-called Roche lobe, and stellar matter flows outward through the inner Lagrangian point, thus forming an accretion disk around the white dwarf. Some fraction of this material ultimately ends up on top of the white dwarf, where it is gradually compressed by more material that is still being accreted. This compression heats the envelope up to the point at which the ignition conditions for driving thermonuclear processes are reached. In the classical novae scenario, an explosive nuclear reaction takes place on just the surface of the white dwarf, reaching peak luminosity (nova outburst) of about 10^5 $L_\odot$. During the nova outburst, a significant fraction of the formerly accreted envelope is ejected, and the nuclear reactions temporarily stop. Since the temperature attained in the white dwarf envelope during a classical nova explosion reaches typical peak values about $T = 2 - 4 \times 10^8$ K, one may expect that the ejected materials show significant nuclear processing, for low mass nuclides up to A ~ 40 (Trimble, 1983). In fact, some local enhancement effects detected on the binary system white dwarf surface, such as those reported from Hubble space telescope observations indicating the presence of phosphorus with abundances 20–900 times the solar value in the classical nova VW Hydri system, suggest that ^{31}P nuclide may be produced through proton capture during classical nova explosions, particularly those belonging to the O-Ne novae class (Sion et al., 1997; Sion and Sparks, 2014).

On the other hand, in the so-called Type-Ia SN, the accretion onto the white dwarf causes a runaway thermonuclear reaction that consumes the entire volume of the companion star, reaching a much higher peak luminosity of about 10^9 $L_{\odot}$ (Trimble, 1982). When one properly combines the nucleosynthesis products resulting from type-II SN models with the material ejected from classic novae and type-Ia SN, which attain higher peak temperatures allowing for thermonuclear reactions relevant to the formation of ^{31}P nuclides, one appreciates a significant enhancement of this element with respect to the value obtained by considering type-II SN ejecta alone, as it can be readily appreciated by comparing Figures 5.5 and 5.6.

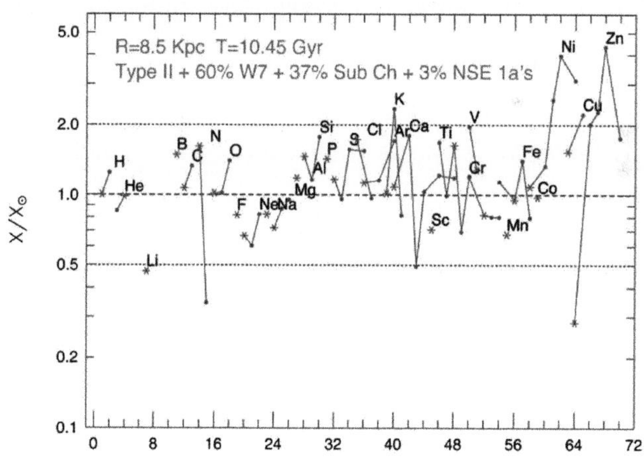

FIGURE 5.6 Integrated nucleosynthesis from the grid of massive stars considered in Figure 5.5, including the nuclear products from type-Ia supernova as well as classical novae compared to the solar abundances. Type-II supernovae produce more α-elements (e.g., O, Mg, Si, S, Ca, and Ti) relative to Fe with respect to the solar abundance ratios, whereas type-Ia supernova produces mostly Fe and little α-elements.

Source: Reprinted figure with permission from Woosley, Heger, and Weaver (2002). Copyright 2002 by the American Physical Society.

5.6 CHEMICAL EVOLUTION OF PHOSPHORUS IN THE GALAXY

As a galaxy evolves the chemical composition of the stars it contains and the gas pervading the ISM progressively change, so that the abundance ratios of different elements in these astrophysical objects provide useful information regarding the underlying processes. In particular, the measurement of elemental abundances in main sequence F-, G-, and K-type stars allow for the study of the chemical evolution of phosphorus and other biogenic elements in our Galaxy using these long-lived stars as suitable tracers. Thus, high-resolution spectroscopy observations of late spectral type individual stars in our Galaxy, nearby galaxies and high-redshift quasars have shown different elemental abundance patterns, which suggest different chemical enrichment histories of these objects (Kobayashi et al., 2006).

The passage of time in galactic chemical evolution is usually tracked in terms of the so-called metallicity ratio [Fe/H] $\equiv \log\frac{N_{Fe}}{N_H} - \log\left(\frac{N_{Fe}}{N_H}\right)_{\odot}$, which systematically increases as iron atoms are progressively released to the ISM by SN explosions. As we mentioned in the previous Section, type-II SN ejected materials enrich the ISM with α-elements such as O, Mg, Si, S, Ca, and Ti atoms relative to iron, while type-Ia SN ejected materials enhance the ISM iron content instead. Accordingly, [X/Fe] versus [Fe/H] and [X/H] versus [Fe/H] plots (see Figure 3.3) are very useful diagnostic tools to study the galactic chemical evolution of a given element X.

During past decades chemical evolution studies of phosphorus remained quite elusive because this element is difficult to detect in the photospheres of relatively cool, late-type stars (see Section 6.1). Fortunately, recent technological advances have allowed for the observation of this biogenic element in an increasing number of the main sequence stars in our Galaxy (Caffau et al., 2011; Cescutti et al., 2012; Jacobson et al., 2014; Roederer et al., 2014; Maas et al., 2017), as well as in other sources elsewhere (Oliva et al., 2001; Crowther et al., 2002; Hubrig et al., 2009). For instance, the average phosphorus abundance in a small sample of horizontal branch B-type stars in the globular clusters NGC 6397 and NGC 6752 appears strongly enhanced by 1.9 dex and 1.6 dex, respectively, as compared to solar abundances (Hubrig et al., 2009), probably due to strong diffusion effects in their atmospheres (see Section 6.1).

In Figure 5.7, we show the [P/H] versus [Fe/H] curve for a sample of FGK stars displaying a broad range of metallicity values. Quite remarkably, we can clearly appreciate a smooth and very well defined building up of over ~ 4 orders of magnitude in phosphorus element with the galactic evolution time as measured in terms of the metallicity ratio (Jacobson et al., 2014).

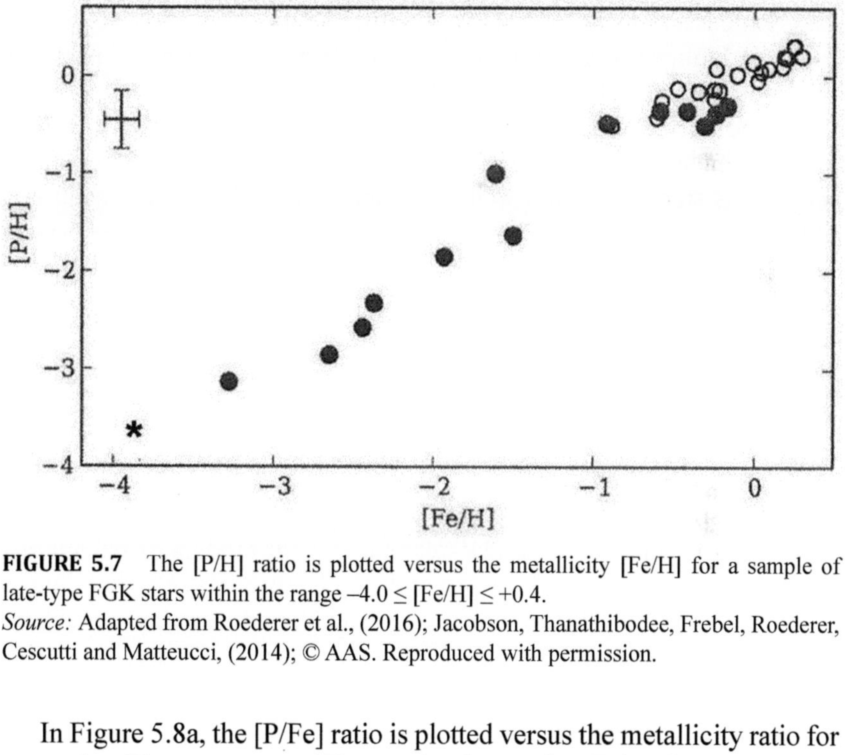

FIGURE 5.7 The [P/H] ratio is plotted versus the metallicity [Fe/H] for a sample of late-type FGK stars within the range $-4.0 \leq$ [Fe/H] $\leq +0.4$.
Source: Adapted from Roederer et al., (2016); Jacobson, Thanathibodee, Frebel, Roederer, Cescutti and Matteucci, (2014); © AAS. Reproduced with permission.

In Figure 5.8a, the [P/Fe] ratio is plotted versus the metallicity ratio for a sample of late-type stars. We can clearly see that [P/Fe] is close to zero for solar metallicity stars, and increases as [Fe/H] decreases backwards in time. In Figure 5.8b, the considered [Fe/H] range is significantly expanded by including poor metallicity stars up to [Fe/H] ~ –4.0. Below [Fe/H] ~ –2.0 the points scatter around the [P/Fe] ~ 0.0 value, which is consistent with a primary origin of phosphorus atoms in SN events related to very old, low metallicity massive stars, probably including population III representatives. In this regard, it is noteworthy the location of the extremely metal-poor ([Fe/H] = –3.8 ± 0.19) red giant halo star BD+44°493 (star symbol in Figures 5.7 and 5.8b), which has been proposed as a candidate second-generation star enriched by metals from a single Population III star only, and exhibiting the low value [P/Fe] = –0.34 ± 0.21 (Roederer et al., 2016). Starting from [Fe/H] ~ –1.5 on, the [P/Fe] ratio starts to increase with increasing metallicity, before returning down to [P/Fe] ~ 0.0 near solar metallicity values. This behavior closely resembles that predicted by the chemical evolution models of Kobayashi et al., (2011), as can be seen in Figures 5.8b and 5.8c. In fact, if one considers stars whose formation time-scale is shorter than the average in the solar neighborhood (solid line model in Figure 5.8c), as is the case in our galaxy bulge (long-dashed line model in Figure 5.8c) and thick disk (dot-dashed line model in Figure 5.8c), the chemical enrichment contribution from these stars appears at a higher metallicity than in the solar neighborhood. Thus, intermediate-mass AGB stars, low-mass AGB stars and Type Ia SN start to contribute at [Fe/H] ~ –2.5, –1.5, and –1.0, respectively, in the solar neighborhood, but at a higher [Fe/H] in the bulge and thick disk models. Accordingly, the abundance ratio [P/Fe] is predicted to be higher in these places because of their higher metallicity.

On the other hand, if the chemical enrichment time-scale is longer than in the solar neighborhood, as occurs in our galaxy halo (short-dashed line model in Figure 5.8c), the contribution from low-mass AGB stars becomes significant, and the [P/Fe] ratio is lower because of the overall lower metallicity (Kobayashi et al., 2011). By closely inspecting Figure 5.8a one realizes that current models of phosphorus production (solid line) do not match well the observed abundances, as they generally predict too low P abundances for the considered galactic disk stars sample. To match the observed [P/Fe] ratios, the current SN explosion phosphorus yields should be increased by factors of 1.5–3, as illustrated by the dashed line in Figure 5.8a (Cescutti et al., 2012; Maas et al., 2017). In addition, by comparing the low metallicity observational data shown in Figure 5.8b with suitable chemical models by Cescutti et al., (2012), it has been concluded that hypernovae (SN explosions of larger than 20 $M_\odot$ stars releasing ~ 10^{45} J) were important sources of phosphorus in the early universe (Jacobson et al., 2014; Maas et al., 2017).

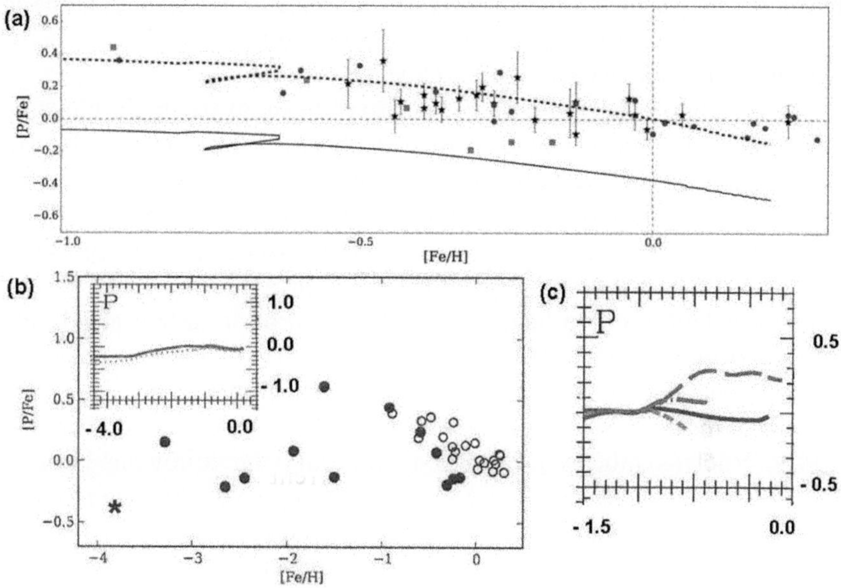

FIGURE 5.8 Phosphorus abundances relative to iron are plotted as a function of the metallicity ratio [Fe/H] for a sample of late-type FGK stars (including several solar twins) within the ranges (a) $-1.0 \leq$ [Fe/H] $\leq +0.4$. (Source: Maas, Pilachowski, and Cescutti (2017); © AAS. Reproduced with permission), and (b) $-4.0 \leq$ [Fe/H] $\leq +0.4$ (Roederer et al., 2016). The dashed (solid) lines in (a) are chemical evolution model yields of [P/Fe] in the solar neighborhood. For the sake of comparison, a [P/Fe] versus [Fe/H] curve obtained from chemical evolution numerical models (Kobayashi et al., 2011) are shown in the inset of (b) and (c). The solid (dotted) line in the inset includes contributions from Type II and Ia SN as well as hypernovae (along with rotation effects).
Adapted from Kobayashi, Karakas, and Umeda (2011); by permission of Oxford University Press.

5.7 SUMMARY AND REVIEW QUESTIONS

By all indications, the first phosphorus atoms were synthesized and released by very massive Population III stars when the universe was about 4×10^8 years old (see Table 2.1). Since then phosphorus has mainly been formed in massive enough stars, say $M \geq 8\ M_{\odot}$, mainly in explosive C- and Ne-burning layers undergoing SN explosion, preferably by proton capture on ^{30}Si nuclei (see Figure 5.3) and proton capture on ^{34}S nuclei (see Table 5.1), and subsidiarily during hydrostatic Ne-burning shells in the pre-SN stage, likely involving α capture on ^{27}Al nuclei. Other types of explosive nucleosynthesis, such as those occurring in classical novae and in type-Ia SN, have also significantly contributed to the current inventory of phosphorus in the Galaxy (see Figures 5.5 and 5.6).

Thus, when compared to the nucleosynthesis of the main biogenic elements C, O, and N, the case of phosphorus is quite remarkable. In the first place, the nucleosynthesis of ^{31}P nuclides can only take place in the minor subset of stars which are massive enough to explosively ignite the C and Ne fuels, in the range of 15–100 $M_{\odot}$. This represents less than 10% of the star population in a typical galaxy. In fact, stars somewhat less massive than 4 $M_{\odot}$ will fail to burn carbon at all, and will probably become white dwarfs, while stars of intermediate-mass (4–8 $M_{\odot}$) may be entirely disrupted by carbon ignition. Therefore, only the most massive stars enter C-burning phase quiescently (Trimble, 1982). Nucleosynthesis processes in this and the following burning stages are characterized by a great variety of nuclear reactions made possible by the progressively higher temperatures reached in the star inner shells, the proliferation of trace elements from previous burning stages, and the fact that some of the key reactions, like carbon and oxygen fusion, liberate free neutrons, protons, and α particles able to readily react with surrounding nuclei. Except for a range of transition masses around 8–11 $M_{\odot}$, each massive star ignites a successive burning state at its center using the ashes of the previous stage as a fuel for the next. Therefore, explosive nucleosynthesis conditions, involving many competing nuclear processes, occurring not only in type-II SN, but in type-Ia SN and classical novae as well, are required to get phosphorus in the amounts demanded by astronomical observations. This implies that, in contrast to the other main biogenic elements, the synthesis of ^{31}P proceeds through rather involved pathways, which are taking place in a minor fraction of the overall stellar population consisting of very massive stars during the last and violent episodes of stellar evolution. These facts, leading to an overall low ^{31}P nuclide nucleosynthesis yield, properly explain why phosphorus is so scarce as compared with the other main biogenic elements.

Albeit scarce, phosphorus is also evenly distributed throughout the universe. Two main factors contribute to account for its reported ubiquity. On the one hand, the explosive conditions leading to phosphorus formation guarantee that this element will be efficiently spread out in rapidly expanding SN remnants progressively diluting into the ISM. On the other hand, the high stability of ^{31}P nuclei prevents them to be consumed by the relatively low-temperature nuclear reactions taking place in low mass stars belonging to Populations II and I, which originally incorporated these atoms from ISM during their protostellar condensation phase. The P nuclei so preserved in low mass stars are subsequently exposed in their external layers due to dredge up stellar convective motions, and finally returned back to the ISM again during the late planetary nebula phase. In this way, substantial amounts of phosphorus atoms are recycled from low mass stars to the ISM, where they will be ready to react with other available atoms to form molecules of increasing complexity.

I would like to conclude this chapter by considering a couple of questions to think about. The first one refers to the importance of the continued contribution of atoms heavier than helium delivered by stars belonging to the Populations III and II over the time. This can be appreciated by realizing that only about 25% of the current inventory of helium in the universe has a primordial origin (see Section 5.1). Bearing in mind that most of the He nuclei produced during the H-burning stage are subsequently consumed during the 3α burning phase in low mass stars, as well as by the higher temperature burning stages from carbon to silicon in high mass stars, we can figure out that the number of stars required to obtain the remaining 75% helium atoms must be really huge. Would you venture to make a quantitative guess?

The second question has to do with backyard astronomy. Indeed, one may wonder as to whether phosphorus nuclei are being synthesized in the cores of some of the most conspicuous stars we can see in a starry night. On the basis of the information given through the Sections 5.4.1 to 5.4.7, and after inspecting the H-R diagram information provided in Section 2.3.2, I guess that P nuclei synthesis may be nowadays going on in Betelgeuse (α Orionis, M2-type, $M = 18–19\ M_\odot$, $L = 4.1 \times 10^4\ L_\odot$) and Antares (α Scorpius, M1-type, $M = 15.5\ M_\odot$, $L = 3.7 \times 10^4\ L_\odot$) stars, and it would be expected to occur in the near future (astronomically speaking) in the stars Deneb (α Cygnus, A2-type, $M = 19\ M_\odot$, $L = 3.2 \times 10^5\ L_\odot$), Rigel (β Orionis, B8-type, $M = 23\ M_\odot$, $L = 7 \times 10^5\ L_\odot$), Spica (α Virginis, B1-type, $M = 10\ M_\odot$, $2.5 \times 10^4\ L_\odot$), or Mintaka (δ Orionis, O-type, $M = 20\ M_\odot$). Would you agree? Why (or why not)?

KEYWORDS

- classical novae
- Coulomb barrier
- explosive nucleosynthesis
- H-, He-, C-, Ne-, O-, and Si-burnings
- metallicity
- phosphorus chemical evolution
- population III stars
- primordial nucleosynthesis
- radiation pressure
- resonant quantum tunneling
- thermonuclear reactions
- types Ia and II supernovae

FIGURE 2.2 The cycling of matter and energy in the interstellar medium of spiral galaxies (background) involves the injection of material from stars stemming from condensing dense molecular clouds (M42), first in the form of continuous wind (young open cluster M45) and later on during briefer episodes which take place in their final stages. The majority of stars (those with masses comprised in the range 1–8 $M_{\odot}$) transition into planetary nebulae (M57), a stage which lasts about 10,000 years, where the star sheds its outer layers, losing almost all of its original mass and becoming a very hot, UV emitting white dwarf that ionizes its surrounding envelope, rendering it pretty colorful (Kwok, 2010). Higher mass stars undergo explosive events, yielding rapidly expanding supernova remnants (M1). The materials from these nebulae flow into the diffuse ISM, forming relatively dense clouds (ρ Oph) in its due time. These clouds eventually collapse into even denser molecular clouds (M16), which generally become gravitationally unstable and form a new generation of stars along with preplanetary disks ultimately evolving into planetary systems (β-Pic). During this stage, most of the material originally released from the original stars is heavily processed, but some pristine matter from molecular clouds still survives, preserved in planetesimals and the nuclei of comets. (The Messier objects M1, M15, M16, M42, M45, and M57, as well as the background galaxy NGC 1232, are reproduced by courtesy of The Hubble Heritage Team – AURA/STScI/NASA, Baltimore, MD).
Source: The artist's conception showing planet formation around the very young type A star β Pictoris was downloaded from http://imagine.gsfc.nasa.gov/Images/bios/roberge/BetaPictoris.jpg; NASA/FUSE/Lynette Cook.

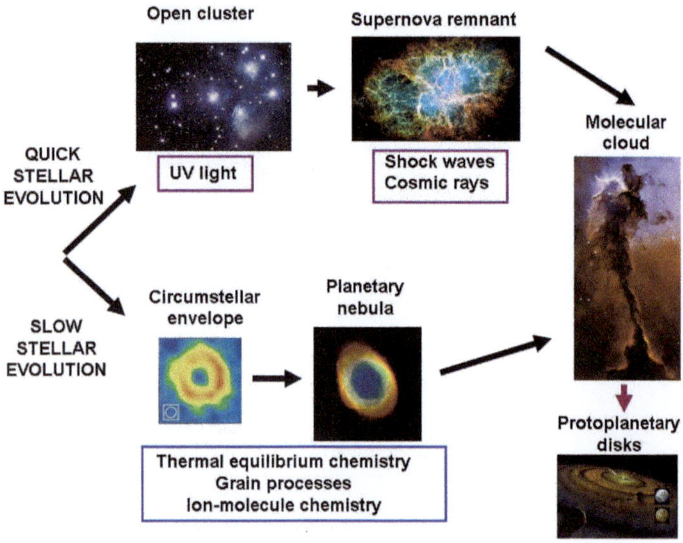

FIGURE 2.3 This diagram indicates the physicochemical mechanisms that dominate molecule and dust-grain production in several astrophysical objects shown in Figure 2.2. Additional information regarding most astrophysical objects in this diagram is given in Section 2.4.

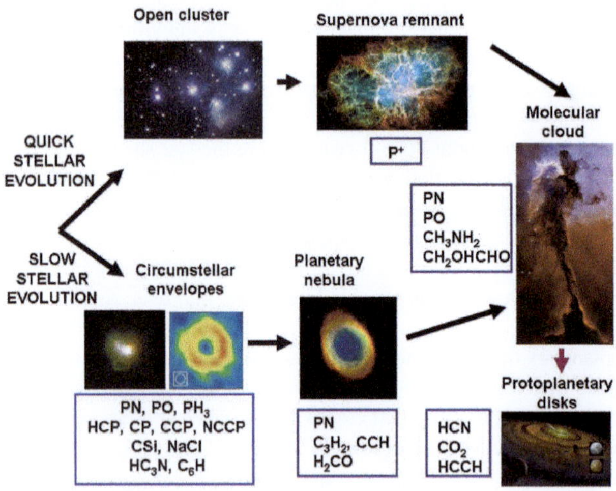

FIGURE 2.4 This diagram illustrates how molecules and dust grains cycle from circumstellar envelopes and supernova remnants through various phases of interstellar matter in the Galaxy. Some representative molecules appearing in different astrophysical objects during the whole ISM life cycle are indicated (Tielens, 2013; Ziurys et al., 2015), including the seven phosphorus compounds detected to date (see Table 1.2). Additional information regarding most astrophysical objects in this diagram is given in Section 2.4.

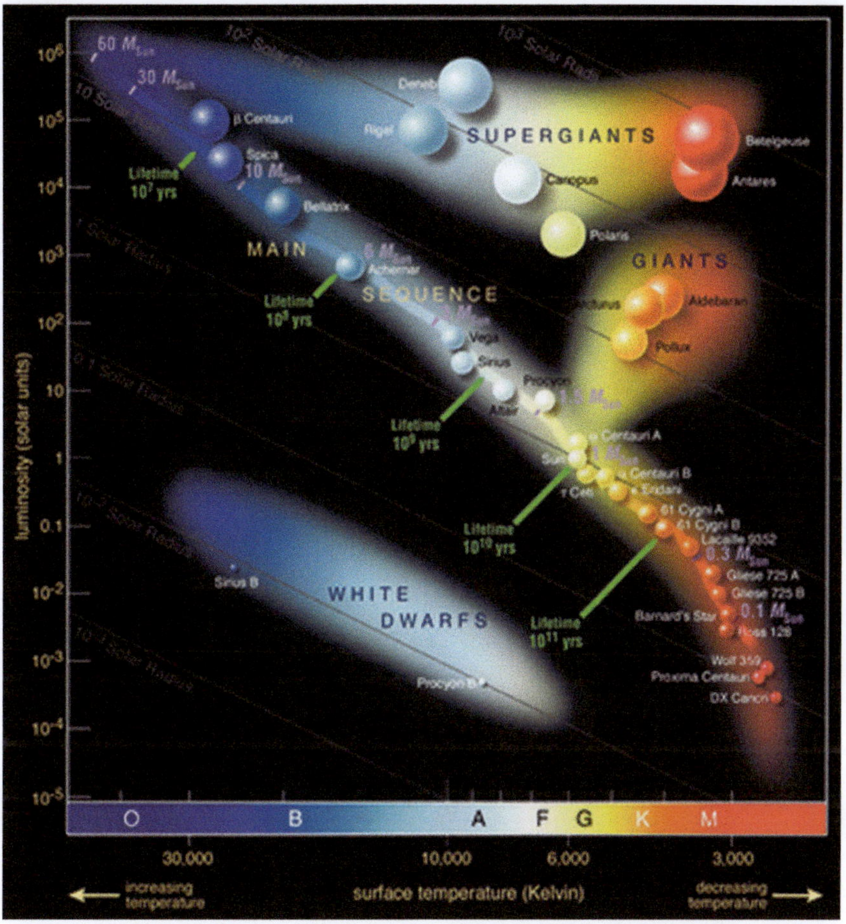

FIGURE 2.6 The Hertzsprung-Russell (HR) diagram displays the luminosity of a star against its surface temperature (spectral class). Stars of greater luminosity are toward the top of the diagram, and stars with higher surface temperature are toward the left side of the diagram. By ESO - https://www.eso.org/public/images/eso0728c/.

FIGURE 2.7 Rho (ρ) Ophiuchi cloud
Reproduced by courtesy of NASA/JPL-Caltech/WISE Team; https://commons.wikimedia.org/wiki/File:Rho_Ophiuchi.jpg.

FIGURE 2.8 Pleiades open cluster
Reproduced by courtesy of The Hubble Heritage Team – AURA/STScI/NASA, Baltimore, MD.

FIGURE 2.9 Orion molecular cloud
Reproduced by courtesy of The Hubble Heritage Team – AURA/STScI/NASA, Baltimore, MD.

FIGURE 2.11 This color composite image of VY Canis Majoris consists of a mixture of HST Wide Field and Planetary Camera 2 images, taken in blue, green, red, and near-infrared light. The image reveals a complex pattern of circumstellar ejecta, with arcs, filaments, and knots of material formed by the massive outflows stemming from several outbursts. (Composite image created from HST data taken by R. Humphreys.
Reproduced by courtesy of The Hubble Heritage Team – AURA/STScI/NASA, Baltimore, MD.

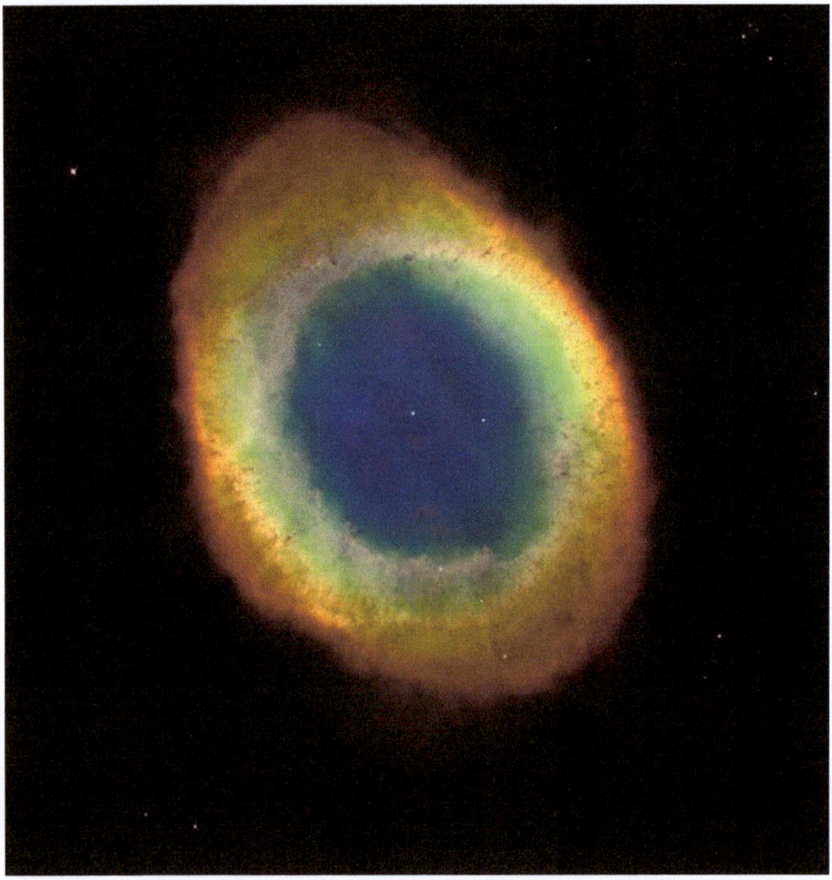

FIGURE 2.12 NASA's Hubble Space Telescope captured this sharp view of the best known Ring Nebula (M57). The photo, in approximately true colors, shows long dark clumps of material in the gas at the edge of the nebula, and the dying central star floating in a hot gas blue haze. The nebula is about a light-year in diameter and is located some 2,000 light-years from earth towards the constellation Lyra. Going outwards from the very hot central star, blue shows emission from helium atoms, green represents ionized oxygen, and red shows ionized nitrogen, radiated from the coolest, farthest gas. The ultraviolet radiation from the remnant central star, with a white-hot 120,000 K surface temperature, gives rise to the gradations of color in the glowing gas.
Reproduced by courtesy of The Hubble Heritage Team – AURA/STScI/NASA, Baltimore, MD.

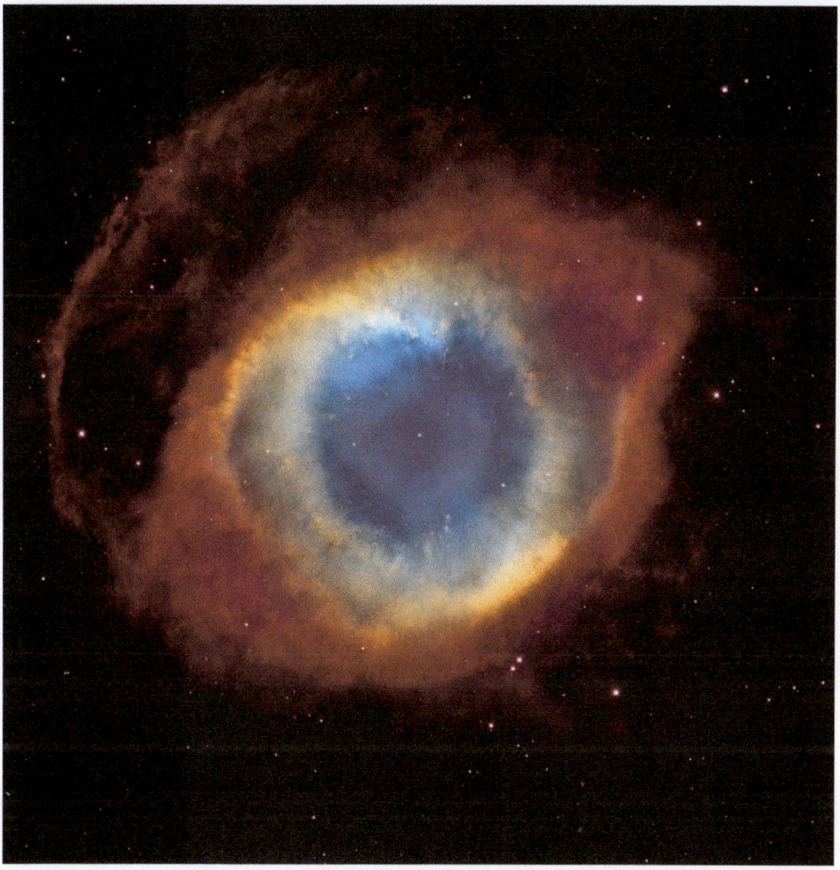

FIGURE 2.13 Covering an apparent area in the sky equal to half the full moon (actually corresponding to 5.6 light-years across) the Helix nebula (NGC 7293) is the closest of all known planetary nebulae, and it lies 140 parsecs (456 light-years) away from Earth in the constellation Aquarius. The star that ejected these gases is at the center of the glowing shell. The bluish-green color comes from ionized oxygen, the pink and red from ionized nitrogen and hydrogen. The small radial blobs in the red shell (each one is about 150 AU across) give the object its alternate name, the Sunflower Nebula. The ejected shell is actually nearly spherical; it looks like a ring because of the substantial thickness of the shell when we look near the shell's rim.
Reproduced by courtesy of The Hubble Heritage Team – AURA/STScI/NASA, Baltimore, MD.

FIGURE 2.14 The Crab Nebula, named for the arm like the appearance of filamentary structure, is the remnant of a supernova observed on July 4, 1054 A. D. in the constellation of Taurus by the imperial astronomer to the Chinese court, Yang Wei-T'e. A thousand years after the explosion these gases are still moving outward at about 1,800 km s^{-1}. The distance of the nebula is about 2,000 parsecs (6,520 light-years), so its present angular dimensions of 4x6 arc minutes correspond to linear dimensions of approximately 2 by 3 parsecs (6.5 by 9.8 light-years). At the center of the nebula, there is a rapidly rotating neutron star. This pulsar, the first one discovered, has a period of 0.033s. The cloud of gases shines with a luminosity of 75,000 times the Sun's powered by the synchrotron radiation stemming from electrons whirling along the powerful magnetic field of the neutron star.
Reproduced by courtesy of The Hubble Heritage Team – AURA/STScI/NASA, Baltimore, MD.

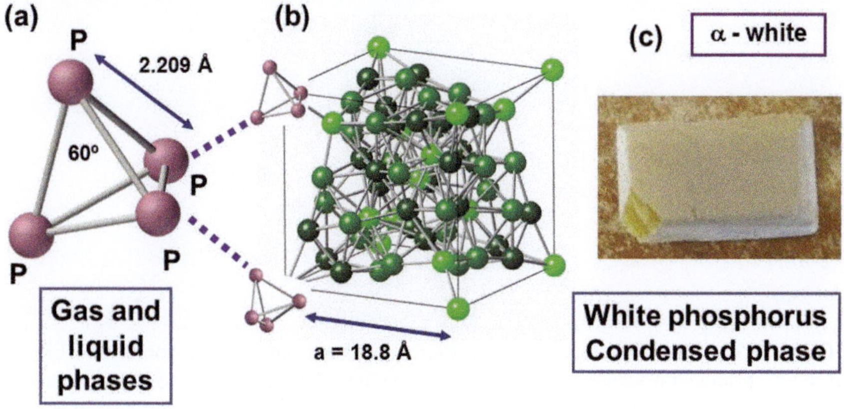

FIGURE 4.1 The α-white phosphorus is a molecular crystal whose structure is based on stable P_4 tetrahedral molecules (a) with a bonding energy of 3.08 eV per atom, a P-P distance of d = 2.209 **Å**, and P-P-P bond angles of 60 ± 0.5°, which arrange in a complex body-centered-cubic unit cell shown in (b), with a large lattice constant of a = 18.8 Å, where each ball in the reference α – Mn structure (shown in green) is replaced by a P_4 molecule. In (c) we show a centimeter-sized crystal of α – white phosphorus.

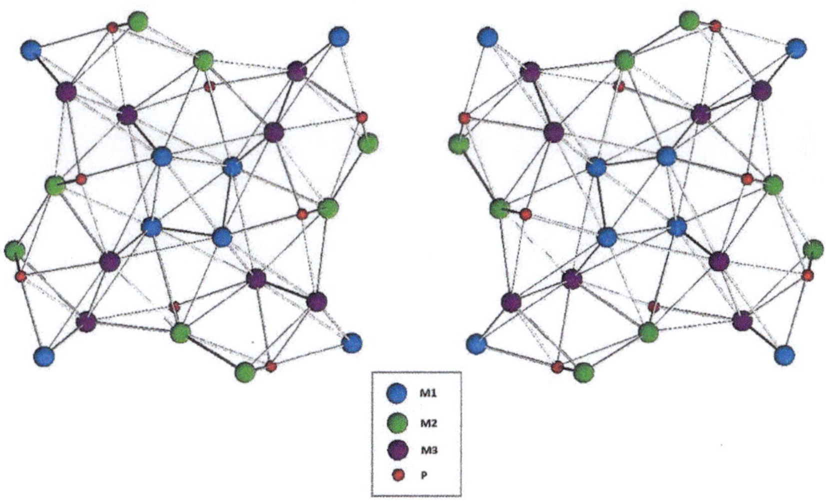

FIGURE 4.9 The regular (on the left) and inverse (on the right) enanthiomorphic crystal structures of schreibersite.
By courtesy of Cameron Pritekel, (2015).

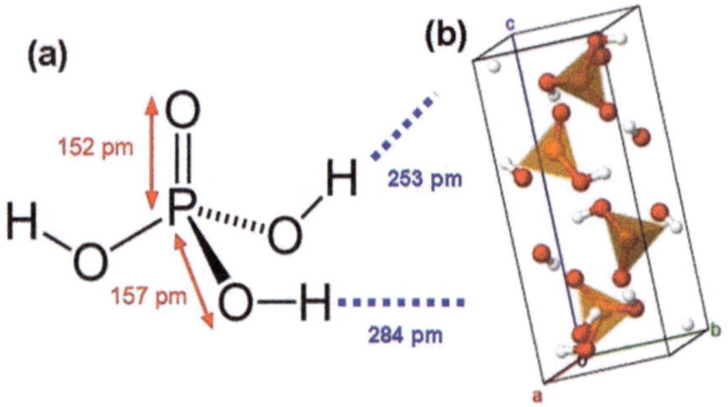

FIGURE 4.13 (a) Molecular structure of the orthophosphoric acid molecule H_3PO_4 exhibiting tetrahedral symmetry; (b) crystal structure of the anhydrous orthophosphoric acid crystal containing four tetrahedral H_3PO_4 molecules in the unit cell (Souhassou et al., 1995). The P, O, and H atoms are shown in orange, red, and white, respectively. The 3D unit cell view has been generated employing the *Jmol* online service provided by IUCr journals.

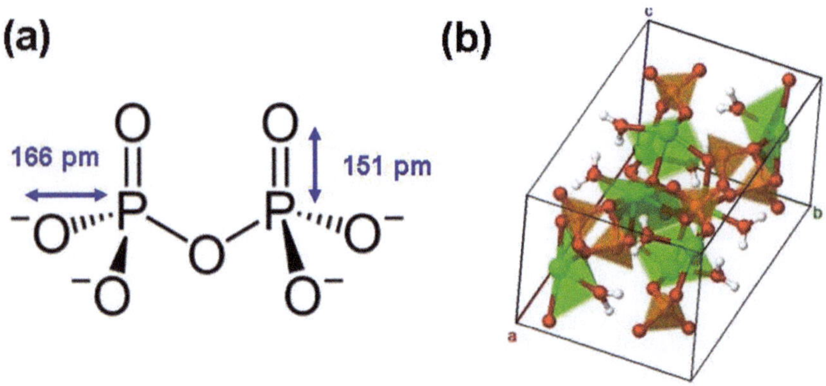

FIGURE 4.14 (a) Molecular structure of the pyrophosphoric acid molecular ion $P_2O_7^{4-}$ (angle and bonds obtained from Jmol); (b) crystal structure of the calcium pyrophosphate dihydrate ($Ca_2P_2O_7{\cdot}2H_2O$) containing four pyrophosphate molecules (in orange) in the unit cell (Gras et al., 2016). The Ca, P, O, and H atoms are shown in green, orange, red, and white, respectively. The 3D unit cell view has been generated employing the *Jmol* online service provided by IUCr journals.

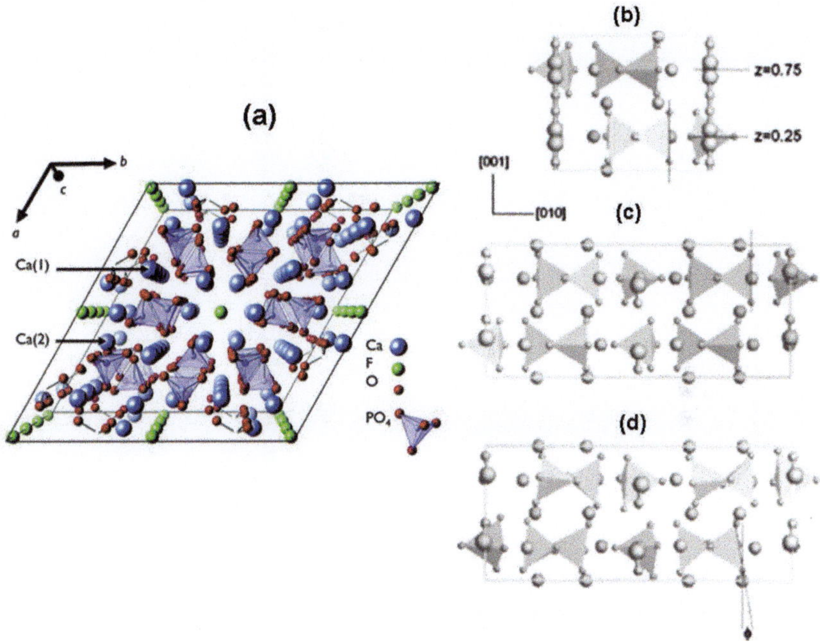

FIGURE 4.17 (a) Structural model for the fluorapatite, with the two calcium sites indicated. Ca, F, and O atoms are in blue, green, and red, respectively, and phosphoryl groups are shown explicitly. (Source: Reproduced from Jay, Rushton, and Grimes, (2012); with permission of The Royal Society of Chemistry). Different structural models of hydroxyapatite: (b) disordered $P6_3/m$ structure, having the superposition of the reflected OH^- and the $O_3 - O_3$ edge parallel to the c axis, (c) superstructure with the $P6_3/m$ arrangement of the Ca and PO_4 framework, but with an ordered alternate arrangement of the OH^- ions, (d) the monoclinic $P2_1/b$ structure having a relaxed Ca-PO_4 arrangement in addition to the alternate orientation of the OH^- ion. (Source: Reprinted figure with permission from Haverty, Tofail, Stanton, and McMonagle, (2005). Copyright (2005) by the American Physical Society.)

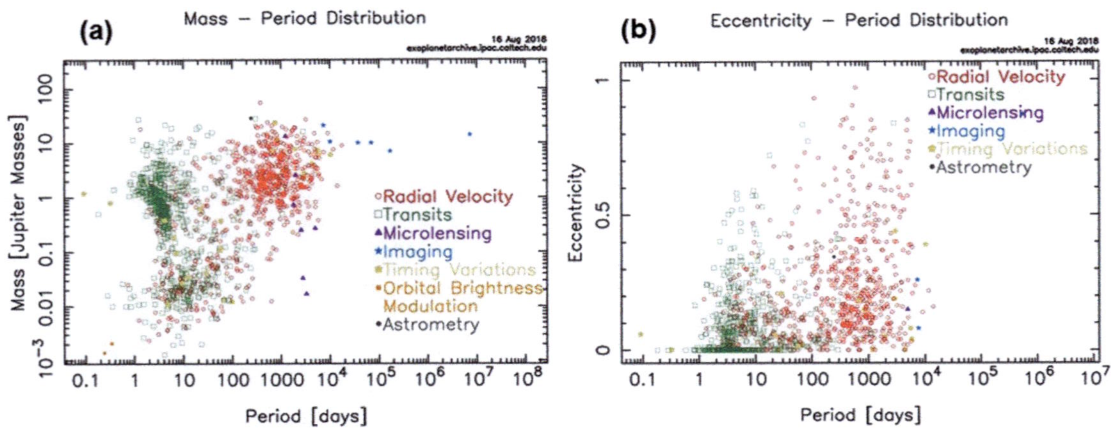

FIGURE 7.4 Graph showing exoplanet (a) masses (measured in terms of Jupiter's mass) and (b) eccentricity against orbital period (in days). The different colors indicate different detection techniques.

Source: NASA Exoplanet Archive, operated by the California Institute of Technology, under contract with the National Aeronautics and Space Administration under the Exoplanet Exploration Program with permission.

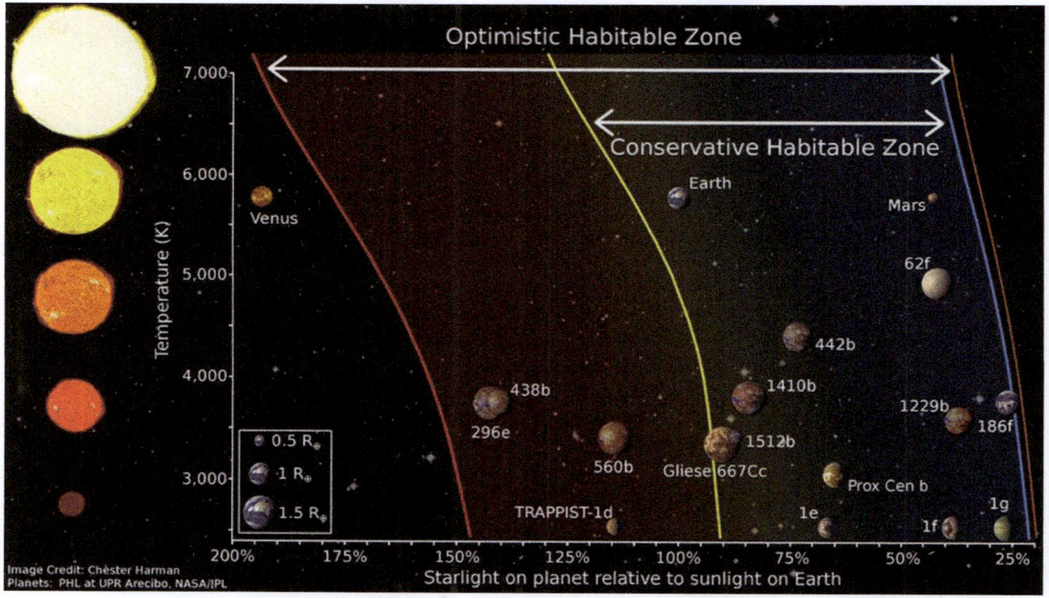

FIGURE 7.5 A diagram depicting the habitable zone (HZ) boundaries, and how the boundaries are affected by stars ranging in spectral type from F to M. For the sake of comparison this plot includes solar system planets (Venus, Earth, and Mars) as well as especially significant exoplanets such as TRAPPIST-1d, Kepler 186f, and our nearest neighbor Proxima Centauri B. This plot shows the limits for both the "conservative habitable zone," which are based on one-dimensional climate model calculations (Kopparapu et al., 2013), and for the "optimistic habitable zone," which are based on observations that Mars once had liquid water at the surface and Venus used to have more water, possibly contained in oceans.

Credit: Chester "Sonny" Harman, using planet images published by the Planetary Habitability Lab at Aricebo, NASA, and JPL.

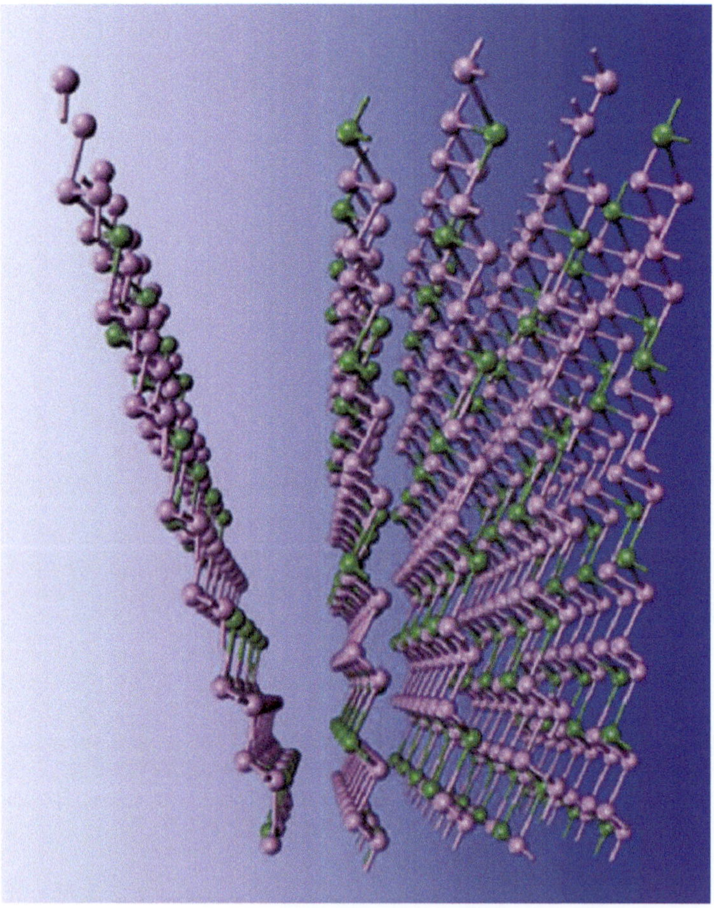

FIGURE 9.6 Atomic structure of GeP_3. Phosphorus (germane) atoms are shown in pink (green).
Reprinted with permission from Jing, Ma, Li, and Heine, (2017); Copyright (2017) American Chemical Society.

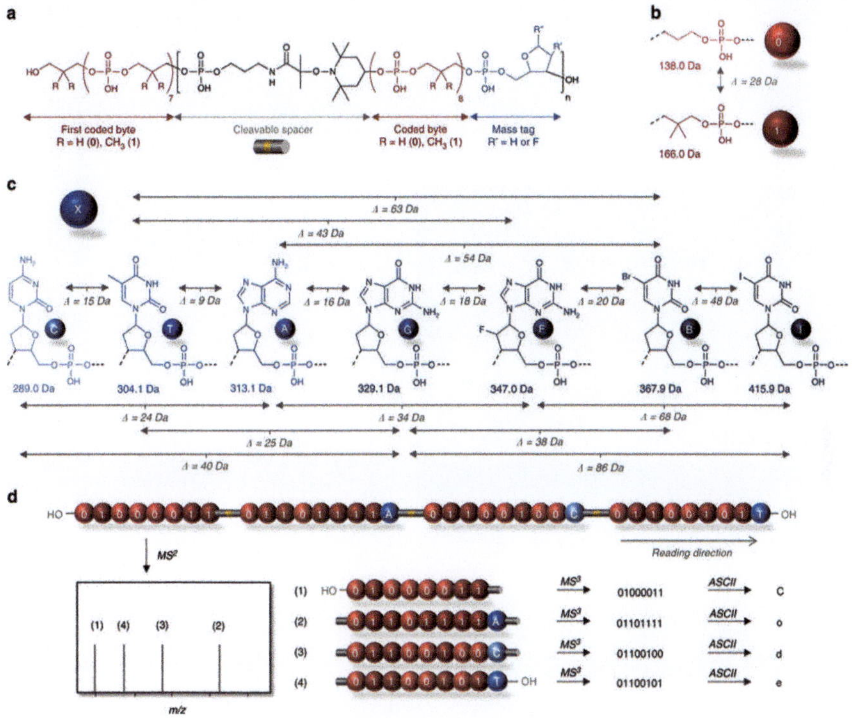

FIGURE 9.8 General concept for the sequencing of long digital polymer chains. (a) Molecular structure of a digital polymer contains n + 1 coded bytes (in red; a byte is a sequence of eight coded monomers that represent 8 bits). Two consecutive bytes are separated by a linker noted in black, which contains an N-O-C bond that can be preferentially cleaved during mass spectroscopy analysis. In order to sort out the bytes after cleavage, n bytes of the sequence are labeled with a mass tag noted in blue. (b) Molecular structure and mass of the two coded synthons that define the binary code in the polymers. (c) Molecular structure and mass of the tags that are used as bytes labels. In order to induce identifiable mass shifts after cleavage, the mass of a byte tag (noted in blue) shall not be a multiple of 28, which is the mass difference between a 0 and a 1 coded unit. In addition, the mass difference between two tags (noted in grey) shall not be a multiple of 28. (d) Schematic representation of the mass spectrometry sequencing of a digital polymer containing 4 bytes of information.
Source: Al Ouahabi, Amalian, Charles and Lutz, (2017); Creative Commons CC BY 4.0 license.

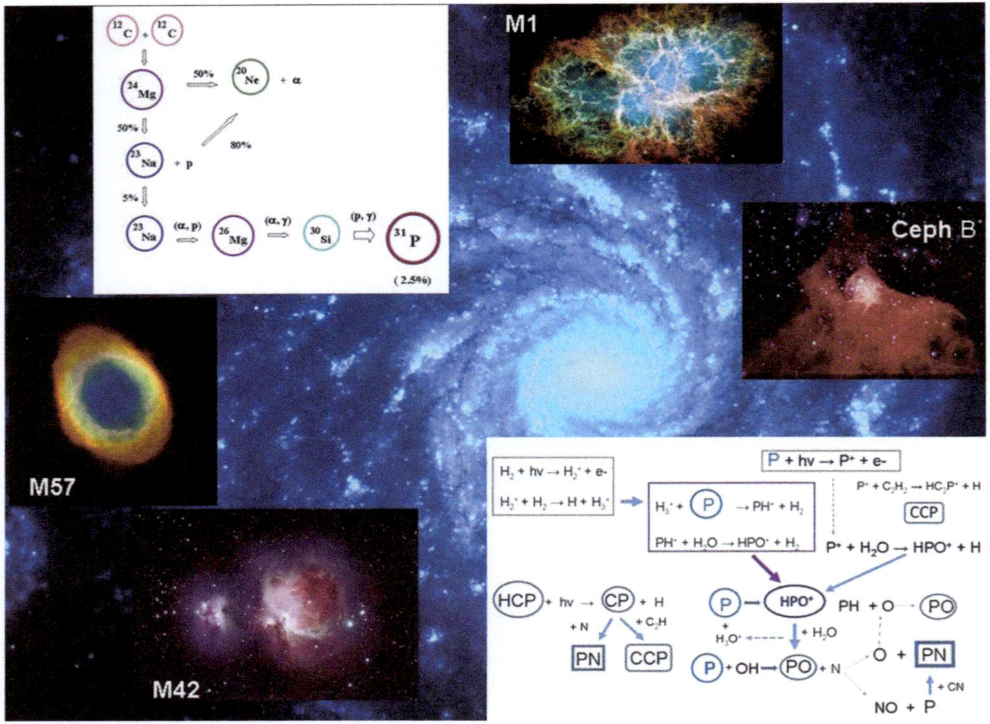

FIGURE 10.3 The cycling of matter and energy considered at different scales, ranging from thermonuclear reactions taking place at a subatomic scale inside stellar cores (upper left box) to chemical reaction pathways occurring at the atomic and molecular scales (lower right box) in the interstellar medium of spiral galaxies.

CHAPTER 6

From Stellar Atmospheres to the Interstellar Medium and Back Again

"Much can still be learned about the physical structures of interstellar clouds and the formation of molecules by a combination of observational, theoretical and experimental studies of the smallest such molecules."

(Ewine F. van Dishoeck & John H. Black, 1986)

In this chapter, we will review the route of phosphorus atoms from stellar photospheres to circumstellar envelopes first, and then all the way up to its dissemination throughout the ISM via planetary nebulae outflows (low mass stars) or supernova (SN) remnants (high mass stars). Thereafter, our journey will continue by considering the slow collapse of diffuse interstellar gas under the action of gravitational forces leading to the formation of dense molecular clouds and the emergence of protostars within them, as it is pictorially sketched in Figure 6.1.

Albeit the molecular phosphorus inventory in the ISM is probably quite incomplete at the time being, a glimpse to the molecules listed in Figure 6.1 suggests that chemical synthesis of phosphorus bearing species quickly and efficiently proceeds in circumstellar regions, and it likely slows down during the planetary nebula stage. Conversely, most molecules formed in circumstellar regions around giant and supergiant stars, such as IRC +10216 and VY CMa, are subsequently destroyed during the energetic events taking place in SN explosions, followed by photodissociation processes in the diffuse ISM regions. Later on, molecular chemical synthesis restarts in dense molecular clouds and stellar forming regions preceding the formation of planetary systems. Unfortunately, the scarce information currently available regarding phosphorus compounds within these regions does not yet allow us to attain a complete picture encompassing all the chemical processes involving P atoms in such diverse astrophysical scenarios, particularly if we take into account the possible presence of phosphorus, not only in the gas phase, but also in still unknown condensed matter forms trapped in interstellar grains.

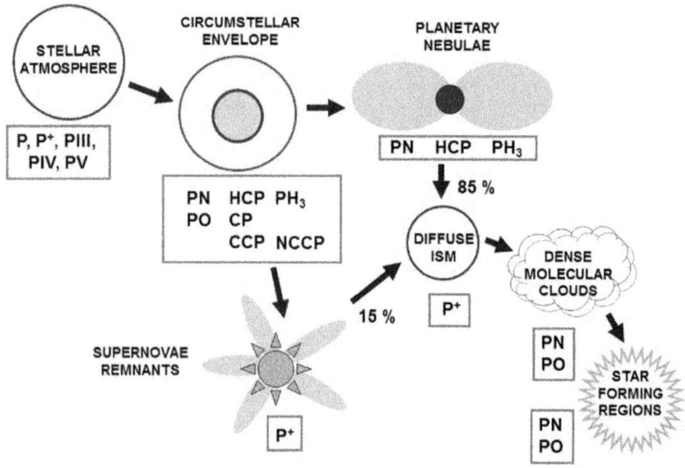

FIGURE 6.1 Sketch illustrating different astrophysical scenarios where phosphorus compounds can be formed and destroyed in the course of the molecular chemical evolution of a galaxy. About 85% of the ISM materials originate with mass loss from low-mass stars that do not become supernovae. Phosphorus bearing species which have been currently detected in some representative objects of each class in our Galaxy are listed in the boxes. Most phosphorus compounds have been detected around circumstellar envelopes.

6.1 PHOSPHORUS IN STELLAR PHOTOSPHERES

Stars are born from collapsing clouds of dust and gas pervading the ISM, hence incorporating all the elements originally present at this medium. As we have learned in Chapter 5, during their life cycle massive stars process these raw materials and enrich them with the contribution of their nucleosynthetic products, including phosphorus nuclei in the case of the more massive ones. On the contrary, the amount of phosphorus should not be modified in the interior of low- and medium-mass stars, where convective motions will progressively dredge-up these and other heavy atoms from the inner regions were they originally condensed during the protostar stage to the outer layers near the star photosphere. Therefore, the presence of heavy atoms in a star's photosphere does not necessarily imply that these atoms were formed in the considered star.

The presence of phosphorus in a stellar atmosphere was first reported by Struve (1930) by inspecting several absorption lines within the range 4,060.7–4,246.6 Å in the spectrum of the B2-type star 88 γ Pegasi, which were properly identified as belonging to doubly ionized (PIII) atoms.* Since then, neutral phosphorus atoms (which only show up spectral lines in the far UV) and the ions PII, PIII, and PV have been detected in the atmospheres of stars belonging to different spectral classes and evolutive stages, including main sequence stars of the O, B, A, F, G, K spectral classes, giant, and supergiants, B type sub-dwarfs (Ohl et al., 2000) (Table 6.1), extremely hot Wolf-Rayet stars that have evolved beyond the asymptotic giant branch (Marcolino et al., 2007), and white dwarfs (Chayer et al., 2005), including dwarf nova stars (Sion et al., 1997).

TABLE 6.1 Phosphorus Abundances in Some Illustrative Stars*

Star	Spectral Class	T_e (K)	Spectral Range λλ (Å)	[P]	[P/P$_\odot$]	References
HD 190429A	O4I	39000	PV 1118–28	5.11	–0.30	Bouret et al., 2005
HD 96715	O4V	43500		5.41	0.0	
CPD-72 1184	B0III	29500	PIII 4057.4	5.4[(a)]	0.0	Tobin and Kaufmann, 1984
PG 0749+658	sdB	24600	PIII 1125.70	4.69	–0.72	Ohl et al., 2000
3 Cen A HD 120709	B5III-IVp	17500		7.0	+1.6	Castelli et al., 1997
Feige 86	Bp	16100	PII 4044–4499	7.1[(b)]	+1.72	Kafando et al., 2016
HR 1512	Bp	15200	PII 4420–6165	7.08[(c)]	+1.67	Lyubimkov et al., 2008
HD 213781	Bp	13300	PII 4044–4499	6.70[(d)]	+1.29	Kafando et al., 2016
Procyon (α CMi) HD 61421	F5IV	6500	PI 9750.75	5.29	–0.12	Kato et al., 1996
			PI 9796.91	5.37	–0.04	

*In the standard spectroscopic notation a neutral atom is denoted by the Roman numeral I, and its successive ionization states by the following numerals II, III, and so on.

TABLE 6.1 *(Continued)*

Star	Spectral Class	T_e (K)	Spectral Range λλ (Å)	[P]	[P/P$_\odot$]	References
HD 13555	F5V	6470	PI 1053.241	5.28	–0.13	Caffau et al., 2016
(η Ari)			PI 1058.447	5.27	–0.14	
HD 22484	F9IV-V	5960		5.33	–0.08	
				5.41	0.0	
HD 1461	G0V	5765	PI 1053.241	5.62	+0.21	Caffau et al., 2016
			PI 1058.447	5.59	+0.18	
Arcturus (α Aur) HD 124897	K0III	4286	PI 10581.569	5.11[(e)]	–0.30	Maas et al., 2017
GD71	DA	32000	PIII 1003.615 PV 1117.985 PV 1128.020	3.43[(f)]	–1.98	Dobbie et al., 2005

*Spectral class, effective temperature (T_e), considered spectral range, and elemental abundance of phosphorus ([P] = log(N_P/N_H) + 12, where N_X is the respective element column abundance, see Section 3.1) in the atmospheres of some representative stars belonging to several spectral classes. The reference phosphorus solar abundance [$P_\odot$] = 5.41 ± 0.03 is taken from Scott et al., 2015 (see Table 3.1) and the originally published data for the ratio P/$P_\odot$ have been properly updated to this value. The stars 3 Cen A, Feige 86, and HR 1512 belong to a chemically peculiar class characterized by the presence of weak He absorption lines (Bp spectral class). 3 Cen A is the prototype of the so-called "phosphorus stars" (Ga-P subclass), due to its high P overabundance. Star HD 213781 was tentatively classified as a Bp main-sequence star, though it may probably be a blue horizontal-branch star. The solar-like star HD 1461 hosts two super-earth planets. GD71 is a white dwarf star with an H dominated photosphere containing minor traces of other elements. Reported error ranges: (a) ±0.1, (b) ±0.2, (c) ±0.29, (d) ±0.2, (e) ±0.15, (f) ±0.1 dex.

Since the ionization potentials of the phosphorus atom are 10.49, 19.77, 30.20, and 51.44 eV for the ionization processes PI → PII, PII → PIII, PIII → PIV, and PIV → PV, respectively (Ralchenko et al., 2010), one should expect to find the higher states of ionization in the hotter early spectral type stars, while PI should be the dominant ionization state in the cooler late-type stars atmospheres. In fact, this general trend can be clearly appreciated in the examples listed in Table 6.1, where we also indicate some representative phosphorus abundances measured in stars belonging to different spectral classes. As we can see, no PI line is available in the visible range in spectra of stars belonging to the spectral types F, G or K. Some PII and PIII lines are observable in the UV and visible spectra of B-type stars and in aged stars belonging to globular clusters in the Galaxy's halo, and some lines of PV, PIV, and PIII can be detected in the UV spectra of hot O-type stars. Both PIII and PV lines have been observed in the UV spectra of white dwarfs.

In most main sequence stars observed to date the obtained phosphorus abundance has values very close to or slightly under abundant than those measured in the Sun's photosphere (Figure 6.2), although also a few cases have been reported of stars showing a significant phosphorus enhancement in their atmospheres in many representatives of the so-called *chemically peculiar* stars (Bidelman, 1960). These stars belong to the A and B spectral types (see Table 6.1), located at the upper main sequence or the horizontal branch region in the H-R diagram and they exhibit some anomalous compositions in their atmospheres. The mechanisms responsible for producing the chemical peculiarities in these stars are still unclear, though it is generally assumed that their peculiar atmospheric abundances are a consequence of atoms diffusion from the stellar interior resulting from competition among gravitational settling, radiative levitation, turbulent convective mixing, and fractionated mass loss (González et al., 1995). In fact, evidence for a significant stratification of phosphorus atoms in the atmosphere of the chemically peculiar B-type HR1512 star, described by a linear increment of P abundance with the average height, in good agreement with the diffusion theory expectations, has been reported (Lyubimkov et al., 2008).

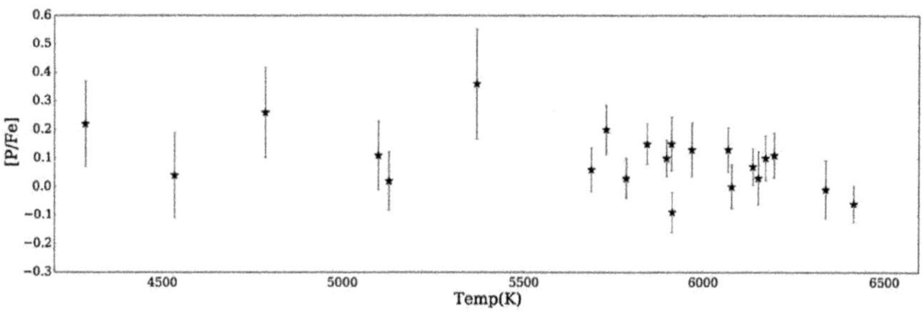

FIGURE 6.2 The difference between the phosphorus abundance (relative to iron) of a sample of late-type FGK stars and that of the Sun ([P/Fe] = log(P/Fe) – log(P/Fe)$_{\odot}$) is plotted as a function of the star photosphere effective temperature. No significant dependence of the phosphorus abundance on stars' temperature is observed.

Source: Maas, Pilachowski, and Cescutti (2017); © AAS. Reproduced with permission.

In Sun-like stars, G2 spectral class with photospheric temperatures about 5800 K, no spectral lines of neutral phosphorus are found in the optical range, and only a few weak PI multiplets are present in the near-infrared. These lines are too weak to be detected in aged, late-type stars with low metallicities, so that absorption lines in the near-UV at 2,135–2,149, 2,152–2,154 and 2,534–2,555 Å multiplets, have recently been used to study metal-poor, main-sequence F, G, and K spectral type stars with metallicities between $-3.3 < [Fe/H] < -0.2$ (Roederer et al., 2014; see Figure 5.8a). In doing so, the weighted mean abundance ratio [P/Fe] = $+0.04 \pm 0.07$ has been reported for a 13 stars sample, hence indicating similar phosphorus abundances to those observed in the Sun, although the star in the sample with higher metallicity ratios shows a decreasing trend of [P/Fe] with increasing [Fe/H] (see Figure 5.8b).

When going from the asymptotic giant branch to the planetary nebula stage, intermediate-mass stars loss their external, hydrogen dominated layers (see Section 6.2.3), exposing their very hot (typically within the range 4×10^4–1.5×10^5 K) central cores which are referred to as Wolf-Rayet stars. Phosphorus has been identified in several Wolf-Rayet stars spectra through the doublet transition of PV at $\lambda = 1,118$ and 1,128 Å. In this way, an oversolar abundance of about 4–25 times solar phosphorus has been reported for the central star ($T_e = 74,600$ K) located in the planetary nebula NGC 5315 (Marcolino et al., 2007). Since these P atoms will remain trapped in the resulting white-dwarf star, this observation provides evidence on the significant amount of phosphorus being sequestered from further chemical evolution, thereby reducing the availability of this biogenic element in the Galaxy (Exercise 5).

6.2 FORMATION OF PHOSPHORUS COMPOUNDS IN STELLAR ENVIRONMENTS

6.2.1 Stellar Atmospheres

The stellar atmospheres physical conditions change throughout the evolution of the stars. In their late stages, low-, and intermediate-mass stars develop strong stellar winds allowing for the formation of some stable molecules when the atmosphere temperature falls below about 3,000 K. Accordingly, molecules found in the photospheres of these stars are unique since they can survive at the stellar atmospheres rather high temperatures (as compared to the usual conditions of terrestrial laboratories), and thus these molecules are recognized as high-temperature molecules from the chemist viewpoint.

Leaving aside hydrogen and the chemically inert helium atom, the next most abundant elements in the cosmic ranking are oxygen and carbon (Figure 1.1), so that these atoms readily combine to form the strongly bounded and highly stable molecule carbon monoxide, $C \equiv O$ (dissociation energy $D = 1{,}079$ kJ mol$^{-1} \approx 11.2$ eV, bond length $d = 1.128$ Å). Therefore, the molecular inventory of late-type stars atmospheres is mainly determined by whether there is oxygen, carbon, or neither of them left over after CO formation (Jaschek and Jaschek, 1995). Presently, red giant stars are thought to follow an evolutionary sequence whereby they originally start as oxygen-rich M-stars, and afterward they gradually transform into the so-called S and then C-stars by the cumulative addition of carbon into their atmosphere, thus progressively changing the chemical equilibrium conditions, atmospheric structure, gas opacity and light flux properties of the star (Piccirillo, 1980). Consequently, the molecular composition of red giants stellar atmospheres can be divided into three broad classes: the commonest oxygen-rich M-type stars class, the less common carbon-rich C-stars class, mainly formed by those stars that have undergone the so-called third-dredge up in the upper part of their AGB ascension, and the rare S-type stars class, which have a C/O ratio of approximately unity.

In common M-stars, the excess oxygen not locked up in CO molecules drives a chemistry dominated by oxygen-bearing molecules such as water, H_2O ($D = 9.51$ eV), and silicon monoxide SiO ($D = 8.31$ eV). Conversely, the chemistry in C-rich stars reflects the excess carbon not locked in CO, and is characterized by the presence of numerous organic compounds, such as acetylene, C_2H_2, or hydrogen cyanide, HCN. In Figure 6.3, we list a number of representative molecules detected in these stars classes, as well as those found in the Sun, which is an O-rich star hotter than a typical M-type star. As we see, most detected molecules are relatively simple diatomic ones, although some triatomic representatives (H_2O, CaOH, HCN, C_2H, SiC_2, and C_3) and a molecule containing four atoms (C_2H_2) have also been reported. Out of 44 molecules listed in Figure 6.3, only seven can be found in the four considered types of stars (central box), five of which are hydrides, the other two being CO and CN ($D = 7.82$ eV). On the other hand, no molecule is exclusively shared by C- and S-type stars, whereas only NH can be observed in both M- and C-type stars, but not in S-type stars. Conversely, M-, and S-type stars share many different compounds, most of them metallic oxides and hydrides also observed in the Sun's photosphere. Carbon stars display a remarkably rich organic chemistry, since C atoms in excess give rise to several stable polyatomic molecules such as HCN, C_3, and C_2H_2 in significant amounts. In comparison, the chemistry of M-type stars is relatively simple: after CO and H_2, SiO is one of the most abundant molecules in their atmospheres, with H_2O molecules also playing an important role, along with a number of metal oxides. Finally, in S-type stars we note the presence of three diatomic metallic sulfides, indicating that the sulfur to silicon ratio plays in S-stars an analogous role to the C/O ratio in C-rich ones. Quite remarkably, no phosphorus bearing molecules have been detected in stellar atmospheres to date (31 August 2019), albeit the possible presence of PN, PS, and PH molecules has been predicted in chemical equilibrium model calculations.

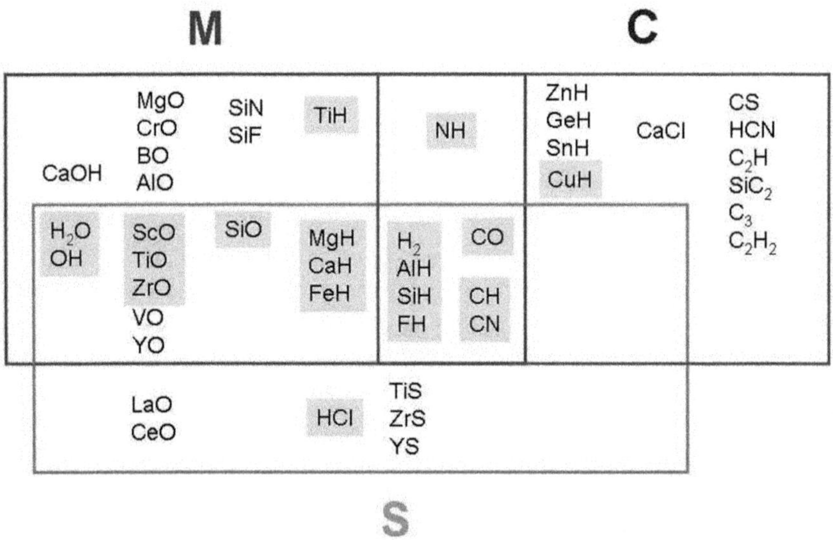

FIGURE 6.3 Molecules observed in the atmospheres of late-type M-, S-, and C-stars. Molecules observed in the Sun are highlighted for the sake of comparison. The ZrO (D = 8.47 eV) absorption molecular bands are a defining feature of S-stars.
Source: Jørgensen, 1997; Tennyson, 2003.

In fact, several models have been introduced during the last decades in order to account for the molecular inventory observed in different stars' classes atmospheres. To a first approximation, the photospheric chemical composition is assumed to be in local thermodynamical equilibrium (LTE) owing to the relatively high densities and temperatures of the gas, so that every elemental volume can be treated as containing an ideal gas at constant pressure, P_g, and temperature, T. The most fundamental problem is to determine the equilibrium partial pressure of the free neutral atom for each element, since it is generally determined by several species that consume different amounts of the element yielding the main related molecular compounds. Thus, the atmospheres of late-type stars are complex chemical systems in which about a hundred elements are mixed giving rise to many possible reactions to form molecules through intertwined networks.

Fortunately enough, when setting up the required chemical equilibrium models, it is not always necessary to incorporate all atomic species at once. For instance, as a first step, the reaction networks among the most abundant atoms H, O, C, and N can generally be determined independently of the other, far less abundant elements. Once the equilibrium partial pressure of each main element is known, determining the abundance of any related molecule is straightforward, and the fraction of atoms that are locked in molecules (the so-called molecular association degree) is then determined from the coupled chemical reactions involving different elements (Tsuji, 1973). The most important molecular data in calculating the required equilibrium constants is the dissociation energy (D), which does not depend on the formation mechanisms of the related compounds. On the contrary, the chemical equilibrium distribution of less abundant atoms sensitively depends on temperature, gas pressure and the detailed chemical composition of star's atmosphere and can only be determined by solving the overall chemical reaction network, which involves the most abundant elements along with all their chemically related molecules. Such an overall description is essential, since if just one major species were omitted in the chemical equilibrium calculations, the resulting equilibrium abundances may not be correct at all. These calculations are very sensitive to the adopted thermochemical parameters in the reaction networks as well. This important point will be illustrated by considering the case of the thioxophosphino (PS) radical in detail.

In Figure 6.4a, we show the numerically obtained partial pressures for different phosphorus related compounds as a function of the photosphere temperature, making use of an earlier chemical equilibrium model for a stellar atmosphere with a solar elemental abundance, including 36 species in the considered chemical networks (Tsuji, 1973). According to this model, almost no P_2 is formed, so that phosphorus atoms are available for the formation of heteroatomic molecules such as PS, PN, PO, or PH. In the lower temperature regime (i.e., T < 1,500 K) the more abundant P-bearing molecules follow the sequence PS > PN > PO > PH. We note that, keeping in mind the expected absence of C-bearing molecules in an O-rich star, this ordering correlates with the dissociation energy values of the corresponding molecules (see Table 4.2), with the exception of PS. As the photosphere temperature is increased above 1,500 K, the PO abundance crossover that of PS at about 2,100 K and attains its partial pressure peak at about 2,500 K, whereas neutral P atoms become the major phosphorus species at higher temperatures. Thus, the thioxophosphino PS radical should be the dominant phosphorus bearing compound in the low-temperature regime, ranging from 1,000–2,000 K, whereas neutral phosphorus predominates above about 2,200 K. The preponderance of PS compound at low photospheric temperatures was confirmed by Milam et al., (2008), who considered an expanded version of the earlier Tsuji's model (Figure 6.4b).

However, by comparing Figures 6.4a and 6.4b we realize that the PO abundance peak ($f \cong 8 \times 10^{-7}$) is shifted toward lower temperatures and located at ~ 1,500 K, and the PN abundance is reduced with respect to the earlier Tsuji's model values over the entire temperature interval. In this way, in the range from 1,000 to 1,400 K the P-bearing molecules abundance obeys the ranking PS > PO > PN > PH. At T ~ 1,500 K, the abundances of PS and PO are almost the same, and in the temperature range, 1,500–2,000 K the PO abundance surpasses that of PS.

In the light of these predictions a systematic search for the possible presence of the radical PS was performed in different astronomical objects, including Orion KL and Sgr B2 molecular clouds, the prestellar core L134N, the carbon star IRC +10216, the VY CMa oxygen-rich star and the protoplanetary nebula OH 231.8+4.2, without success (Ohishi, 1988). These negative results suggested that either the thermochemical data adopted in the numerical calculations may be inappropriate, or the considered chemical networks were incomplete.

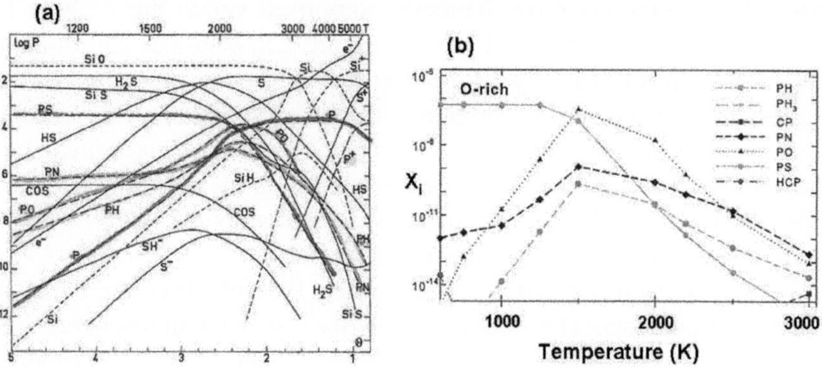

FIGURE 6.4 (a) Chemical equilibrium model abundances (expressed in terms of their partial pressures logarithm; $\log P_g = 3.0$) of compounds based on elements belonging to the 14, 15, and 16 groups of the periodic table for a solar composition ([P] = 5.43) stellar atmosphere as a function of its temperature in the interval 1,000–5,040 K. (Source: Tsuji, (1973); © AAS. Reproduced with permission). Note the predicted absence of both P_2 and CP molecules over the entire considered temperature range. (b) Calculated fractional abundances, X_i, relative to the total hydrogen content (i.e., $H_2 + H + H^+$) of P-bearing molecules as a function of the photosphere temperature for a model star with solar composition and a particle number density of $n \sim 10^{11}$ cm^{-3} as a function of the temperature in the interval 800–3,000 K. (Source: Milam, Halfen, Tenenbaum, Apponi, Woolf, and Ziurys (2008); © AAS. Reproduced with permission).

The formation of PS molecule is assumed to run via the endothermic reaction $2PH_3 + 2H_2S \rightarrow 2PS + 5H_2$, which proceeds toward the right with increasing temperature, so that the PS abundance should increase as the stellar atmosphere temperature rises, at variance with the trend displayed by the abundance curves plotted in both panels of Figure 6.4, but in good agreement with the PS fractional abundance curve obtained in the recent calculations shown in Figure 6.5b, which increases with the temperature, as required. In fact, it has been recently pointed out that information given in the 4th edition of the NIST-JANAF Thermochemical tables contained erroneous data regarding the thermochemistry for the PS radical (Lodders, 2004; Viana and Pimentel, 2007). In particular, the indicated PS dissociation energy value D = 543 kJ/mol = 5.62 eV (used in the computations shown in Figure 6.4) is too high, and the substantially smaller value D = 438 ± 10 kJ/mol = 4.5 ± 0.1 eV, should be used instead (see Table 4.3). In doing so, the conditions under which the PS formation reaction takes place is constrained to a remarkably small region in the temperature-pressure phase diagram, comprised within the 1,100 < T < 1,400 K and –6 < logP < –4 ranges (Figure 6.5a), so that the formation of PS molecules can only take place at low enough pressures. In fact, according to LeChatelier's principle, the reaction yielding PS molecules proceeds toward the right with decreasing pressure, because there are 4 gas molecules on the left and 7 gas molecules on the right. In the light of these results, one should not expect significant amounts of PS molecules to be formed in most of the stars atmospheres.

In any event, if formed at all the reactive PS radical could probably be destroyed by chemical processes taking place in outer atmospheric layers of the star, where the photodissociation of water molecules produces highly reactive hydroxyl radicals, which allow for the endothermic reactions $PS + OH \rightarrow PO + HS$, and $PS + OH \rightarrow PH + SO$, consuming PS molecules (Ohishi et al., 1988). In that case, we can expect PO and/or PH species to be spectroscopically detected instead of PS in outer star regions. Indeed, as we will describe in the next section, PO molecule has been reported to be present in the circumstellar envelope around the oxygen-rich, late-type star VY CMa (Tenenbaum et al., 2007).

By inspecting Figures 6.5a and 6.5b, we see that, under equilibrium conditions, P_4O_6 gas is the dominant phosphorus compound at the lower temperatures, ranging from 300 to 1,100 K, depending on the pressure value, and this phosphorus oxide is replaced as temperature increases by either PH_3 or P_2. Indeed, the conversion of P_4O_6 to PH_3 occurs at relatively high pressures by the reaction $P_4O_6 + 12H_2 \rightarrow 4PH_3 + 6H_2O$, whereas the conversion of P_4O_6 to P_2 takes place at lower pressures, generally in CO-dominated atmospheres (see the CH_4 = CO boundary dash-dotted line in Figure 6.5a), via $P_4O_6 + 6H_2 \rightarrow 2P_2 + 6H_2O$. The relative importance of both reactions depends on the location in the phase diagram relative to the PH_3 - P_4O_6 - P_2 triple point located at T ~ 1,100 K and $\log P_T \sim -1.2$ bars, where all three gases have equal abundances. In any event, according to the results shown in Figure 6.5b, the P-bearing molecular inventory of a typical M-type star photosphere is expected to be dominated by relatively reduced compounds, such as PH_2 or P_2 over a wide temperature range, and only at temperatures below 1,100 K tetraphosphorus hexaoxide (P_4O_6) molecule is expected to appear. In addition, the relative abundance of PS molecules is clearly subsidiary, as compared to that corresponding to the above-mentioned compounds, hence accounting for the unsuccessful detection of the thioxophosphino radical in M-type stars to date.

In Y dwarfs, with T ~ 500 K, phosphine should exhibit a strong feature in the mid-infrared at 4.3 μm, where it is predicted to be the dominant source of photospheric opacity. T dwarfs, with T < 1,300 K, will mostly be dominated by other molecules (e.g., H_2, NH_3, CH_4, CO, H_2O, and CO_2) but phosphine can still have a significant contribution to the shape of the spectrum. Thus, PH_3 is predicted to be the dominant phosphorus molecule in cool T dwarfs such as Gliese 229B, where it is expected to carry all the phosphorus in the atmosphere with approximately 0.6 ppm. As the temperature rises, as it occurs in hotter environments such as HD2009458b and L dwarfs, PH_3 is readily converted into P_4O_6, and the phosphine abundance appreciably decreases to about 50 ppb (Visscher et al., 2006) (Table 6.2).

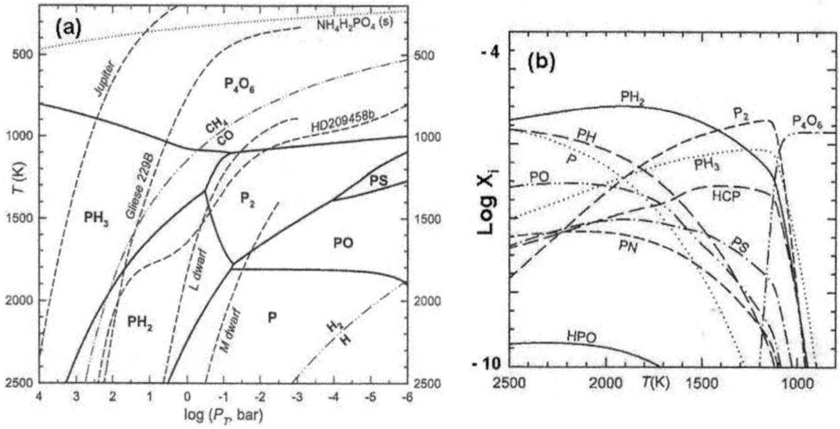

FIGURE 6.5 (a) Phosphorus chemistry as a function of temperature and pressure in a protosolar composition atmosphere (Lodders, 2003; [P] = 5.54 ± 0.04). The solid lines indicate the borders where major P-bearing gases have equal abundances. The condensation curve for ammonium monophosphide ($NH_4H_2PO_4$) is shown in the upper low-temperature region (dotted line), and the H_2 = H equal abundance boundary is shown in the lower right corner (dash-dotted line). Atmospheric T(P) profiles for representative substellar objects (dashed lines) are given for the sake of reference. (b) Fractional abundances, X_i, of phosphorus-bearing compounds in an L-type brown-dwarf as a function of its temperature in the interval 800–2,500 K. Note that the temperature increases from right to left in Figure 6.5b, at variance with the criteria adopted in Figures 6.4a and 6.4b. (Source: Visscher, Lodders, and Fegley (2006); © AAS. Reproduced with permission).

At variance with the results obtained for O-rich stellar atmospheres, the most abundant phosphorus bearing compound in a late S-type star of a similar temperature is the PN molecule, as can be seen in the Table 6.2, where we show molecular abundances calculations performed by Sauval (1978) considering a chemical reaction network including 83 elements and 1,600 molecular compounds. In this case, the abundance ranking reads PN > PS > PH > PH_2 >> PH_3, where we note the expected absence of PO molecules. Nevertheless, keeping in mind the previous comments regarding the PS radical, it is probable that the calculated abundance of this compound may be somewhat overestimated.

TABLE 6.2 S-Type Star Molecular Abundances*

log [n(...)/n(ZrO)]	Compound	P-Bearing
10	H_2	
7	CO, N_2	
6	SiO, SiS	
4	HCl, CS	PN
3	HF, AlCl, H_2O	PS
2	CN, AlF	PH
1	NH, NH_2, NH_3, NaCl	PH_2
0	HCO	
–1	NS, OCS	
–2	H_2CO	PH_3

*The obtained column abundances (relative to ZrO) for different compounds obtained from a detailed model network calculation with T_e = 2,000 K and O/C ratio within the range 0.9–1.1

Source: Modified from Sauval (1978).

Regarding C-rich stars, a thermochemical equilibrium model for the C-type star IRC+10216 was considered by Turner et al., (1990) on the basis of the approach earlier introduced by Tsuji (1973). Their model included 36 atoms and 240 molecules in total and the resulting abundances for the P-bearing compounds indicated that the triatomic HCP molecule is the dominant compound over the interval $600 \leq T \leq 1,500$ K, whereas neutral phosphorus is the predominant form above $T \geq 1,800$ K. Note that the presence of HCP compound is also expected in the atmospheres of O-rich stars (see Figure 6.5b), although exhibiting fractional abundances well below those of P_2 or PH_3 molecules in this case. In the intermediate temperature range 1,600–1,800 K a number of P-bearing molecules reached peak concentration values which are several orders of magnitude lower than those attained by HCP and P species, following the ranking PN ≅ CP > PH > PS > P_2. These earlier results were confirmed by Milam et al., (2008), though the peak abundances of the P-bearing molecules mentioned before are shifted toward the somewhat higher value T ~ 2,000 K (Figure 6.6).

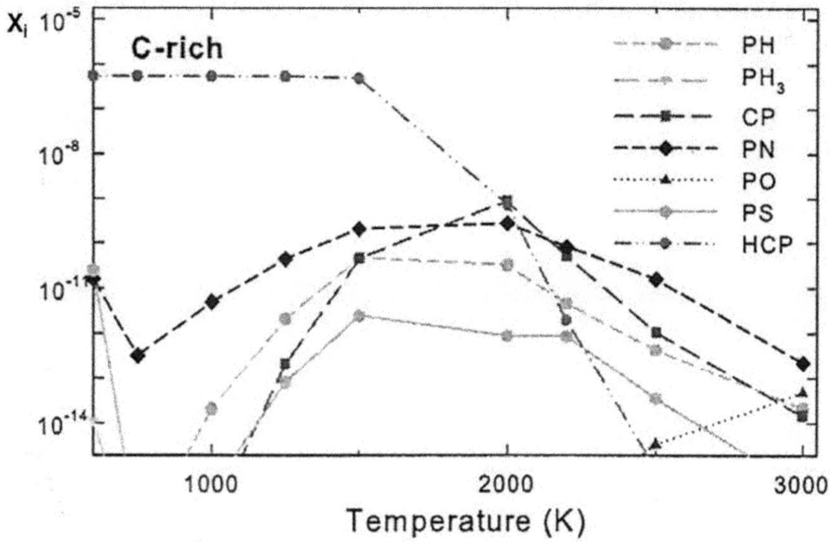

FIGURE 6.6 Chemical equilibrium abundances of phosphorus compounds for a C-rich star atmosphere with n ~ 10^{11} cm^{-3} and C/O ~ 1.5, over the range 600–3,000 K.
Source: Milam, Halfen, Tenenbaum, Apponi, Woolf, and Ziurys (2008); © AAS. Reproduced with permission.

6.2.2 *Circumstellar Shells*

During the red giant phase, stars begin to lose their outer atmospheres, creating a relatively cool circumstellar envelope (Kwok, 2004). When the atmospheric gas of an evolved star expands to form a colder envelope, most heavy elements condense in small grains, generally referred to as dust. For instance, the condensation temperature of schreibersite $(Fe,Ni)_3P$ (see Section 4.3.2) is estimated to be ~ 1,250 K (Lodders, 2003) and, at even lower temperatures, equilibrium condensation of ammonium monophosphate $NH_4H_2PO_4$ (see Figure 6.5a) is expected to occur via the thermochemical reaction (Visscher et al., 2006)

$$10\,H_2O + 4\,NH_3 + P_4O_6 \rightarrow 4\,NH_4H_2PO_4 + 4\,H_2,$$

which consumes previously formed P_4O_6 molecules. However, neither P_4O_{10} nor elemental P condenses under equilibrium conditions over the temperature-pressure phase diagram shown in Figure 6.5a. Dust grains and the remaining gas phase elements then take part in an active photochemistry driven by stellar UV photons, along with ion-neutral and neutral-neutral reactions involving reactive molecules and radicals.

It is thought that maybe around 50% of evolved stars have O-rich, as opposed to C-rich, circumstellar envelopes (Kwok, 2004). C-rich envelopes are preferentially found on the red giant/supergiant branches, as well as on the AGB stars, where their convective atmospheres dip into the He-burning shell and bring up afresh carbon atoms to the stellar surface. Thus, it is believed that this "third dredge-up" occurring during the late AGB stage converts an O-rich star, such as VY CMa, to a C-rich one, such as IRC+10216, which can be properly regarded as typical representatives of their respective families.

In Figure 6.7, we compare their respective chemical inventories. By inspecting this figure, the following conclusions can be drawn:

1. The most abundant molecules in both sources are CO, HCN, and SiS, followed by CS in IRC +10216 and SiO in VY CMa.
2. The molecules which are common to both stars include six carbon- and two silicon-bearing compounds, along with PN and NaCl molecules containing third-row elements.
3. About 72% (33%) of the compounds present in IRC +10216 (VY CMa) contain carbon, whereas 12% (50%) of these compounds contain oxygen, respectively. Therefore, the C/O ratio does not completely control the relative chemistries in these two objects (Tenenbaum et al., 2010).
4. While species with as many as eleven atoms (HC_9N) have been detected in IRC +10216, only diatomic and triatomic molecules are found in VY CMa. Several species in IRC +10216 contain single C-C bonds, and are formed by neutral-neutral and ion-neutral reactions involving photodissociation products of certain "parent" molecules (Agúndez et al., 2008).
5. Sulfur chemistry appears to play an important role in VY CMa, with six S-bearing molecules observed, along with the ubiquitous

presence of SO_2 and SO. Sulfur's predominance may be a condensation effect, since three of the main dust constituents in VY CMa contain silicon (amorphous silicate, Mg_2SiO_4, and $MgSiO_3$), whereas none include sulfur. Therefore, excess S atoms may be available for gas-phase molecule formation (Tenenbaum et al., 2010).

6. Phosphorus chemistry is apparently richer in IRC + 10216 (where six different compounds have been detected, including molecules containing up to 4 atoms) than in VY CMa, where only two diatomic molecules, PN, and PO, have been observed to date.

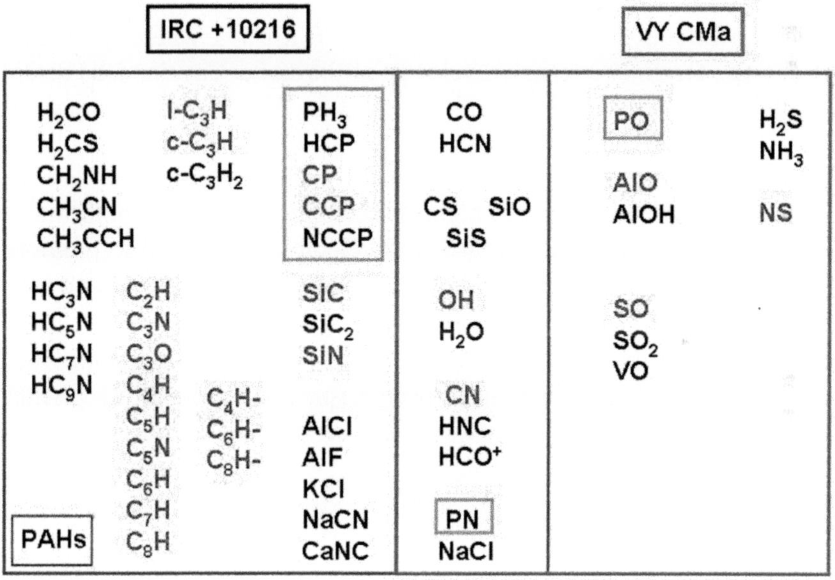

FIGURE 6.7 Molecular species observed toward IRC+10216 and VY CMa stars envelopes. Free radicals (in gray) are present in both circumstellar objects, but are mainly observed in the C-rich envelope. P-bearing molecules are boxed, and we note that PN molecule is present in both stars. PAHs stand for polycyclic aromatic hydrocarbons.
Source: Agúndez and Cernicharo, 2006; Ziurys et al., 2007; Halfen et al., 2008; Ziurys, 2008; Tenenbaum et al., 2010; Tielens, 2013; Cernicharo et al., 2019.

The chemistry of phosphorus-bearing compounds in the circumstellar envelopes of both oxygen-rich and carbon-rich evolved stars have received increasing attention during the last two decades. In these stars, a molecular photosphere extends beyond the conventional optical photosphere and below an outer dust cocoon.

Current models available in the literature use the same general approach to describe these circumstellar regions by considering that certain molecular species are formed at local thermodynamic equilibrium conditions near the stellar photosphere: the so-called parent molecules. As parent molecules flow into the outer envelope, they interact with ambient UV photons, and the subsequent photochemistry creates radicals and ions which react among them. Most chemical models for circumstellar envelopes considered to date focus on the chemistry taking place in a steady, spherically-symmetric, gas outflow from the star's surface, with constant mass-loss rate (within the range $1\times10^{-5}-4\times10^{-4}$ $M_{\odot}$ yr^{-1}), and uniform expansion velocity (in the interval $6\leq v_{\infty}\leq 30$ km s^{-1}), where v_{∞} denotes the average terminal expansion velocity of the stellar envelope, as determined from the measured spectral lines profiles. However, it has recently been pointed out that the supergiant star VY CMa also exhibits highly directional, non-spherical outflows, which require an additional term in the density profile in terms of a conical section depicting such outflows.

In general, high mass-loss rates and velocity values tend to promote circumstellar chemistry. The resulting chemical abundances are calculated as a function of the radial distance from the star, beginning at an inner radius in the range $r_0\approx 2-50\times10^{12}$ m (13–334 AU), and continuing outwards until all molecules have been destroyed by photodissociation (Willacy and Millar, 1997). The radial number density is given by:

$$n(r)=\frac{\dot{M}}{4\pi r^2 v_{\infty} m}, \tag{1}$$

where, $\dot{M}$ is the mass-loss rate and m is the considered molecule mass. A similar function, $\rho(r)\propto r^{-2}$, describes the density distribution around the star, so that there exists a density threshold value defining the scale of the so-called photodissociation region, where the interactions among the energetic photons flux emitted by the star's photosphere govern the resulting chemistry. The photodissociation field is given by the expression:

$$\beta(r)=\beta_0 e^{-A\left(\frac{d}{r}\right)^{\gamma}}, \tag{2}$$

where, β_0 gives the unshielded (interstellar) photodissociation rate, A is the opacity factor, d measures the shielding length, and $\gamma < 1$. Finally, the gas temperature distribution is generally given by a radially decreasing function of the form:

$$T(r) = T_0 \left(\frac{r_0}{r} \right)^{\alpha}, \tag{3}$$

where, T_0 is the temperature at the origin of the circumstellar region, r_0 is the inner radius of the circumstellar shell, and the exponent $\alpha < 1$ determines the temperature decreasing rate profile. For the sake of illustration in Tables 6.3 and 6.4, we list the model parameters used to describe the circumstellar shells around three AGB evolved stars and two supergiant oxygen-rich stars, where both PO and PN molecules have been detected.

TABLE 6.3 Modeling Parameters for Circumstellar Shells*

Star	v_∞ (km s^{-1})	r_0 (10^{12} m)	T_0 (K)	α
TX Cam	13.5	1.3	125	0.75
R Cas	7.0	1.4	1050	0.91
IK Tau	16.0	2.0	960	0.98
VY CMa	20.0	25.0	230	0.62
NML Cyg	24.0	50.0	270	0.50

*Terminal expansion velocity, inner radius, the temperature at the inner radius and thermal profile exponent for several circumstellar envelopes.

Modified from Ziurys et al., 2018.

TABLE 6.4 Model Parameters Employed in the Analysis of the O-Rich Stars R Cas, TX Cam, IK Tau, and NML Cyg, Arranged According to Increasing Mass Loss Rate

Source	R_* (10^{11} m)	T* (K)	d (pc)	Mass Loss (M_0 yr^{-1})
R Cas	2.0	1800	172	4×10^{-7}
TX Cam	2.0	2600	390	4×10^{-6}
IK Tau	2.1	2200	250	5×10^{-6}
NML Cyg	26.0	2500	1740	2×10^{-4}

Modified from Ziurys et al., 2018.

Earlier thermodynamic calculations, on the basis that the dominant form of phosphorus in the inner circumstellar envelope was PS and HCP for O-rich and C-rich stars, respectively (see Figures 6.4 and 6.6), reproduced the observed CP abundance reasonably well in C-rich stars, but yielded a too low abundance of PO in the case of O-rich stars (MacKay and Charnley, 2001). As we mentioned previously, the assumption of PS being the dominant form in these stars' photosphere was flawed, and it was subsequently realized that phosphorus hydrides were favored at temperatures of about 2,500 K (Figure 6.5b). Thus, in the O-rich model of Willacy and Millar (1997), LTE processes produce a single important parent molecule: phosphine, PH_3, which subsequently photodissociates to create PH_2, PH, P, and P^+ species. PO radicals ($f \sim 3 \times 10^{-10}$ relative to H_2) are formed by the reaction of oxygen atoms with either PH or PH_2 molecules in a shell located at radii of $r \sim 10^{14}$–10^{15} m (670–6,700 AU) from the star. Phosphine and P^+ atoms are the most abundant species in the inner and outer regions of the circumstellar envelope, respectively, with $f \sim 3 \times 10^{-8}$ relative to H_2. However, phosphine molecules have only been detected to date in circumstellar regions around C-rich stars (see Figure 6.7), with an estimated fractional abundance $f \sim 10^{-8}$ relative to H_2 (Agúndez et al., 2014; Tenenbaum and Ziurys, 2008), instead of those around O-rich stars, as predicted by theoretical models, thus indicating that the chemistry of phosphine in circumstellar envelopes is not yet well understood.

The carbon-phosphorus chemistry in a C-rich envelope, such as that shrouding the IRC +10216 star, is based on the parent species HCP, which photodissociates according to the sequence given by Eq. (4) releasing the reactive species CP and P (MacKay and Charnley, 2001). This scenario has been confirmed in a recent model including 24 elements and 259 related molecules in the equilibrium region located at about 2–3-star radii of a C-rich star. The model predicts that nearly all phosphorus is in the molecular form HCP in the gas phase, with a relative abundance of 2.7×10^{-7} relative to H_2 at $r \leq 10\ R_*$ (Figure 6.8a). The other P-bearing compounds, including PN, have abundances below 10^{-10} (Agúndez et al., 2007). For O-rich envelopes, the major phosphorus reservoirs in the vicinity of the star are PO, prevalent at ~2 R_*, with $f \sim 10^{-7}$, and PS attaining an equivalent abundance at $\sim 3\ R_*$. Beyond a distance of $\sim 4\ R_*$, P_4O_6 becomes the only significant phosphorus carrier (Figure 6.8b). We note that phosphine is only a minority compound according to this model, at variance with the previously mentioned earlier computations reported by Willacy and Millar (1997).

$$\begin{aligned} HCP + \nu &\rightarrow CP + H, \\ CP + \nu &\rightarrow P + C, \\ P + \nu &\rightarrow P^+ + e^-, \end{aligned} \tag{4}$$

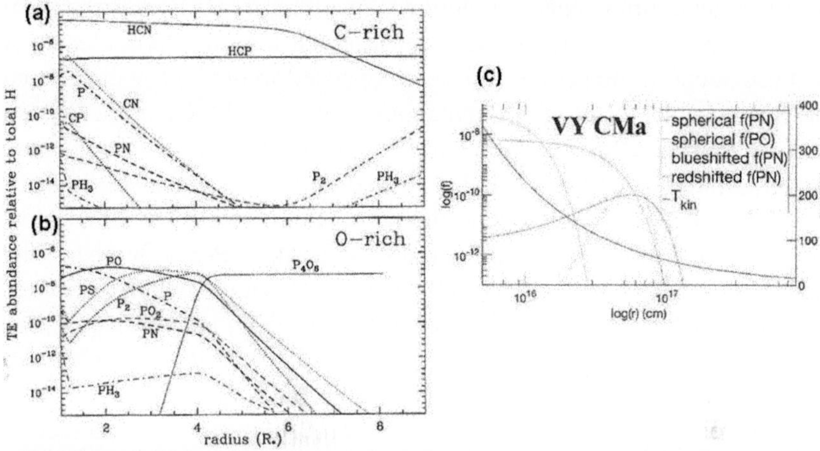

FIGURE 6.8 Radial abundance plot measuring the thermochemical equilibrium model abundances of P-bearing compounds in the innermost region of a circumstellar envelope as a function of the distance from the star (measured in stellar radii) for a C-rich (a) and an O-rich (b) model stars. (Source: Agúndez, Cernicharo, and Guélin, (2007); © AAS. Reproduced with permission). In (c) the radial abundance of two P-bearing molecules in the supergiant star, VY CMa envelope is plotted in terms of a spherical component for the PO molecule and three contributions for the PN molecule. (Source: Ziurys, Schmidt, and Bernal, (2018); © AAS. Reproduced with permission).

The abundance profiles obtained from this O-rich star model do not satisfactorily agree with recent measurements of PO molecule in VY CMa with an observed fractional abundance of $f \sim (5 \pm 3) \times 10^{-8}$ relative to H_2 (Ziurys et al., 2018), a figure about one order of magnitude below the $f \sim 3 \times 10^{-7}$ value estimated from the thermochemical calculations shown in Figure 6.8b, although the observed angular extent of this molecule lines is approximately 1" (Ziurys et al., 2007), corresponding to about 50 stellar radii, which is a significantly larger distance scale than that considered in Figure 6.8b. On the other hand, the predicted PN abundance in the thermochemical model is $f \sim 10^{-10}$, about one order of magnitude lower than the observed abundance of $(7 \pm 3) \times 10^{-9}$ in VY CMa (Ziurys et al., 2018). The reported discrepancies between model results and spectroscopic observations may be related to the fact that, being a 25 $M_{\odot}$ supergiant with $T_e \sim 2{,}800$ K, VY CMa is more active and luminous than the typical AGB stars considered in the Agúndez and co-workers model. In particular, it is now well known that the star exhibits asymmetric outflows that are quite probably causing shocks. This has been explicitly considered in the model shown in Figure 6.8c, which takes into account the presence of more energetic, non- spherical outflows, resulting in a different chemical environment. In this way, it has been observed that the PN molecule is most prevalent in the spherical flow, as opposed to the highly directional winds, by about a factor of 70. Analogous models for TX Cam, IK Tau and R Cas AGB stars clearly suggest that PN is a parent molecule formed near the stellar photosphere, while PO is formed farther out in the envelope, as can be readily seen by inspecting the radial abundance distribution plots shown in Figure 6.9. In fact, these plots suggest that PO is formed in the outer circumstellar envelope where photodissociation begins to occur. According to the work by MacKay and Charnley (2001) this molecule may be formed through the exothermic reaction P + OH → PO + H, although the maximum amount of PO derived in their model was of $f \sim 10^{-10}$, well below the observed abundances. A more recent study considers molecular abundances in the gas phase under LTE conditions first, and then allows a shock wave to propagate through the material. This model yields a PO abundance of $f \sim 10^{-7}$, which quickly drops to 2–3 $\times 10^{-10}$, as the shock impacts the chemistry (Gobrecht et al., 2016). The differences between these model predictions and the observations suggest that shocks are not the main chemical pathway to producing PO molecules.

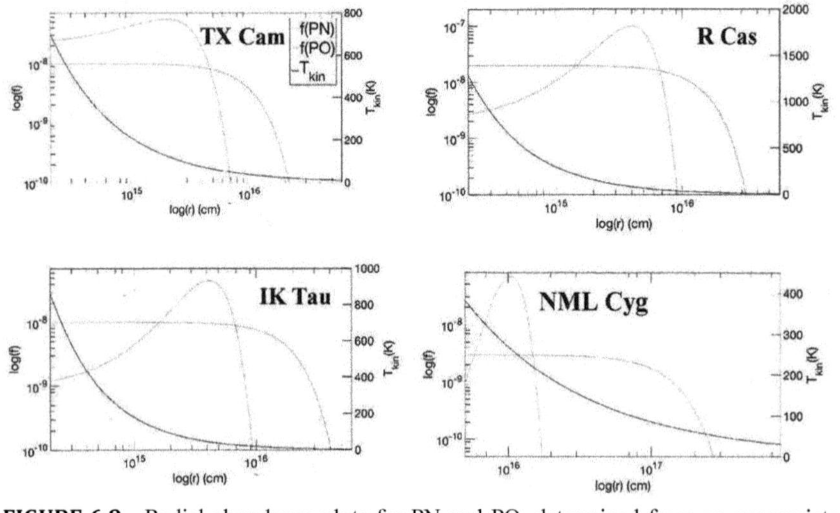

FIGURE 6.9 Radial abundance plots for PN and PO, determined from an appropriate modeling. Kinetic temperature as a function of distance is based on the temperature profile parameters given in Table 6.3.
Adapted from Ziurys, Schmidt and Bernal, (2018); © AAS. Reproduced with permission.

On the other hand, to fit with the observations in IRC +10216 one must consider that the HCP abundance decreases steeply between 5 and 20 stellar radii, so that is already depleted at a few stellar radii from the photosphere, either because it condenses on grains in the dust formation region (at 5–10 stellar radii) or because it is destroyed by gas-phase reactions. The second decrease in abundance occurs farther out, at $\sim 3\times10^{14}\ m \cong 460\ R_*$, due to photodissociation processes. Indeed, the reaction sequence (Eqn. (4)) predicts a hollow-shell distribution for CP, as shown in Figure 6.10, in agreement with spectroscopic observations showing that molecules in IRC +10216 are either concentrated around the star (e.g., HCN) or distributed in a hollow shell of apparent radius 10–20" (e.g., CN, l-C_3H, C_4H). Nevertheless, the CP and PN abundances predicted by this model in the inner circumstellar region are too low to explain the column abundances derived from the observations. Indeed, both compounds are more efficiently formed in the outer envelope ($r > 3\times10^{14}$ m), namely, CP from the photodissociation of HCP and PN through the neutral-neutral reactions.

$$CP + N \rightarrow PN + C,$$
$$P + CN \rightarrow PN + C. \quad (5)$$

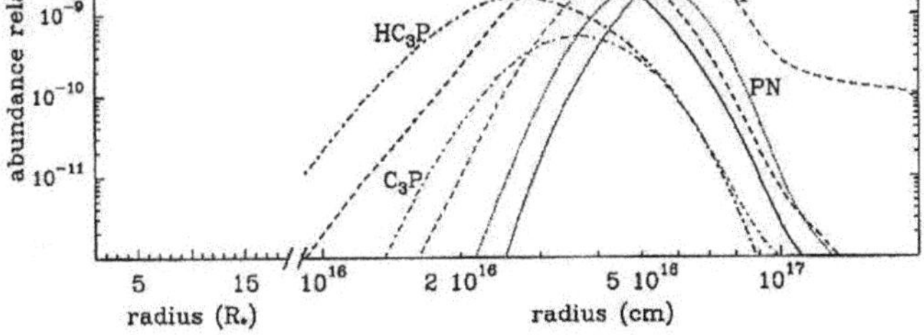

FIGURE 6.10 Abundances of phosphorus compounds in the outer envelope of IRC +10216. Angular distance is given in the top axis for an assumed distance of the star of 150 pc *Source*: Agúndez, Cernicharo, and Guélin, (2007); © AAS. Reproduced with permission.

Noteworthy, albeit the molecules HC_3P and C_3P are predicted to be present with significant relative abundances at distances of about 3×10^{14} m from the IRC +10216 star (Figure 6.10), none of them have been detected to date. Instead, the radical CCP (not predicted by the model) was found with a fractional abundance of $f \sim 10^{-9}$ relative to H_2, hence suggesting a 5:1 ratio of CP to CCP (Halfen et al., 2008). A plausible formation mechanism for the CCP radical in IRC +10216 involves the ion-neutral reaction where acetylene reacts with ionized phosphorus atoms (present with an abundance near $f \sim 10^{-8}$ in IRC +10216) to produce $HCCP^+$, which then recombines with an electron to form CCP. However, there is little evidence of ion-molecule type chemistry in ICR +10216, so that the alternative neutral-neutral reaction mechanism, based on the relatively abundant CP radical as a parent species reacting with other C-bearing radicals (prevalent in the outer envelope) has been proposed (Halfen et al., 2008).

$$\begin{aligned} P^+ + C_2H_2 &\rightarrow HCCP^+ + H, \\ HCCP^+ + e^- &\rightarrow CCP + H, \end{aligned} \tag{6}$$

$$\begin{aligned} CP + CCH &\rightarrow CCP + CH, \\ CP + C_3H &\rightarrow CCP + CCH. \end{aligned} \tag{7}$$

In VY CMa, both PO and PN disappear earlier in the outflow than molecules such as HCN and HCO^+ (Ziurys et al., 2007). In fact, in Figure 6.9 we see that PO (PN) molecule is rapidly destroyed in all cases at distances ranging from 70–450 (1,000–1,400) stellar radii, depending on the considered star. Similarly, according to the model results shown in Figure 6.8b, PO is the most abundant species until four stellar radii, where it is converted to P_4O_6. The confined nature of phosphorus chemistry, which is also observed in IRC +10216 (Agúndez et al., 2007), maybe indicating the P-bearing molecules are condensing into dust grains in both sources, suggesting that phosphorus acts as a refractory element (Exercise 13). Indeed, condensation calculations suggest that schreibersite $(Fe,Ni)_3P$, is the most prominent solid-state form of phosphorus in both oxygen- and carbon-rich environments (Tenenbaum et al., 2007). In O-rich stars, the dust forming in the wind includes silicates (e.g., forsterite Mg_2SiO_4) and metal oxides (e.g., alumina Al_2O_3). If condensation processes in evolved stars outer envelopes really incorporate most available phosphorus into refractory grains, searches for new P-bearing molecules may have the greatest opportunity for success in protoplanetary nebulae (PN), where phosphorus would be released from grains through the action of strong shock waves (MacKay and Charnley, 2001).

6.2.3 Planetary Nebulae

Asymptotic red giant branch stars evolve into the final stage of their life cycle: the planetary nebulae phase, which lasts about 10^4 years. By then, most of the original stellar mass is lost and forms a shell around an extremely hot (~ 10^5 K) residual core referred to as a Wolf-Rayet star. This star emits a strong UV radiation flux that readily ionizes the surrounding material, so that one could expect that most molecules formed in the circumstellar region would be destroyed via photodissociation during this phase. Notwithstanding this, spectroscopic observations have shown a significant number of molecules in different planetary nebulae (Table 6.5). The presence of such a diversity of molecular compounds in the gas phase of these nebulae, including molecules containing 4 or even 5 atoms, suggests that more organic species may be discovered in these sources and other similar planetary nebulae (Ziurys, 2008). In addition, the measured abundances of three representative molecules, CO, CS, and HCO^+, detected in a number of planetary nebulae spanning ages ranging from 1,000 to 12,000 years, do not appear to vary significantly with the age (Edwards et al., 2014). A possible explanation for these molecules endurance may be offered by the presence of finger-like dust clumps with densities as high as $n = 10^5$ cm^{-3} which have been observed in Spitzer images of IR emission (see Section 2.4.6). Such clumps, mainly composed of molecules mixed with dust, would provide a self-shielding protection against photodissociation processes.

In Figure 6.11, we list the molecular inventory of the protoplanetary nebula CRL 2688 (Egg nebula), whose age is estimated to be ~ 350 years (Zhang et al., 2013). The protoplanetary nebula is a rapid (~ 1,000 years) transition phase between the AGB and full-fledged planetary nebula phases (Kwok, 2003).

TABLE 6.5 Molecules Detected in Planetary Nebulae*

Source	Compound	Age (yr)	References
Helix	H_2, CO		Young et al., 1999;
NGC 7293	**HCN, HNC, CN, HCO^+**	12,000	Bachiller et al., 1997; Tenenbaum et al., 2009
	CCH, H_2CO, c-C_3H_2		
NGC 6853	CO, **CS, HCO^+**	10,000	Edwards et al., 2014
Dumbbell			
Ring	CO,		Tenenbaum et al., 2009;
NGC 6720	**CCH, CS, HCO^+**	7,000	Edwards et al., 2014
	HCN, HNC, CN, c-C_3H_2, H_2CO		
M2–48	CO, SO_2, N_2H^+		Edwards et al., 2014;
	CS, HCO^+	4,800	Edwards and Ziurys, 2014
	CN, SO, SiO, HNC		
NGC 6537	CO, N_2H^+		Edwards et al., 2014;
Red spider	**CS, CN, SO, HCO^+**	1,600	Edwards and Ziurys, 2013
	CCH, HNC, H_2CO		
K4–47	CO,	900	Edwards et al., 2014;
	CS, HCO^+, CCH		Schmidt and Ziurys, 2017
NGC 7027	H_2, OH, CO, CO^+, CH, CH^+, H_2O, N_2H^+, SiS, HCS^+, HC_3N	700	Zhang et al., 2008; Edwards et al., 2014
	HCN, CS, HCO^+,CN CCH,C_3H_2		

*Molecules observed in planetary nebulae at different stages of their evolution cycle. The nebulae are listed by decreasing estimated age. Bold-faced molecules have been also detected in diffuse ISM clouds (see Section 6.2.4).

By inspecting Figure 6.11 we see that the P-bearing molecules, HCP and PH_3, respectively containing 3 and 4 atoms, are present in a C-rich giant star, and a protoplanetary nebula, hence indicating that these phosphorus compounds can be spotted along the evolutionary track described by these astrophysical objects. In CRL 2688 the abundances of both PN and HCP are about one order of magnitude higher than in IRC +10216, but the PN/HCP ratio is roughly the same (~ 0.01–0.02). This may be suggesting that the chemistry of these two species mimics that of IRC +10216 under LTE conditions (Milam et al., 2008).

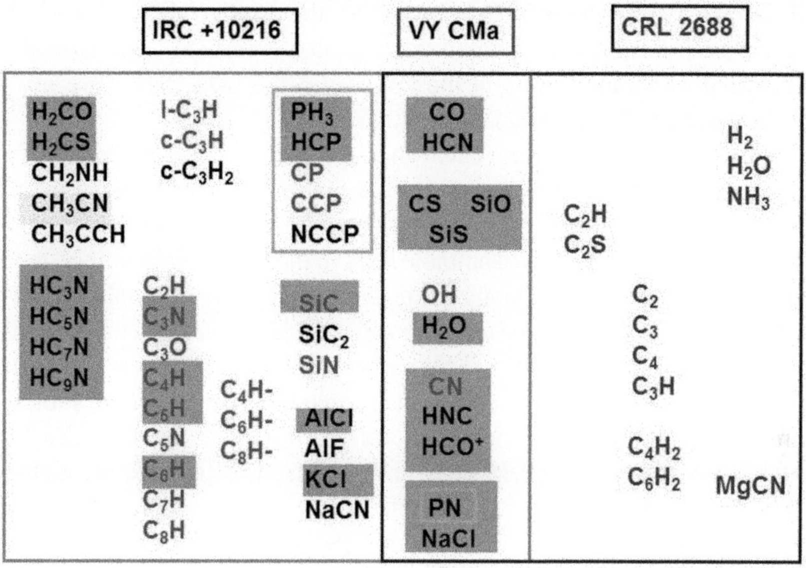

FIGURE 6.11 Reported and tentatively detected molecules in the protoplanetary nebula CRL 2688 (highlighted) are compared with those molecules which have also been detected in the C-rich star IRC+10216 (left panel) and the O-rich star VY CMa (central panel). In the right panel, we list the molecules which have been observed in the protoplanetary nebula CRL 2688 but not in the other two sources.

6.2.4 *Diffuse Clouds*

The gas and dust from planetary nebulae and SN remnants eventually disperse into the diffuse ISM (typical densities within the range 1–10 cm^{-3}), enriching this material with heavy elements (see Figure 6.1). It is estimated that planetary nebulae supply about an order of magnitude more interstellar mass than SN in our Galaxy (Stanghellini et al., 2000). One way of studying the ISM composition is to perform very high-resolution spectroscopy on the starlight absorption lines caused by the intervening clouds of gas and dust. Thus column abundances can be determined and expressed as ratios relative to the column density of atomic hydrogen, the most abundant element in the ISM. These ratios can then be compared to those corresponding to the elemental cosmic abundance. In doing so, elements not observed or underobserved in interstellar gas must be trapped in a nonatomic/ionic phase, namely, as molecular species or dust grains

(Exercise 13). In this way, by comparing the relative abundance of P and O atoms, it has been concluded that they have probably followed similar chemical evolution tracks in the ISM from the time they were released from the stars (Lebouteiller et al., 2005).

A number of studies revealed that the solar system abundances seem to be enhanced in heavy elements by a factor of 1.5 to 3 above the mean for stars of comparable age (see Sections 3.2.2 and 5.6), so that one should consider the composition of suitable stars in the solar neighborhood in order to obtain representative relative abundance values. For the sake of illustration, earlier depletions measured for different biogenic elements toward the sightline of ζ Oph (a bright O9.5V early-type star at a distance of about 138 pc from the Sun) relatively cold diffuse interstellar cloud, where substantial grain growth has already occurred, are –0.20 dex for C, –0.14 dex for O, +0.07 dex for N and +0.35 dex for S, referred to the average elemental abundances of a suitable stars sample in the galactic disk (Snow and Witt, 1996). P is depleted by –0.2 ±0.1 dex (Savage and Sembach, 1996), the same value than that reported for carbon. More recent measurements give –0.2 dex for C, –0.1 dex for N, –2.5 dex for O, –0.7 dex for S and –0.65 dex for P (Ritchey et al., 2018), a figure similar to that previously reported (–0.5 ± 0.2 dex) in cooler and denser clouds (Jenkins, 2009; Lebouteiller et al., 2005; Dufton et al., 1986). Albeit these updated measurements indicate that phosphorus is significantly more depleted from the gas phase than the remaining main biogenic elements, the measured phosphorus depletions are weaker than expected, attending to its condensation temperature value. This seems to suggest that P atoms are either somehow prevented from being fully incorporated into the initial grains that condense in the outer atmospheres of late-type stars, or are incorporated primarily into grain mantles that are subsequently destroyed or disrupted by shocks after the grains are incorporated into the ISM (Ritchey et al., 2018).

In fact, depletion on grains depends on both condensation temperature and chemical reactivity of a given element. The condensation temperature is defined as the temperature at which a 50% of the element is depleted by capture on the solid or liquid phase, and it fits the general picture that grains come, in part, from atmospheres of red giant stars and are injected into the ISM by slow mass outflow. Observational data indicate that the abundance of atoms in interstellar dust not only depends on their condensation temperature, but it is heavily influenced by chemistry in the ISM (Cardelli, 1994; York, 1994). Indeed, it has been suggested that phosphorus may be chemically blocked from depleting in the outer atmospheres of late-type stars because it can form the stable molecule PN (see Table 6.2 and Figure 6.6). Upon entering the ISM, the PN molecules would be subject to the interstellar radiation field and would dissociate, allowing the P atoms to deplete along with other refractory elements in interstellar clouds. Therefore we must consider two complementary environments as possible repository places of phosphorus in the ISM: the gas-phase and the solid grains (see Section 6.3).

The UV radiation field from ambient starlight is thought to be quite high in diffuse clouds, and earlier models suggested that all polyatomic molecules would be dissociated there (van Dishoeck and Black, 1986). Radio astronomical observations, however, have demonstrated that these clouds have an unexpectedly rich molecular content including CN, CS, SO, SiO, C_3, HCN, HNC, HCO+, CCH, H_2CO, and c-C_3H_2 compounds (Maier et al., 2001; Listz et al., 2006). These molecules remained undetected for years because the densities in diffuse clouds are so low that rotational levels cannot be excited via atomic collisions, as they are in denser planetary nebulae. Thus, the molecular transitions had to be observed in absorption against background quasars.

By comparing the molecules listed above with the species given in Table 6.4, we realize that both diffuse clouds and planetary nebulae appear to contain similar sets of molecules. This remarkable coincidence suggests that the molecular content of diffuse clouds may be remnant material which must come from the winds of planetary nebulae, since the density of diffuse ISM clouds is too low for efficient polyatomic molecule production (Ziurys, 2008). Indeed, many molecular ratios appear to be similar in both astrophysical objects. For instance, the CO/HCO^+ ratio is about 10^3 in both types of sources, while the CN/HCN ratio is ~ 4–9 in the planetary nebula and ~ 7 in diffuse clouds (Ziurys et al., 2015). On the other hand, the observed molecular abundances in diffuse clouds are a factor 10–100 lower than those measured in the oldest planetary nebulae (e.g., Helix), which is consistent with a substantial partial destruction from the expected photodissociation processes mentioned above.

6.2.5 Molecular Clouds and Star-Forming Regions

Diffuse clouds with typical densities $n = 10^2$ cm^{-3} and temperatures ~ 20 K gravitationally collapse over a period of about 10^6 years into denser clouds, which then undergo a free-fall collapse keeping the temperature nearly constant at T ~ 10 K until a final density within the range 10^4–10^6 cm^{-3} is reached. In massive enough clouds a proto-star is formed in the center, followed by a warm-up of the resulting proto-star envelope. In this way, a robust gas-phase organic chemistry takes place, under physical conditions far from thermodynamic equilibrium, at densities ranging from 10^3–10^7 cm^{-3} and temperatures within the interval T = 50–300 K. Quite interestingly, all basic organic functional groups such as simple alcohols (CH_3OH, CH_3CH_2OH), carboxylic acids (HCOOH, CH_3COOH), ethers (CH_3OCH_3), esters (CH_3OCHO), aldehydes (CH_3CHO), ketones ($((CH_3)_2CO)$), amines, and imines (CH_3NH_2, CH_2NH), as well as amides (NH_2CHO, CH_3CONH_2) have been detected in dense molecular clouds, which also contain simple sugars ($CHOCH_2OH$, see Figure 1.5). Accordingly, the chemistry of dense clouds is highly organic in nature, a feature relevant to the origin of life, since in due course most of these dense clouds will collapse into protostellar nebulae, and the formed organic material can coalesce into small planetary systems bodies such as planetesimals, meteorites or comets present in their related preplanetary disks, as we will describe in Chapter 7.

Of the seven P-bearing species currently detected in the ISM, the only species containing phosphorus identified in molecular clouds to date have been PN (Ziurys, 1987; Turner and Bally, 1987) and PO (Rivilla et al., 2016; Lefloch et al., 2016). These molecules have been observed in several star-forming regions such as Sgr(B2), W51 e1/e2, W3(OH) and L1157 (see Section 2.3.2). The reported abundances of PN and PO in the high-mass star-forming regions W51 and W3(OH) are similar (~ 10^{-10}), the PO abundance being higher by a factor ~ 2. The measured abundances in the solar-type star-forming region L1157 are [PN] = 9×10^{-10} and [PO] = 2.5×10^{-9}, leading to the ratio [PO]/[PN] = 2.8. These ratios (estimated assuming LTE conditions) are similar to those found in the O-rich circumstellar envelopes of the evolved stars VY CMa and IK Tau, suggesting that phosphorus seems to be equally distributed in the form of PO and PN in both circumstellar and interstellar material. In fact, chemical modeling indicated that these two molecules may be chemically related and are formed via gas-phase ion-molecular and neutral-neutral reactions during the cold free-fall collapse of dense molecular clouds. Then the molecules could freeze out onto grains at the end of the collapse, and subsequently evaporate during the warm-up phase once the temperature reaches ~ 35 K (Rivilla et al., 2016). During the shock phase, the key player for the formation and destruction of PO and PN molecules seems to be atomic nitrogen, which in turn is mainly released from ammonia previously condensed onto grains. Thus, it has been estimated that the reaction N + PO → PN + O could contribute to ~ 85% to the total destruction of PO (Aota and Aikawa, 2012; Lefloch et al., 2016).

Searches in dense clouds for other P-bearing molecules such as PS, HPO, HCP, and PH_3 have given negative results (Turner et al., 1990). The lack of P-containing molecules in molecular clouds is surprising. The solar N/P ratio is relatively low (~ 263, see Table 3.1), and N-bearing molecules are extremely abundant in both interstellar and circumstellar gas. Moreover, N-bearing analogs exist for almost all known P-containing compounds observed in circumstellar envelopes, such as CN, HCN, NO, NH_3, or CCN (Halfen et al., 2014). In cold dense clouds, the low abundance of phosphorus nitride ($[PN]/[P] < 10^{-4}$) suggests that most P is condensed on dust grains, so that the transfer of phosphorus from gas to the solid and back to the gas phase largely involves carbon atoms via the HCP linear molecule. As a consequence, phosphorus is believed to mainly reside in adsorbed HCP molecules, which are released to the gas by photodesorption in warm media and subsequently readily photodissociated (Turner et al., 1990). Thus, P ≡ N is only a trace repository of phosphorus in the ISM. Accordingly, its elemental abundance is considered to be depleted by a factor of 100 with respect to its solar value in quiescent molecular clouds (Lefloch et al., 2016; Jiménez-Serra et al., 2018). Phosphorus seems also strongly depleted from the gas phase in warm star-forming regions, where $[PN]/[P] \sim 10^{-3}$, and where neither HPO nor HCP or PH_3 has been detected (Agúndez et al., 2007; Turner et al., 1990). Thus, in order to explain the absence of P-bearing molecules other than PN and PO in dense clouds it has been suggested that phosphorus may be present in more complex molecules, such as PH_2CN, CH_3PH_2, PH_2CHO or $CH_3CH_2PH_2$ which are very difficult to be spectroscopically detected (Halfen et al., 2014; Ziurys et al., 2015; Turner et al., 2016).

6.3 THE CHEMISTRY OF PHOSPHORUS IN THE ISM

6.3.1 *Gas-Phase Chemistry*

At the low temperatures (~ 10 K) and very low densities (1–10^2 cm^{-3}) conditions prevailing in diffuse clouds a molecule typically collides with other only once every four months on average. In dense clouds (n ~ 10^5 cm^{-3}) this rate increases to about one collision every three hours (Exercise 14). As we see, the density in these ISM regions is much lower than the better vacuum attained in current high-vacuum chambers in laboratories (~ 10^{10} cm^{-3}), which in turn is about nine orders of magnitude lower than the air density at sea level on earth (~ 10^{19} cm^{-3}). In addition, collisions involving non-charged atoms or molecules (the so-called neutral-neutral reactions) are often endothermic and/or possess appreciable activation energies, so that they can only take place in those particular environments where shock waves occur. Thus, the formation of interstellar molecules in the gas phase is mainly driven by collisions including one charged species, which electrically polarizes its partner (ion-molecule reactions), hence increasing the chances for collisions to occur by a high enough kinetic energy. These reactions are certainly more frequent in dense clouds where the gas density (~ 10^5–10^6 cm^{-3}) is significantly higher than in diffuse clouds. The required ions are produced by the pervasive flux of high energy (~ 100 MeV) cosmic rays in dense clouds and by UV photons in diffuse clouds, respectively.

Since hydrogen is the overwhelming element in the ISM, reactions of both H^+ and H_2^+ ions with the most abundant O, C, and N neutral atoms and H_2 neutral molecules are particularly important in initiating the gas-phase chemistry. Thus, the H_2^+ ion rapidly reacts with H_2 to form H_3^+, which readily transfers by itself a proton to many other molecular species, specially to CO, CO_2, and N_2, leading to the formation of HCO^+, $HOCO^+$, and HN_2^+, respectively. Accordingly, HCO^+ is the most abundant ion in most ISM regions (see Table 6.5 and Figures 6.7 and 6.11), and it has even been observed in several distant galaxies (Seaquist et al., 1998). The H_3^+ ion can also react with oxygen atoms leading to the formation of water molecules through the reactions sequence (Draine, 2011):

$$H_3^+ + O \rightarrow OH^+ + H_{2,}\ k = 8.4 \times 10^{-10}\ cm^3s^{-1}$$
$$OH^+ + H_2 \rightarrow H_2O^+ + H,\ k = 1.0 \times 10^{-9}\ cm^3s^{-1}$$
$$H_2O^+ + H_2 \rightarrow H_3O^+ + H,\ k = 6.4 \times 10^{-10}\ cm^3s^{-1}$$
$$H_3O^+ + e^- \rightarrow H_2O + H \qquad (8)$$

where, the last reaction has a branching ratio of 25%. On the other hand, He^+ reacts with CO to form C^+ and O. In turn, the C^+ ion reacts only feebly with H_2, but avidly with methane, CH_4, acetylene, C_2H_2, and many other molecules to launch sequences that build up many organic compounds (Herschbach, 1999).

All the atoms considered in the reactions above belong to the first and second rows of the periodic table. Broadly speaking one may expect that the formation of molecules bearing elements from the third row would require higher temperatures than those needed for compounds involving second-row atoms, since these elements are significantly more abundant. Therefore, the formation of compounds with elements of the third row is likely to involve the breaking of a stronger bond of a stable molecule among second and first row elements in order to build a much weaker one, generally resulting in an endothermic reaction. Indeed, earlier laboratory studies on the gas-phase chemistry of phosphorus indicated that the ion-molecule chemistry of P-bearing molecules significantly differs from that of the closely related N-bearing molecules under the low-temperature conditions prevailing at the ISM because, unlike NH_n^+ ions, PH_n^+ ions react endothermically with H_2. Thus, PH, PH_2, and PH_3 compounds are expected to be very scarce in the gas phase (Thorne et al., 1983; Adams et al., 1990), at variance with the ubiquitous presence of ammonia and its derivatives in the ISM (see Figure 1.3). In fact, PH_n^+ ions have not been observed in the ISM as yet (Table 6.6).

TABLE 6.6 Laboratory Data on Reactions for Ions Containing Phosphorus*

Reaction P^+	**Rate Coefficient k (10^{-9} cm^{3}s^{-1})**	**Reaction PH^+**	**Rate Coefficient k (10^{-9} cm^{3}s^{-1})**
$+ NH_3 \rightarrow PNH_2^+ + H$	1.80–2.06	$+ NH_3 \rightarrow PNH_3^+ + H$	1.20–2.1
$+ CH_4 \rightarrow PCH_2^+ + H$	0.94–0.96	$NH_3 \rightarrow PNH_2^+ + H_2$	0.90–2.1
$+ H_2O \rightarrow HPO^+ + H$	0.51–0.55	$+ H_2O \rightarrow HPO^+ + H_2$	0.76–1.2
$+ O_2 \rightarrow PO^+ + O$	0.51–0.56	$+ CH_4 \rightarrow PCH_3^+ + H_2$	0.60–1.1
$+ CO_2 \rightarrow PO^+ + CO$	0.46–0.49	$+ O_2 \rightarrow PO^+ + OH$	0.49–0.54

*On the left (right) columns we list the reactions of P^+ (PH^+) ions with different molecules arranged by their rate coefficient value. P^+ ions do not react with H_2, N_2, CO, and HCN molecules. PH^+ ions do not react with H_2, N_2, CO, and CO_2 molecules.
Source: Thorne et al., 1984; Adams et al., 1990.

A possible route to PH^+ is the exothermic proton transfer reaction $H_3^+ + P \rightarrow PH^+ + H_2$. Since PH^+ reacts only very slowly with H_2, this ion will survive in interstellar clouds, where it would be available to react with other species. In Table 6.6, we list the measured rate coefficients for reactions involving different simple molecules with P^+ and PH^+ ions. As we see, reactions with ammonia show the largest k values, followed by those with methane (for the P^+ ion) and water (for both the P^+ and PH^+ ions). Now, the fractional abundances of methane and ammonia are significantly lower than that of water in dense molecular clouds. For example, ammonium column density values $f(NH_3) \sim 10^{-8}$ cm^{-2} and $f(NH_3) = 1.1 \times 10^{-7}$ cm^{-2}, have been respectively measured in SgrB2 and the Orion ridge, as compared to that of water $f(H_2O) = 4.3 \times 10^{-6}$ cm^{-2}. Accordingly, the possible reactions of P^+ and PH^+ ions with methane and ammonia listed in Table 6.6 were not included in the model reaction network considered in the earlier model by Thorne and co-workers (1984), which considered reactions involving the P-bearing compounds P, P^+, PO, PO^+, PH^+, HPO^+, and H_2PO^+ among them and with CO ($f \sim 5 \times 10^{-5}$), O_2 ($f \sim 1 \times 10^{-5}$), H_2O ($f \sim 2 \times 10^{-6}$) and CO_2 ($f \sim 1 \times 10^{-7}$) molecules, in a dense cloud with $n(H_2) = 10^5$ cm^{-3}. Not surprisingly the resulting model predicted that the most abundant interstellar phosphorus compounds would be those species containing P - O bonds, because both P^+ and PH^+ ions readily react with H_2O according to the reactions (see Table 6.6).

$$P^+ + H_2O \rightarrow HPO^+ + H, \; PH^+ + H_2O \rightarrow HPO^+ + H_2 \qquad (9)$$

followed by:

$$HPO^+ + H_2O \rightarrow H_3O^+ + PO, \; k \approx 1 \times 10^{-9} \text{ cm}^3\text{s}^{-1} \qquad (10)$$

which, upon dissociative recombination of the hydronium H_3O^+ ion (last reaction in Eq. (8)), can be expressed as the net reaction $P^+ + H_2O + e^- \rightarrow PO + H_2$. Thus, Thorne et al., (1984) predicted that about 85% of the available phosphorus remained in atomic form, 14% resides in PO (with $[PO]/[H_2] = 1.4 \times 10^{-10}$), and ~ 1% exists in all other P species, preferentially PO^+, HPO^+ (which have not yet been detected in the ISM) and P^+. In addition, it was argued that species containing P–N and P–C bonds would be rare because they require reaction of P^+ and PH^+ ions with NH_3 and CH_4 molecules, respectively, which are appreciably less abundant than H_2O

ones in the gas phase. On this basis, the very low ratio $[PN]/[H_2] < 10^{-14}$ was estimated for densities typical of dense molecular clouds.

However, at variance with these theoretical predictions, the first phosphorus bearing species found in the ISM was the $P \equiv N$ molecule, with fractional abundances ranging from 1×10^{-11} to 4×10^{-10} with respect to H_2 (Ziurys, 1987; Turner et al., 1987). This value is about four orders of magnitude higher than the predicted one, and it is comparable to the figure originally derived for the PO molecule. In order to account for this unexpected finding, Millar et al., (1987) added to the Thorne et al., ion-neutral reaction scheme given by Eq. (9) and (10) the neutral-neutral reactions given by Eq. (11) (calculated to be exothermic) with an estimated total rate coefficient $k = 3.4 \times 10^{-11}$ cm^{3}s^{-1} at $T = 50$ K, thereby efficiently destroying PO molecules and yielding PN ones.

$$PO + N \rightarrow PN + O, PO + N \rightarrow P + NO \tag{11}$$

Thus, the formation of PN in detectable abundances could start with the gas phase reactions $N + CP \rightarrow PN + C$ and $P + CN \rightarrow PN + C$, and then proceed through the reactions given by Eq. (11). Assuming that these reactions have no activation barrier, PN was thus predicted to be the most abundant P-bearing species under steady-state conditions (after neutral P), containing about 20% of available gas-phase phosphorus, with a fractional abundance of 1.2×10^{-9} relative to H_2, followed by HPO with 1.4×10^{-10}, and PO with 5.7×10^{-11}, which is slightly higher than the observational upper limit $[PO] < 3.2 \times 10^{-11}$ in Orion-KL (Ziurys, 1987). Once formed the stable PN molecule is depleted only via ion-molecule reactions. In this way, PN formation in dense clouds could be explained in terms of a relatively low-temperature gas-phase synthesis route, which was efficient due to two main reasons, namely:

1. Water molecules necessary to run reactions Eqs. (8) and (9) are abundant enough; and
2. The destruction routes for PN are inefficient since reactions with H_3^+, HCO^+, and other protonated species form HPN^+, which then reacts with electrons to give either PN or PH, which, in turn, is readily converted into PN through the very exothermic reaction $PH + N \rightarrow PN + H$.

Since earlier detections of PN were reported in hot and turbulent high-mass star-forming cores (with kinetic temperatures higher than 50 K), and later on this molecule was also observed in other star-forming regions (see Table 1.2), but not in several quiet dense clouds (Turner et al., 1990), the very possibility that this molecule was produced by processes related to either high-temperature gas-phase reactions or grain disruption was originally seriously considered, via thermal desorption of phosphine (PH_3) from grain mantles at temperatures above ~100 K, followed by rapid gas-phase reactions which transform it into PN, PO, or atomic P in a time scale of about 10^4 years (Turner and Bally, 1987; Charnley and Millar, 1994). However, recent observations indicate the presence of PN molecules with column densities ~10^{11}–10^{12} cm^{-2} (slightly lower than the values derived in energetic star-forming regions) in relatively cold and quiescent gas clouds (with kinetic temperatures in the range ~20–60 K), hence challenging the chemical models that explain the formation of PN (Fontani et al., 2016). Therefore, due to the limited number of observations available and the limited range of physical conditions of the observed regions with detected P-bearing molecules, the possible formation routes for PN and PO molecules are currently strongly debated. In particular, three plausible routes have been proposed for the formation of PO and PN in star-forming regions:

1. Shock-induced desorption of P-bearing species (e.g., PH_3) from dust grains and subsequent gas-phase chemical synthesis at low temperatures (Lefloch et al., 2016; Rivilla et al., 2018; Fontani et al., 2019);
2. High-temperature gas-phase chemistry after thermal desorption of PH_3 from ices (Charnley and Millar, 1994); and
3. Gas-phase formation during the cold collapse phase and subsequent thermal desorption by protostellar heating (Rivilla et al., 2016).

As we see, along with the gas phase, all the considered scenarios explicitly consider the presence of grains playing a significant role in the ISM phosphorus chemistry.

6.3.2 *Solid Phase Chemistry*

There is now ample evidence for the presence of substantial amounts of submicron-sized dust particles in the ISM. Interstellar light extinction observations indicate that the total grain mass relative to total hydrogen mass is larger than about 0.8%. What is this dust made of? In current dust models, refractory elements such as Ca, Fe, Ti, Al, Cr or Ni, are expected to be the most depleted in the gas phase. Oxygen is the principal element in silicates, forming the core composition of an important grains population. Carbon is the main constituent of a distinct population of grains, such as amorphous carbon, graphite, or polycyclic aromatic hydrocarbons (PAH). Thus, carbon and oxygen play prominent roles, along with more refractory elements and, consequently, the depletions of these elements are important parameters in constraining ISM dust models.

Since H atoms contribution to the grain mass is quite subsidiary (e.g., polyethylene molecules $(CH_2)_n$ are 86% carbon by mass), and He and Ne are chemically inert, the only way to attain the observed dust to hydrogen mass ratio is to build the grains out of the most abundant condensable elements, namely, O, C, Mg, S, Si, and Fe. Indeed, absorption-line spectroscopy of C, Mg, Si, and Fe shows that these elements are underabundant in the gas phase ISM, with about 66% of C and 90% or more of Mg, Si, and Fe presumed to be incorporated in dust grains in typical diffuse interstellar clouds (Draine, 2011).

Accordingly, the main materials that are considered in dust grain models are:

1. Silicates containing Mg, Fe, Si, and O atoms in different proportions;
2. Oxides of Si, Mg, and Fe (e.g., SiO_2, MgO, Fe_3O_4);
3. Solid carbon phases (graphite, diamond, and amorphous carbon);
4. Hydrocarbon molecules (e.g., PAHs);
5. Carbides, particularly SiC;
6. Metallic phases containing Fe along with Ti and/or Cr.

To date, little attention has been paid to the very possibility of grains containing P-bearing condensed compounds, such as P_4O_6, apatite, or schreibersite. I will consider this issue in more detail in Section 10.2.3.

The most clearly demonstrated ubiquitous solid material in interstellar space was originally referred to as "astronomical silicate." The term silicate refers to the presence of absorption and/or emission features at about 9.7 μm and 18 μm which are characteristic of Si–O and O–Si–O stretching in rocky materials, respectively. The term astronomical refers to the fact that, until now, no naturally found or laboratory-produced "pure silicate" bears exactly the same spectral signature as that observed in space, although they closely resemble the spectra of pyroxene ($Mg_xFe_{1-x}SiO_3$) and olivine ($Mg_{2x}Fe_{2-2x}SiO_4$) representatives. Greenberg redefined the term "astronomical silicate" by interpreting the 9.7 μm and 18 μm interstellar features not in terms of absorption by a pure silicate, but rather in terms of silicate core/organic refractory mantle particles (Greenberg and Li, 1996). According to this model, most particles found in the ISM will show a multiple shell structure consisting of a silicate core, generally embedding small carbon aggregates, and a thin ice mantle containing a small inventory of volatile molecules. Molecular species observed in the mantles of interstellar grains include CO, H_2O, NH_3, H_2CO, OCN^-, NH_4^+, CH_3OH, OCS, CO_2, and CH_4. The possible presence of H_2S, $(H_2CO)_n$, and S_2, as well as hexamethylenetetramine ($C_6H_{12}N_4$), ethers, alcohols, ketones, amides, and other diverse compounds related to polyoxymethylene, has been strongly inferred from laboratory spectra of realistic interstellar ice analogues (Bernstein et al., 1995; Schutte et al., 1993) and models of grain mantle evolution (Greenberg and Mendoza-Gomez, 1993). The possible presence of P-bearing chemical species in these mantles has been recently considered in order to investigate to what extent the C – P bond strength can lead to the formation of methyl phosphine (CH_3PH_2) and other higher-order organophosphorus compounds (particularly alkyl phosphonic acids, see Section 7.3.3) in ices of phosphine and methane upon interaction with energetic electrons in the track of galactic cosmic ray particles penetrating ice-coated grains in cold molecular clouds (Turner et al., 2016, 2018).

In fact, dust grains are subjected to dramatic changes at chemical, morphological, and structural levels during their lifetime, from the condensation of dusty cores within outflows of evolved stars to the final destruction in shocks or their aggregation in planetesimals during the formation of new generations of stars and planetary systems. In this way, dust grains undergo several growth (molecules adsorption and icy mantle accretion) and erosion (photoprocessing, sputtering) episodes as they move within first diffuse and then denser nebular media. Ice mantles on dust grains are formed from the accretion of gas-phase molecules during the gravitational collapse of these clouds. Thus, interstellar grains provide a surface on which accreted species can meet and react and to which they can donate excess reaction energy, an essential step in order to complete the chemical reaction process. The primary process initiating dust surface chemistry is the collision of a molecule from the ISM gas phase with the surface. Grain surface chemistry is then governed by the accretion rate, which sets the overall time scale for the process. The accretion rate is a measure of how often molecules will stick to the dust surface, and it depends on the collision energy, the temperature of the grain surface, and the chemical nature of the surface itself. For instance, silicates are highly polar and hence attract other polar molecules and atoms. For typical interstellar dense clouds conditions ($n = 10^4$ cm^{-3}, $T = 10$ K), a flux of about 10^5 hydrogen atoms cm^{-2} s^{-1} on a single grain corresponds to one H atom accreting every couple of months (for a 5 nm-sized grain) to one H atom every day (for a 100 nm-sized grain). For the sake of comparison surface chemistry in earthly based laboratories proceeds at much higher fluxes, as an ultrahigh vacuum background pressure of 10^{-13} bar corresponds to a flux of 10^{11} atoms cm^{-2} s^{-1} (Tielens, 2013).

Adsorption of molecules to the surface can result in dissociation of these molecules, producing reactive atomic species. The efficiency of surface chemical reactions depends on the mobility of the reactants. The surface migration rate, which governs the reaction network, is set by the interaction energy of the species with the surface, necessary to break the bonds of highly stable molecules. Grain surface chemistry involves predominantly hydrogenation and oxidation reactions. The most important reaction in the condensed phase of ISM is the formation of the neutral H_2 molecules from the union between two H atoms on the grain surface which then desorb back into the gas phase, releasing thermal energy which heats the grain (Exercise 15). Thus, the grain acts as a catalyst. Heterogeneous catalysis is also proposed for the formation of the ice mantles around the grain's core. Co-adsorption of H, O, and N atoms leads to the formation of water and ammonia ices. Adsorption of CO molecules onto the ice surface then provides a carbon source to initiate organic synthesis according to the following sequence of reactions:

$$CO + H \rightarrow HCO + H \rightarrow H_2CO + H \rightarrow H_3CO + H \rightarrow CH_3OH \qquad (12)$$

Which account for the formation of formaldehyde and methanol, hence illustrating the potential for hydrogenation of species on the grain surfaces (Shaw, 2006; Tielens, 2013). Other important ices formed on grain surfaces are H_2O, CO_2, and CH_4. To form water molecules, accreted oxygen atoms migrate and react to form molecular oxygen first. Once a small amount of

O_2 is present, further accreted O will react to form ozone, O_3 molecules. Atomic H will then preferentially react with O_3, forming OH, which immediately reacts with H_2 present on the surface to form H_2O. During water formation, some of the oxygen atoms can be lost through reactions with CO forming CO_2. However, in view of the low efficiency of this reaction, other routes to CO_2 on interstellar grains have been explored, such as the reaction $OH + CO \rightarrow CO_2 + H$ (Tielens, 2013). In summary, this sort of surface reactions increases the chemical complexity of the mantles which, when their temperature rises, release their frozen molecular material to the gas phase again, a process named thermal desorption. Since icy grain mantles appear to be dominated by simple hydrides, such as H_2O, CH_4, and NH_3, one would expect the presence of PH_3 in the mantles.

In fact, according to Charnley and Millar (1994), the most abundant P-bearing compound on grain mantles is phosphine, which is formed through hydrogenation of P atoms previously adsorbed on the grain mantles, and subsequently evaporates in the shock regions, thereby triggering the gas phase phosphorus chemistry. Thus, the released PH_3 molecules firstly react with H atoms following the sequence (Yamaguchi et al., 2011):

$$PH_3 + H \rightarrow PH_2 + H_{2,}$$

$$PH_2 + H \rightarrow PH + H_2 \tag{13}$$

Although these reactions have activation-energy barriers, they can proceed in warm enough conditions (i.e., gas temperature greater than 100 K, Charnley, and Millar, 1994). Then, PH reacts with O atoms to form PO, which further reacts with N atoms to produce PN according to Eq. (11).

Quite interestingly the presence of both PO and PN molecules with similar fractional abundances ~ 10^{-10} relative to H_2 has recently been reported in star-forming regions (Rivilla et al., 2016), the PO abundance being higher by a factor two (PO/PN = 1.8–3). In order to account for this relative abundance values, chemical models coupling both P-bearing compounds have been proposed on the basis of the gas-phase ion-molecule reactions given by Eqs. (14) and (15), which couple to each other through the neutral-neutral reaction given by Eq. (16) along with Eq. (11).

$$P + H_3O^+ \rightarrow HPO^+ + H_{2,}$$

$$HPO^+ + e^- \rightarrow PH + O,$$

$$\rightarrow PO + H \tag{14}$$

and

$$P^+ + H_2 \rightarrow PH_2^+,$$
$$PH_2^+ + e^- \rightarrow PH + H \tag{15}$$
$$O + PH \rightarrow PO + H \tag{16}$$

All these reactions would take place during the initial cold collapse stage of the molecular cloud. The produced molecules then freeze out onto grains at the end of the collapse phase and subsequently PO, and PN desorb simultaneously during the warm-up period once the cloud temperature reaches ~ 35 K. Once the temperature reaches ~ 100 K, water ice evaporates, and the abundance of protonated water correspondingly increases allowing for the reaction given by Eq. (17) to proceed.

$$PN + H_3O^+ \rightarrow HPN^+ + H_2O \tag{17}$$

In turn, HPN^+ has two equally probable channels of dissociative recombination:

$$HPN^+ + e^- \rightarrow PN + H,$$
$$\rightarrow PH + N \tag{18}$$

Since the abundance of atomic oxygen is higher than that of nitrogen by almost one order of magnitude, PH is preferably converted to PO rather than PN. In this way, PN is gradually destroyed, while the PO is additionally produced, thus significantly increasing the PO/PN ratio (Rivilla et al., 2016).

To investigate the main chemical route for PN formation, (that is, surface-chemistry versus gas-phase chemistry), and the dominant desorption mechanism, (namely, thermal versus shock processes), recently performed observations of PN towards a sample of nine massive dense cores in different evolutionary stages have shown that the excitation temperatures of the observed PN molecules were in the range 5–30 K. By comparing the obtained results with those derived from molecules tracing different chemical and physical conditions (SiO, SO, CH_3OH, and N_2H^+), it is concluded that, in six out of the nine targets, PN molecules may be released by sputtering of dust grains due to shocks. This finding is further supported by a statistically significant positive trend between the PN abundance and the abundances of SiO molecules in a sample of star-forming regions including 33 sources in total. This clearly shows that the production of PN in these regions is linked to the presence of shocked gas, and rules out alternative scenarios based on thermal evaporation from iced grain mantles (Fontani et al., 2019). Notwithstanding this, the origin of PN is probably not unique, since it can be formed in protostellar shocks, but also through alternative pathways in colder and more quiescent gas (Mininni et al., 2018).

In order to further ascertain this important issue the chemistry of P-bearing molecules under diverse, energetic conditions (UV photons and cosmic rays radiation fields and/or shock waves) has recently been carefully analyzed by considering detailed models including both grain surface chemistry and gas-phase reactions, along with gas-grain reactions as well. The considered chemical network contains 3695 reactions in total, involving 401 species, 265 of which are in the gas phase (Jiménez-Serra et al., 2018). In all the studied cases, the most abundant phosphorus compounds during the cloud free-fall collapse stage are atomic phosphorus and PN, with maximum abundances of $\sim 5 \times 10^{-10}$–10^{-9} at timescales of $\sim 5 \times 10^{6}$ years. After this time these species' gas-phase abundances start to decrease due to their freezing out onto dust grains. During the warming up of the central envelope by the central proto-star the resulting chemistry depends on both the depletion degree of P-bearing molecules onto the grains and the maximum temperature reached in this phase. For T ~ 50 K the only P-bearing compound that is released into the gas phase from the mantles is atomic P, which leads to the formation of PO via the reaction O + PH $\rightarrow$ PO + H, while the formation of PN depends on the resulting abundance of PO through the coupled reaction N + PO $\rightarrow$ PN + O, approaching an equilibrium state as time goes on. For T ~ 100 K and above, the abundance of phosphine in the gas phase shows an initial jump once the ices are thermally desorbed from the grains and become the most abundant P-bearing species for the first few 10^{5} years. After that, it gets destroyed via proton transfer with mainly H_3O^+, while PO and PN molecules are progressively formed, with PO/PN > 1 ratio in almost all considered scenarios. However, in all UV photon-illuminated models, the abundance of PN always remains above that of PO molecules, hence indicating the significant role played by UV photon radiation effects on the chemistry of phosphorus in these clouds. Increasing the ionization rates of cosmic rays, on the other hand, enhances the relative abundance of species such as PN and PO (with PN/PO ~ 1–2) with respect to those predicted by the pure collapse models due to the non-thermal desorption of these molecules by cosmic ray induced secondary UV photons and the rapid photodissociation of phosphine by this secondary UV field. Finally, when considering the chemical evolution of P-bearing compounds under the influence of shock waves (with speeds of about 20 km s^{-1}) it is observed that PH_3 and PN present enhancements larger than ×100 and ×10, respectively, with respect to their values in the pre-shock cloud, although for higher shock speeds (40 km s^{-1}) phosphine is efficiently destroyed due to the endothermic reaction $H + PH_3 \rightarrow PH_2 + H_2$. Nevertheless, in these models the ratio PN/PO > 1 is generally obtained, which is at variance with astronomical observations (Lefloch et al., 2016; Rivilla et al., 2018). In order to reconcile the model calculations with the reported empirical results Jiménez-Serra and co-workers (2018) have introduced a new formation reaction for PO molecule by including the reaction $P + OH \rightarrow PO + H$ ($k = 6.1 \times 10^{-11}$ cm^3 s^{-1}) in their chemical network. In doing so, the abundance of PO in non-shocked models is very moderately increased by a factor of ~ 2–3, whereas in shocked regions it can be increased by several orders of magnitude.

As we mentioned before, phosphorus nitride molecules have also been detected in circumstellar envelopes around both C-rich and O-rich stars, as well as in planetary nebulae (see Figures 6.7 and 6.11). Accordingly, it would also be possible that a significant amount of PN molecules was originally synthesized in the expanding atmospheres of evolving stars, then diluting in the surrounding ISM as a survivor species after the planetary nebula stage. In line with this scenario, the C≡P• radical was the second phosphorus compound detected in the ISM, in the circumstellar region around the late-type, evolved star IRC+10216 (Turner et al., 1990). Its most probable parent molecule according to detailed photochemical models is HCP which, in turn, almost completely condenses onto grains, with perhaps 4% remaining free to form CP. In this sense, phosphorus behaves in circumstellar envelopes like a highly refractory element. These results clearly indicate that both PN and CP are only minor repositories of phosphorus, which remains mainly locked in some condensed form in the ISM (Turner et al., 1990).

To conclude, a few words are in order to account for the possible route formation of the recently detected CCP molecule. To this end, the reaction of P^+ ion with acetylene (C_2H_2) provides a barrier-free exothermic channel leading to the ion $PCCH^+$ with a rate coefficient k = 3.4×10^{-11} cm^{3}s^{-1} at T = 50 K (Adams et al., 1990), which could lead to the CCP molecule through subsequent dissociative recombination. Similar reaction mechanisms may be proposed for other organophosphorus compounds. Thus, in the case of C_3P, the following scheme might be devised (Del Rio et al., 1996):

$$P^+ + C_3H_2 \rightarrow PC_3H^+ + H,$$

$$PC_3H^+ + e^- \rightarrow C_3P + H \qquad (19)$$

6.4 SUMMARY AND REVIEW QUESTIONS

A significant fraction of the atoms present in the stellar atmospheres of low mass stars is expelled through stellar winds, the formation of circumstellar envelopes, and the stripping of their outer envelopes during planetary nebula stage. In addition, high mass stars eject a significant amount of their original mass during SN explosions. This chemically enriched SN remnant leftover contributes to the interstellar medium (ISM), where diffuse gas clouds laced with microscopic dust grains undergo photochemical processes. In the ISM these elements interact with each other through ion-molecule reaction networks in the gas phase, as well as heterogeneous catalytic processes on grain surfaces. In the regions located at the denser and colder molecular clouds, chemical reactions further proceed with the synthesis of complex organic molecules containing up to 12 chained atoms, cyclic benzene-related compounds, and C_{60} cage molecules. Within some of these clouds, certain regions collapse under the action of gravitational forces, triggered by shock density waves, giving rise to the formation of protoplanetary disks where a new generation of stars and planets are born. These bodies incorporate in their own composition most of the materials which have been previously processed at the different stages of chemical evolution in the ISM.

According to this scenario, the presence of phosphorus atoms in the external atmospheres of low and intermediate-mass main-sequence stars should not be interpreted as indicating that this element is actually formed in those stars at the observing time. Rather it indicates this element was formed in previous generations of stars, expelled to the ISM and subsequently incorporated in the gas out of which that star was formed.

Assuming a solar value for the elemental phosphorus abundance, and neglecting the possible presence of yet undetected P-bearing compounds in the sources listed in Table 1.3, we have concluded (Exercise 13) that about 10% of the available phosphorus is in the form of gas-phase molecules in the circumstellar envelope around IRC +10216. Similar values, ranging from 10% to 14% have been obtained for the relative gas-phase abundance of phosphorus in the envelopes of several giants and supergiant stars, while larger values, of about 25% and 40%, are inferred for the giant star R Cas and the protoplanetary nebula CRL 2688, respectively. The higher amount of gas-phase phosphorus in the latter object may result from the energetic outflows present in the nebula, which could be destroying grains and releasing P-bearing compounds (Milam et al., 2008).

These results indicate that there are major gas-phase carriers of phosphorus, such as HCP, in the protoplanetary CRL 2688, and to a lesser extent in the AGB star R Cas (mainly, PO, and PN) as well, so that condensation onto grains may be less important for this element in these two sources. On the contrary, a significant fraction of interstellar phosphorus mainly resides in a condensed form in the remaining circumstellar envelopes, particularly around the C-stars IRC + 10216 and IK Tau, and the supergiant O-rich star VY CMa, where this element may be trapped in grains, although the chemical form in which phosphorus is present in these grains is still uncertain.

To date, the tetrahedral PH_3 and the linear NCCP molecules are the more complex P-bearing species detected in the regions between the stars. Nonetheless, it is reasonable to expect that the complexity of phosphorus chemistry extends beyond tetratomic molecules. Meteorites are known to contain phosphonic acids (see Section 4.5.3), such as the n-butyl phosphonic acid $(CH_3CH_2CH_2CH_2)_3PO_3$, which contains the C-C-C-C-P backbone. These results suggest that longer carbon-phosphorus chains might be found in dense molecular clouds currently undergoing star formation processes (Halfen et al., 2014).

KEYWORDS

- atomic depletion
- circumstellar shell
- condensation temperature
- diffuse clouds
- interstellar grains
- ion-neutral reactions
- local thermodynamic equilibrium
- M-, S-, and C-stars
- molecular clouds
- photodissociation chemistry
- planetary nebulae
- protoplanetary nebulae
- protostar
- solid-phase chemistry
- stellar atmosphere
- stellar formation regions
- thioxophosphino radical

CHAPTER 7

Phosphorus Compounds in Planetary Systems

"The nebula from which our solar system originated was seeded with minerals formed in the cauldrons of dozens of dying earlier-generation stars."

(Timothy J. McCoy, 2010)

The fate of the relatively simple phosphorus bearing species present in circumstellar regions and planetary nebulae, once they are incorporated into the ISM diffuse clouds first, and subsequently inside dense molecular clouds, is still unclear. Nonetheless, observations clearly indicate that certain phosphorus-containing solid-phase condensates, including schreibersite and phosphate minerals, as well as relatively complex organophosphorus molecules, can be found in solar system minor bodies, as it is illustrated on the right side of Figure 7.1.

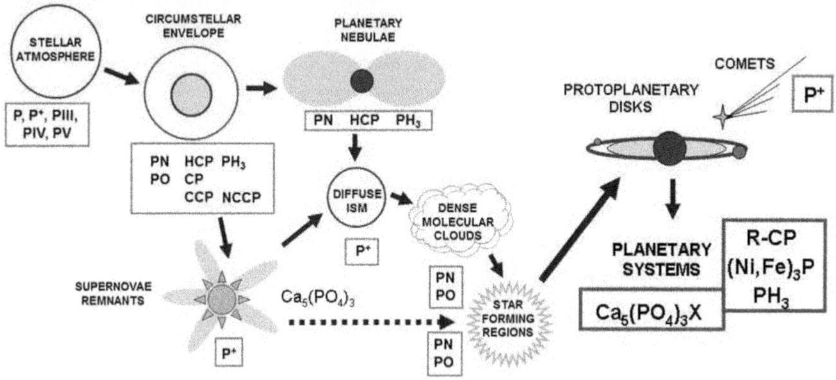

FIGURE 7.1 The phosphorus chemical evolution chart sketched in Figure 6.1 (on the left) is extended to display the phosphorus compounds found in environments related to protostars, protoplanetary disks, and planetary systems. Small apatite crystals found in interstellar grains embedded in some meteorites have been traced back to condensation processes which likely took place in the remnants of a supernova triggering the formation of our solar system.

In this chapter we will consider the physical and chemical processes leading to the possible formation, processing, or simply exposition of these phosphorus compounds during their journey from the ISM to planetary systems, starting when dense molecular clouds collapse under their own gravitation giving rise to star-forming regions where protostars and protoplanetary disks evolve to give birth to full-fledged planetary systems in their due time.

7.1 THE FORMATION OF THE SOLAR SYSTEM

The solar system comprises the Sun, eight planets, five dwarf planets (Ceres, Pluto-Charon, Haumea, Makemake, and Eris), several thousands of asteroids, about a hundred short-period comets, some thousands of long-period comets, and myriads of pristine cometary nuclei located at the outer solar system regions. The four inner planets, Mercury, Venus, Earth, and Mars, are mainly rocky. The four outer planets, Jupiter, Saturn, Uranus, and Neptune, are giant gaseous. Asteroids and comets are small rocky bodies that orbit the Sun, and are thought to be remnants of the planet formation process. Asteroids typically orbit in the region between Mars and Jupiter, although there are some that orbit within the inner solar system and some that pass quite close to the earth (called near-earth objects). It is currently understood that about 85% of the asteroids in the inner main belt (close to Mars' orbit) originated from the disruption of a few (five or six) precursor bodies. Comets originated in the outer regions of the solar system, and they often have very eccentric orbits, passing close to the Sun and then disappearing back in the distance. When close to the Sun, comets' outer layers vaporize, producing a small atmosphere (called coma) that progressively blows away to form a characteristic tail. There are two types of comets: short-period comets orbit in the same plane as the planets, and have periods below a few hundreds of years. Long-period comets come from all different directions and have periods of thousands or tens of thousands years. Accordingly, it is assumed that there exist two different reservoirs of comets in the solar system, respectively located in the Kuiper belt, beyond the orbit of Neptune, and the Oort cloud, extending out to many thousands of AU from the Sun.

The bodies of our solar system formed about 4,650 million years ago, together with the Sun, from a mass of gas, mainly hydrogen, mixed with small quantities of heavier elements, and some additional dust. Gravitational contraction of most of this mass formed the Sun at the center, whereas about 1% of the original material remained in the protoplanetary disk around it, becoming the stuff for forming the planets, satellites, and other minor bodies. The action of gravity combined with magnetic forces and viscous effects redistributed the mass and initial angular momentum within the disk to attain a final state in which about 99.9% of the mass resides in the Sun, while 2% of the solar system total angular momentum is related to the disk (Sánchez-Lavega, 2011). Inside this disk, the gas and dust material of the original cloud contributed to the formation of giant planets and rocky bodies such as asteroids, meteorites, and the terrestrial planets.

In this way, the mineral evolution of solar system bodies begun during the protostar nebula stage. It is currently assumed that the gas nebula from which our solar system originated was seeded with minute minerals, typically up to a few micrometers in size: the so-called presolar grains. These grains were formed during the late stages of earlier-generation stars, when temperatures in the expanding envelopes of red giants or supernova (SN) ejecta fell sufficiently to allow for the condensation of the most refractory atoms and less volatile molecules. These small grains entered the ISM, where many were destroyed by SN shocks and sputtering by the stellar wind. Thus, grains coming from a diversity of stars were incorporated into the dense molecular cloud from which our solar system formed, and they accreted to form planetesimals. These microscopic grains are identified in meteorite samples by their anomalous isotopic compositions, which are characterized by an excess of deuterium to hydrogen or relatively small $^{12}C/^{13}C$ and $^{14}N/^{15}N$ ratios, and contain only about a dozen mineral phases that represent the starting point of mineral evolution in our planetary system. They include diamond, graphite, carbides (SiC, TiC, ZrC, MoC, FeC, $(Fe,Ni,Co)_3C$), nitrides (TiN, α-Si_3N_4), oxides (TiO_2, Al_2O_3, $MgAl_2O_4$, $CaAl_{12}O_{19}$), and silicates (Mg_2SiO_4, $MgSiO_3$) along with Fe-Ni alloy nanoparticles included within a presolar graphite matrix, and silicate glass with embedded metal and sulfides (Hazen et al., 2008; McCoy, 2010). Quite remarkably, the list above has recently been extended with the addition of the phosphate apatite (see Section 4.5.4), identified in presolar grains extracted from the Orgueil and Murchison carbonaceous chondrites, and the Dimmit and Cold Bokkeveld ordinary chondrite meteorites (Jungck and Niederer, 2017). In fact, microsized apatite grains were found in a number of Orgueil meteorite samples, respectively containing 65%, 35%, and 10% of this mineral. On the other hand, apatite was recognized as a companion mineral of graphite layers in the other three meteorites. By all indications, the observed apatite stems from phosphorus produced in a SN explosion, which condensed forming apatite layers grown around a graphite mineral core as the wave of explosion gases pushed outwards interstellar dust originally located around the SN progenitor star (Jungck and Niederer, 2017). This finding provides evidence suggesting the formation of highly oxidized phosphorus compounds during the SN aftermath events, as indicated in Figure 7.1.

Gravitational clumping into a protoplanetary disk, along with Sun formation and the resultant heating in the solar nebula, produced the first greater solids to condense within the solar system 4,565 billion years ago (Connelly et al., 2008), namely, the so-called calcium-aluminum inclusions and the chondrules that characterize the chondritic meteorites (see Section 7.3.3). Currently, there is a general consensus about how the terrestrial planets formed from the gravitational accretion of small growing clumps of solid matter, from micrometer grains to kilometer-sized planetesimals to finally build the planet body. Earth formed via accretion from the solar nebula 4.53–4.47 billion years ago, and differentiated over a 10–150 million years time period. Thus, the formation of earth commenced not earlier than 10 million years after the solar system formation (Kleine et al., 2004). The Hadean marks the first geological eon in earth's history and spans the period from the end of the accretion to the beginning of the Archean eon 3.8 billion years ago. During the Hadean impactors pummeled proto-earth. The most dramatic event during this period occurred within the first 100 million years of earth's history. According to the most widely accepted hypothesis, earth collided with a Mars-sized object, thus leading to the formation of the Moon (Canop et al., 2001). Earth's surface was left in a molten state by the impact heat, and the lack of an atmosphere immediately after impact allowed for cooling within 150 million years yielding a solid basaltic crust. Steam escaping from the crust and gases released from volcanoes generated the prebiotic atmosphere. The bombardment of earth by impactors continued after the Moon-forming impact, resulting in the delivery of a late veneer of material containing highly siderophile elements.

Regarding the formation of giant gas planets, there are two alternative possibilities. The first is based on the formation of an embryo protoplanet at sufficiently large distances from the Sun, a giant planet core (containing a few earth planet masses) that grew by a similar accretion mechanism and was followed by the capture of surrounding gas material from the nebula (mainly hydrogen) that finally formed a large shell, which is the main gaseous body of the planet. The second hypothesis is based on the so-called gravitational disk instability, a mechanism acting on a dense enough disk whose ring-like structure is fragmented into massive gaseous clumps that form the giant protoplanets. These processes are fast compared to the current age of the planetary system, and planet formation was reached in about 5–10 million years. The major satellites of the planets formed around them by a somewhat similar accretion mechanism, whereas other satellites as the Moon formed following large impacts with the rest of the sparse material within the nebula. Minor satellites were incorporated into planetary orbits from the capture of the rest of the disk debris. During the following 2,000 million years, a large quantity of the leftover debris impacted on the planets and satellites, forming surface craters and eroding the gaseous envelopes (primordial atmospheres) formed around the planets and those major satellites with intense enough gravitational fields (Sánchez-Lavega, 2011).

7.2 PROPERTIES OF EXOPLANETARY SYSTEMS

7.2.1 *Protostars and Protoplanetary Disks*

As we mentioned in Section 2.4.8, the edge-on disk surrounding the nearby young star β Pictoris can be regarded as a very illustrative example of what a young planetary system looks like. This debris disk is mainly composed of dust and gas produced by collisions between – and evaporation of – planetesimals analogs of solar system's comets and asteroids. In this regard, one must keep in mind that β Pic disk is exceptionally dusty, most likely due to recent major collisions among unseen planets and asteroid-sized objects embedded within the disk. In particular, a bright dust and gas lobe on the southwestern side of the disk may be the result of a giant collision with a Mars-sized object which was pulverized. In addition, carbon is extremely overabundant relative to every other measured element, a property which apparently helps to keep the gas disk in keplerian rotation (Roberge et al., 2006). Spectroscopic observations of β Pictoris reveal a high rate of transits of small evaporating bodies, named exocomets. In Figure 7.2, the elemental composition of β Pic disk is compared to those corresponding to the Sun, a carbonaceous chondrite meteorite, and comet Halley dust. By inspecting this figure, one sees that the composition of β Pic disk is dissimilar to the composition of all three comparison bodies. In particular, C is overabundant relative to the other measured elements, and the C/O and C/Fe ratios are 18 and 16 times the solar values, respectively. The possible presence of phosphorus was searched looking for absorption lines from PI and PII, but none were detected (Roberge et al., 2006).

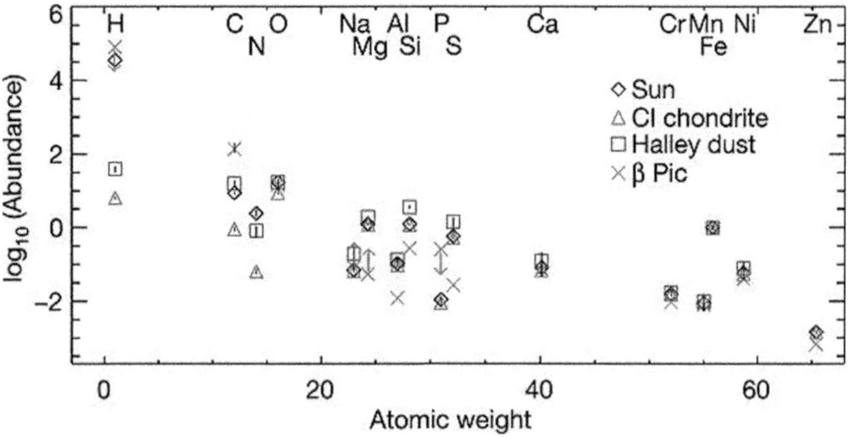

FIGURE 7.2 The elemental abundances of β Pic midplane gas are shown with crosses, the Sun with diamonds, a carbonaceous chondrite meteorite with triangles, and comet Halley dust with squares. The ordinate axis shows the logarithmic elemental abundance, normalized to Fe. The vertical lines on top of the data points show the standard deviations (± 1σ) of the abundances, where available. Upper and lower limits are indicated with arrows.

In Figure 7.3, we list the molecules observed in both protoplanetary disks and the ISM to date. Some refractory compounds, such as SiO, SiC or SiS, simple volatiles, including CO, H_2S or HCN, and more complex organic molecules, such as CH_3CN and HC_3N cyanides, NH_2CHO amide or $HCOOCH_3$ acid, have been detected in solar nebula analogues, indicating that they either survive disk formation or are re-formed in-situ in proto-planetary disks around young stars. This implies that complex organics accompany simpler volatiles in protoplanetary disks and that the rich organic chemistry of our solar nebula was not unique (Örberg et al., 2015). For instance, double deuterated formaldehyde (D_2CO), deuterated methanol (CH_2DOH), methyl mercaptan (CH_3SH), and the prebiotic precursor molecule formamide (NH_2CHO) have been detected in the nearby (400 pc) protostellar system HH 212, which hosts a "hamburger"-shaped dusty disk with a radius of ~60 AU, deeply embedded in an infalling-rotating flattened envelope (Lee et al., 2017).

PROTOPLANETARY DISKS & ISM

SiO	C_2H	CN	H_2S	H_2CO
SiS	NH_2^+	CO	SO_2	D_2CO
	SiC_2	CO^+	HCN	HC_3N
	HCO	CS	HNC	CH_3OH
		SO	HCO^+	CH_2DOH
HNCS				CH_3CN
l-C_3H_2				CH_3SH
$HOHCS^+$				NH_2CHO
				$HCOOCH_3$

FIGURE 7.3 Molecules observed in both protoplanetary disks and the ISM.
Source: Ehrenfreund et al., 2002; Lee et al., 2017.

7.2.2 Exoplanets

The properties of the about 4,100 exoplanets discovered to date (5 September 2019) are quite different from what we would have expected on the basis of the knowledge of our solar system only. In Figure 7.4a, exoplanet masses are plotted against their orbital periods around the parent star (for reference, 1 earth mass is 0.003 Jupiter mass). Those planets detected via the transit method (in green) are quite closer to their star. Red is for planets detected via the radial velocity method, which can detect bodies with a large range of orbital periods, and according to the third Kepler's law, to large radii as well. For the sake of illustration, around a solar-like star, periods between 10^3 and 10^4 days correspond to orbital distances of between 2 and 9 AU. The first planet around a Sun-like star was discovered by using the radial velocity method (Mayor and Queloz, 1995). It was found to have a mass of 0.47 Jupiter masses, but to have an orbital period of only 4.2 days. Hence, this is a gas giant planet that is orbiting its parent star with a much smaller orbital radius than that of Mercury (period of 88 days). It is what is known as a hot Jupiter planet, and many more have since been discovered. The diagonal arrangement of the dots indicates that this method is more sensitive to high-mass planets close to their parent star. Blue is for planets detect via microlensing, and it is more sensitive to bodies located at modest distances from their parent star. Finally, the pink dots are for exoplanets detected via direct imaging, all of which are massive bodies, above the reference Jupiter mass, orbiting at large distances. An intriguing feature of the results shown in Figure 7.4a is that we find planets of all masses in all regions of parameter space where we have the ability to detect planets. Thus, we observe both low-mass (below 0.01 Jupiter mass) and high-mass (above 10 Jupiter mass) planets with periods of less than 10 days (corresponding to separations of less than 0.1 AU). There are also planets with similar masses but with periods in excess of 10^3 days. Based on the knowledge of our solar system, and the earlier theories of planet formation, we would have expected massive gas giant planets in the outer parts and small rocky planets in the inner regions of these exoplanetary systems, at variance with the reported observations. We would have also expected that exoplanets would typically display quite circular orbits. On the contrary, Figure 7.4b shows that exoplanets have a wide range of eccentricities, and some of the observed orbits are remarkably eccentric ($e > 0.9$). This figure also shows that as the orbital period decreases, the range of eccentricity is also reduced. This is because as the planet gets closer to its parent star, it is able to tidally interact with the star, exchanging angular momentum, so that the orbit becomes more circular.

By combining transit observations and radial velocity measurements, one can determine both the mass and the radius of the exoplanet. In this way, it has been concluded that there are some exoplanets with compositions similar to that of the earth (say, 50% iron and 50% $MgSiO_3$), but there are also bodies that appear to be more than 50% water. This naturally leads to the consideration of the possible habitability of certain exoplanets. The so-called habitable zone (HZ) is the region around a star where a planet, with sufficient atmospheric pressure, could support liquid water on its surface. There are a number of factors determining the precise location of the HZ for a given star. One factor is the planet's albedo, which determines how much of the incoming radiation is simply reflected back to space. Another is the thickness and composition of the planet`s atmosphere, including the possible presence of greenhouse gases in the atmosphere, which can contribute to warm the planet more efficiently for a given radiative energy budget. Other additional factors, such as orbital properties, the existence of geochemical cycles, or changes in the stellar light input, can also influence the range of the HZ, but typically we would expect it to be about 0.9 and 1.4 AU around a solar-like star (Domagal-Goldman and Wright, 2016).

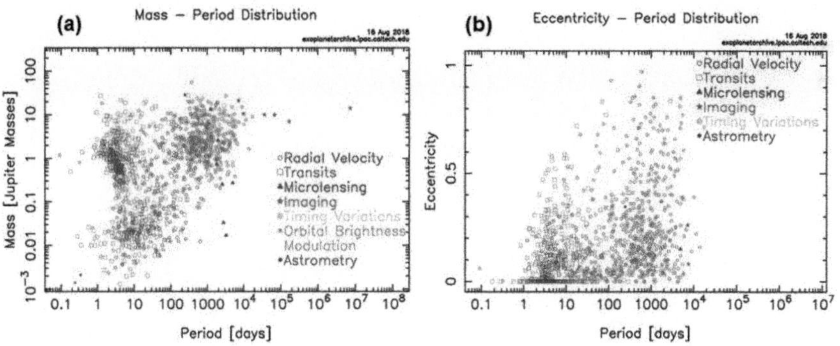

FIGURE 7.4 (See color insert.) Graph showing exoplanet (a) masses (measured in terms of Jupiter's mass) and (b) eccentricity against orbital period (in days). The different colors indicate different detection techniques.

Source: NASA Exoplanet Archive, operated by the California Institute of Technology, under contract with the National Aeronautics and Space Administration under the Exoplanet Exploration Program with permission.

Due to the progressive fusion of H onto He nuclei in their cores (see Section 5.4), main sequence stars get brighter over their lifetime. As a result, the inner and outer HZ boundaries are pushed outward as the star evolves. For the solar system, the 4.6-billion-year continuously HZ has been estimated to extend from 0.95 to 1.15 AU (Kasting et al., 1993). Analogously, for brighter stars, the HZ is shifted outwards because planets must be farther away from them to maintain temperatures necessary for liquid water, and for dimmer stars, the HZ is shifted inwards.

In Figure 7.5, we show the variation of the HZ location and width as a function of the parent star spectral type. For stars cooler than the Sun, it is closer to the star and narrower than in a Sun-like star. Additionally, for stars with masses less than 90% that of the Sun, it falls inside the so-called tidal lock radius, and so planets in the HZ of such stars would have one side always facing the host star.

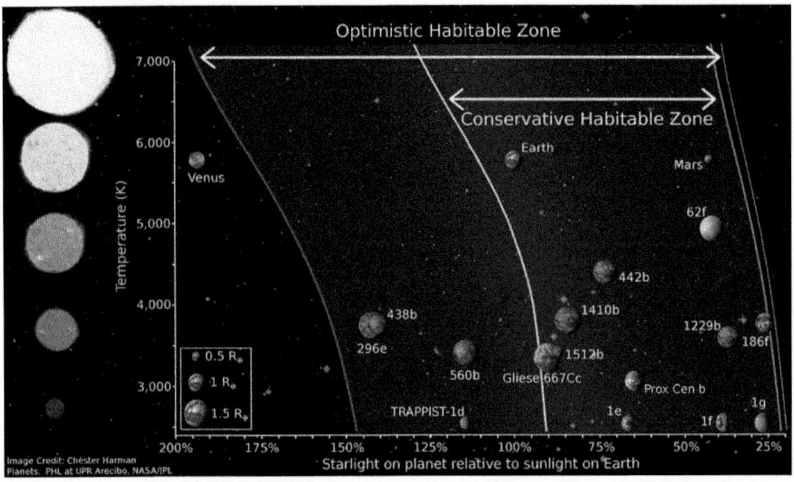

FIGURE 7.5 (See color insert.) A diagram depicting the habitable zone (HZ) boundaries, and how the boundaries are affected by stars ranging in spectral type from F to M. For the sake of comparison this plot includes solar system planets (Venus, Earth, and Mars) as well as especially significant exoplanets such as TRAPPIST-1d, Kepler 186f, and our nearest neighbor Proxima Centauri B. This plot shows the limits for both the "conservative habitable zone," which are based on one-dimensional climate model calculations (Kopparapu et al., 2013), and for the "optimistic habitable zone," which are based on observations that Mars once had liquid water at the surface and Venus used to have more water, possibly contained in oceans.
Credit: Chester "Sonny" Harman, using planet images published by the Planetary Habitability Lab at Aricebo, NASA, and JPL.

As an exoplanet transits in front of its host star, some of the light from the star is absorbed by the atoms and molecules in the planet's atmosphere, causing the planet to seem bigger from a remote observer viewpoint. Plotting the planet's observed size as a function of the wavelength of the light produces a transmission spectrum. Measuring the tiny variations in this transmission spectrum one obtains clues about the properties of the planet's atmosphere. In this way, chemical species composed of relatively light elements, such as H, C, O, Na, and K, have been detected in the atmospheres of several hot giant exoplanets (Vidal-Madjar et al., 2004; Madhusudhan et al., 2016). Recently, the presence of water molecules has been detected in three exoplanets (HD 102195b, WASP-19b, and

WASP-121b) to date, and (Kreidberg et al., 2014), CO, and CH_4 (detected in HD 102195b), as well as molecules bearing heavier metallic atoms, such as TiO (detected in WASP-19b and WASP-33b), has been reported as well (Sedaghati et al., 2017). No phosphorus bearing compounds have been detected as yet in exoplanets (http://www.astrochymist.org/astrochymist_exoplanets.html).

One of the great shocks in the search for exoplanets has been the dawning realization that most of the planets in our Galaxy are not attached to stars. These so-called rogue planets are assumed to be ejected from protoplanetary disks during the early stages of their formation and wander around the Galaxy, most likely in orbits around the galactic center (Summers and Trefil, 2017). Extrapolating to the extragalactic domain, it is natural to hypothesize that planets are common in external galaxies as well. Quite remarkably, by studying the microlensing properties close to the event horizon of a supermassive black hole, some hints for the possible existence of a population of unbound planets between stars with masses ranging from Moon to Jupiter masses have been reported in the gravitationally lensed quasar RXJ 1131–1231 (Dai and Guerras, 2018).

7.3 DISTRIBUTION OF PHOSPHORUS COMPOUNDS IN SOLAR SYSTEM BODIES

7.3.1 Comets

Comets formed in the cold, outer regions of the solar protoplanetary nebula, at distances of at least 5 AU from the Sun, and they are stored in reservoirs as far as 30–10,000 AU from the Sun. Having condensed directly out of the pre-solar nebula, comets are thought to contain relatively pristine interstellar materials that have been moderately processed in the protosolar nebula, and have remained cold ever since. Therefore, their frozen surfaces provide a repository of the primitive preplanetary nebula gas and dust composition about 4.6 billion years ago. Indeed, the presence of highly volatile frozen molecules, such as CO, H_2O, NH_3, CH_4 or N_2 in samples returned from comets, indicates that they formed at very low temperatures, as low as about 30 K. Moreover the remarkable similarity between such volatile ices in interstellar material and in comets strengthens the link between comets and the pristine interstellar material of the solar nebula. However, it is now clear that some solid components (crystalline silicates, Ca-Al inclusion fragments) in comets may well have formed in the inner solar nebula and then diffused outward due to the pressure radiation from the star (Li, 2009). In this way, the chemical inventory of comets provides another window into the early phases of the solar system (Tielens, 2013).

In the last decade, important progress has been made regarding the chemical composition of Kuiper belt objects. Thus, the presence of H_2O, CO_2, CO, CH_3OH, CH_4, and NH_3 has been confirmed in several of these objects (Caselli and Ceccarelli, 2012). Therefore, the most abundant species in comets, with the exception of H_2S and H_2CO (see Table 7.1), are also observed in Kuiper belt objects. In addition, ethane (C_2H_6), thought to be the result of CH_4 photolysis caused by solar wind and cosmic rays, has been detected in the dwarf planet Makemake.

As comets approach the Sun in their orbital journey, their surface ices start to sublimate and to create a gas and dust coma that has been extensively studied through the rotational spectra in the submillimeter and the fluorescence bands in the visible and the IR during the last decades. Radio spectroscopic observations of comets during their visit to the inner solar system have allowed us to detect a wide variety of molecules in their coma. This information has been complemented with detailed analysis of coma and nucleus composition performed during spacecraft flyby to the comets 1P/Halley (1986), 26P/Grigg-Skjellerup (1992), 19P/Borrelly (2001), 81P/Wild2 (2004), and 9P/Tempel 1 (2011), along with the laboratory study of grain samples returned from comet Wild 2 by the Stardust spacecraft (2006) and the recent in-situ measurements performed during Rosetta's landing on comet 67P/Churymov-Gerasimenko in 2015. In this way, over 57 molecules have been identified in comets, as it is shown in Figure 7.6. In Table 7.1, we list the relative abundances of the more frequent molecules observed in comets.

Among organic species we find several molecules of biological interest, such as hydrogen formaldehyde (H_2CO), cyanide (HCN), acetonitrile (CH_3CN), formamide ($HCONH_2$), acetamide (CH_3CONH_2), ethylene glycol ($CH_2(OH)CH_2OH$), acetone (CH_3COCH_3), or the simplest amino acid glycine (CH_2NH_2COOH). In fact, glycine was found in dust samples from comet Wild 2, brought back by the Stardust mission (Elsila et al., 2009), together with the precursor molecules methylamine (CH_3NH_2) and ethylamine ($CH_3CH_2NH_2$). These compounds have been also detected in

TABLE 7.1 Organic Compounds in Comets*

Molecule	Relative Abundance	Hyakutake	Hale-Bopp
H_2O	100	100	100
CO	23	6–30	23
CO_2	6	<7	6
CH_3OH	2.4	2	2.4
H_2S	1.5	0.8	1.5
H_2CO	1.1	0.2–1	1.1
NH_3	0.7	0.5	0.7
CH_4	0.6	0.7	0.6
OCS	0.4	0.1	0.3
C_2H_6	0.3	0.4	0.1
SO	0.3	0.1	0.2–0.8
HCN	0.25	0.1	0.25
CS_2	0.2	0.1	0.2
CS	0.2	-	-
SO_2	0.2	-	0.1
C_2H_2	0.1	0.5	0.1
HCOOH	0.1	-	0.08
HNCO	0.1	0.07	0.06
$HCOOCH_3$	0.08	-	0.08
HNC	0.04	0.01	0.04
CH_3CHO	0.02	-	0.02
NH_2CHO	0.02	-	0.01
CH_3CN	0.02	0.01	0.02
HC_3N	0.02	-	0.02
H_2CS	0.02	-	0.02
NS	0.02	-	-

*Relative abundances of molecules observed in comets with respect to water. The first column gives an average value. No phosphorus compounds have been detected in comets to date.

Source: Llorca (2005); Crosivier (2004); Bockelée-Morvan et al. (2000).

the coma of comet P67/Churyumov-Gerasimenko by the Rosetta orbiter spectrometer for ion and neutral analysis (ROSINA) mass spectrometer on numerous occasions while the comet was approaching perihelion (Altwegg et al., 2016).

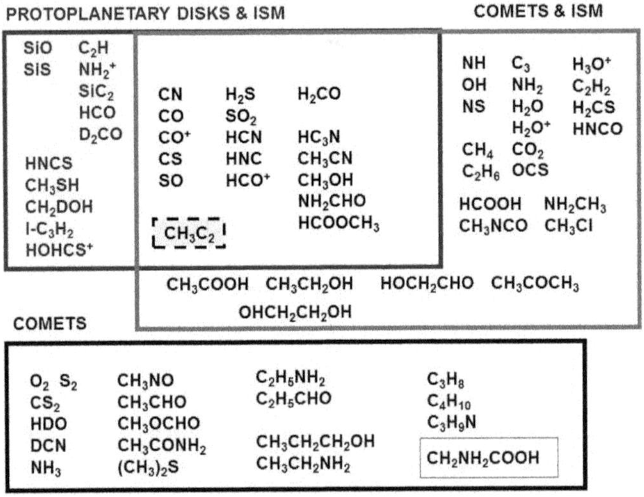

FIGURE 7.6 In the upper panel, we list the molecules observed in protoplanetary disks and the ISM (on the left), the comets and the ISM (on the right) and in the three astrophysical objects (central intersection rectangle). The molecule CH_3C_2, highlighted in the central part, has been detected in both protoplanetary disks and comets, but not yet in the ISM. In the lower panel, we list those molecules which have been detected only in comets to date. The simplest amino acid glycine (CH_2NH_2COOH) is highlighted.
Source: Ehrenfreund et al., 2002; Tennyson, 2003; Crovisier et al., 2004; Elsila et al., 2009; Caselli and Ceccarelli, 2012; Goesmann, 2015; Örberg et al., 2015; Altwegg, 2016, 2017; Fayolle, 2017; Li et al., 2017.

Molecular oxygen was detected in the coma of 67P/Churymov-Gerasimenko by ROSINA. It had a high volume mixing of about 1–10%. Primordial O_2 within the nucleus of comet 67P is compatible with the instrumental observations (Heritier et al., 2018). In addition, glycine was detected around the nucleus of comet 67P/Churymov-Gerasimenko for the first time in October 2014, when the comet was at 3 AU from the Sun, just before Lander delivery. The following glycine observation was during a close flyby (15 km) of the comet on 28 March 2015 at a heliocentric distance of 2 AU, probably associated with dust. Glycine is not a very volatile species whose sublimation temperature is slightly below 150°C, and probably very little of it sublimates from the nucleus surface due to cold temperature. However, the dark small grains released in the coma may attain high enough temperatures to this end when properly heated by incoming sunlight (Altwegg et al., 2016).

The recent in-situ analysis of the composition of the frozen surface of comet 67P/Churyumov-Gerasimenko by the cometary sampling and composition (COSAC) experiments aboard Rossetta's Philae Lander has revealed a significant number of complex organic molecules. Many of the species detected by COSAC have not been observed in the comae of previously studied comets, while some organic molecules observed in the coma of comets are not detected by COSAC (Goesmann et al., 2015).

By inspecting Figure 7.6, one sees that most of the relatively simple molecular species observed in comets are well-known interstellar molecules as well, and one may reasonably expect that more complex molecules already detected in comets but not observed in the ISM yet, will be ultimately found in the ISM in their due time. For instance, methyl isocyanate (CH_3NCO) is the third most abundant species detected on the 67P comet surface, with 1.3% abundance relative to water, following water and formamide (NH_2CHO, 1.8%). But formamide molecule has been only recently detected in the ISM, its search being spurred by its previous detection in this comet's surface (Cernicharo et al., 2016; Halfen et al., 2105). Thus, a good correlation has been found between the type of molecules detected in the comae of comets and those of warm molecular clouds. In fact, overall abundances in comets and in interstellar or circumstellar environments point toward a shared chemical origin. However, there are some differences in detail indicating that the model of a comet as an aggregate of interstellar grains should be complemented with substantial mixing of grains coming from the inner preplanetary nebula as well (Tielens, 2013).

The dust impact analyzers carried by the spacecraft Vega 1, Vega 2, and Giotto provided the first direct, though not quantitative, measurements of the physical and chemical properties of cometary dust. These probes indicated that most of the analyzed particles (up to about 80%) are rich in the light, biogenic elements H, C, N, and O, suggesting the validity of those models describing cometary dust as a source of organic-rich material (Kissel et al., 1986a, 1986b). Organic matter in comets can be divided into two components. The first is volatile, stored in an icy phase dominated by H_2O, CO_2, and CO, and is released from the nucleus into the cometary atmosphere when the comet approaches the Sun. The second component is a refractory phase that remains solid either on the nucleus and/or on the dust particles. Part of the volatile fraction is released directly from the nucleus when the ices sublimate, while other compounds are ejected from previously released dust particles during the most active sublimation stage.

For instance, the distribution of the amino acid glycine molecules detected in the atmosphere of the comet P67/ Churymov-Gerasimenko is consistent with a release from the heated dust particles present in the coma, rather than a direct sublimation from the nucleus itself, as we previously mentioned (Altwegg et al., 2016). The nature of the solid refractory component has eluded a proper characterization for a very long time, though it can be generally split into three groups: one is similar to the insoluble organic matter substance observed in carbonaceous chondrites, the second is highly aromatic matter contained in non-globule-like features, and the third a volatile aliphatic organic found as a halo (Levasseur-Regourd et al., 2018).

The analysis of some obtained mass spectra in Halley's flyby indicated the presence of a minor feature corresponding to the charge-mass ratio m/z = 31, which may correspond to elemental phosphorus. The positive-ion mass spectra obtained during the comet 81P/Wild 2 flyby also exhibited a high signal at m/z = 31 (Kissel et al., 2004). Nevertheless, this assignment was regarded as uncertain due to the difficulty in resolving phosphorus from other organic ions of equal mass ratio, like CH_2OH^+ (Kissel and Krueger, 1987) or H_3CO^+ (Heritier et al., 2017). Phosphorus was not found in any of the 23 particles collected by Stardust, although this element was detected in the residue of two of the seven craters left in the collecting aluminum foils during the passage of the Stardust spacecraft through the coma of comet Wild 2 (Flynn et al., 2006). The new generation analyzer on board the Stardust probe allowed for studying the mass spectra of negative ions of cometary particles for the first time. However, the obtained negative-ion spectra are almost featureless in the range m/z = 60–80, where the possible presence of PO_2 and PO_3 anions should be expected (Kissel et al., 2004; Standford et al., 2006). The possible presence of PO_2, PO_3, and PO_4 ions was considered in the analysis of organic compounds at the surface of comet 67P (Wright et al., 2015) and it was concluded that, if present, these compounds could only be there in relatively minor amounts (Ian P. Wright, private communication 21 September 2015). The first definitive detection of phosphorus in a comet (P67/Churymov-Gerasimenko) was reported on the basis of a clear peak observed in a mass spectra, located at 30.973 Dalton, the exact mass of the P^+ ion (Altwegg et al., 2016). However, it is uncertain which parent molecule produces these ionized atoms, since no specific P-bearing chemical species have been spectroscopically detected in comets to date, and no clear signatures of PO, PN, CP or HCP molecules have been found so far in the studied data. A possible candidate may be phosphine ice, which has been radio spectroscopically searched in the gaseous coma of a few comets, although recent observations of the Oort cloud comets ISON (C/2012 S1) and Lovejoy (C/2013 R1) were unsuccessful, providing upper limits that were not significant enough to conclude whether or not PH_3 is present in these pristine solar system bodies (Agúndez et al., 2014c; Turner et al., 2018).

7.3.2 *Interplanetary Dust Particles (IDPs)*

In particularly clear nights a faint brightness of the sky, called the zodiacal light, can be seen upwards from the eastern horizon before sunrise or above the western horizon after sunset. This glow is produced by scattering sunlight from dust particles with sizes ranging from 1 to 100 μm that are distributed along a flattened disk centered on the Sun, the Zodiacal cloud, and are referred to as interplanetary dust particles (IDPs). The most important sources of IDPs are comets. As a comet approaches the Sun ice components sublimate and dust particles are emitted from the nucleus and accelerated away by radiation pressure which originates from the interaction of sunlight and matter, giving rise to its characteristic tail. For the sake of illustration, the space probes that flew past Halley's comet in 1986 found that its nucleus was injecting approximately 3,000 kg of dust per second into the interplanetary medium (McDonnell, 1986). Dust particles released from the nucleus comet are devoid of the volatile molecules that may have filled the cavities inside the particles, so that IDPs look like fluffy aggregates. Other sources of IDPs are main-belt asteroids and planetary rings. The relative proportions of the cometary and asteroidal contributions to the Zodiacal cloud may vary with time as a result of major disruptions in the main asteroidal belt, on the one hand, and due to the variable frequency of comets approaching the Sun, on the other hand. In addition, both gravity and radiation pressure from the Sun limit the age of the IDPs: either the particles are decelerated and get close to the Sun, where they vaporize, or if the radiation pressure overcomes gravity, they are blown out of the solar system. Consequently, IDPs are not a remnant of preplanetary nebula, but mainly a result of the much more recent evolution of comets and asteroids (Llorca, 2005).

A detailed modeling of the gravitational focusing effect indicates that all near-earth dust collections are highly biased towards the main-belt asteroidal component, due to its lower geocentric velocity as compared to that of comets (Flynn, 1994). On the other hand, most of the cometary component probably comes from short-period comets formed in the Kuiper belt (Brownlee, 1994).

The meteoritic origin chondritic IDPs are generally divided into two groups: those dominated by anhydrous minerals and those dominated by hydrated minerals, and their major refractory element composition is similar to chondritic meteorites. However, the carbon contents, as well as their contents in volatile elements, are significantly higher in most chondritic IDPs than in ordinary chondrites. Thus, most (or all) the chondritic IDPs collected from the earth's stratosphere differ in mineralogy, chemistry, reflection spectrum, porosity, and grain-size from bulk ordinary chondrites, and no significant population among these chondritic IDPs matches either the bulk or separated matrix material of the ordinary chondritic meteorites (Flynn, 1994).

In most IDPs that are studied in detail, phosphorus abundances consistent with carbonaceous chondritic ones are often measured. Sometimes higher abundances of P are seen, while in other IDPs it is undetected, although the latter could be explained by the detection limits. According to the data reported in the study on nineteen anhydrous IDPs of probable cometary origin by Thomas et al., (1993), the average phosphorus content of these particles is of the order of 0.3% in weight and some differences are observed in the content of this element depending on the different IDPs analyzed, ranging from 0.1 to 0.7% in weight, approximately. As a comparison, the bulk composition of the Orgueil CI carbonaceous chondrite contains 0.22% of P in weight (Jarosewich, 1990). Arndt and co-workers (1996) examined the abundances of 28 chemical elements from Na to Zr (excluding the noble gases) in 89 IDPs. Phosphorus was detected in 32 samples (~ 36% of the considered particles), and its elemental abundance ranges from 0.25 to 10.6, normalized to that of iron, with an average value of 1.54. Thus, P ranks the 11th position among the most abundant elements present in IDPs, a figure significantly higher than that corresponding to its cosmic abundance (18th, see Section 3.1). In cases of high total P content, one micron or larger phosphate grains are usually found, generally Mg and/or Ca phosphate (Brownlee, 1996, private communication). In addition, mass spectral evidence of PO_2 and PO_3 phosphate anions in two fluffy IDPs of probable cometary origin was reported (Radicati di Brozolo et al., 1986), and mineralogical evidence of a phosphate of Na, Mg, and Ca, was found in an anhydrous chondritic IDP (Zolensky, 1996, private communication). Interestingly, similar phosphates of mixed cation composition had been previously reported in certain iron meteorites.

7.3.3 Meteorites

Meteorites are generally classified in four broad categories: stony chondrites, stony achondrites, stony-iron, and iron meteorites. Chondrites include a variety of stony meteorites that formed early in the history of the solar nebula from the accretion of fine-grained nebular material into primitive planetesimals. Their most characteristic features are chondrules, which are small spherical objects, typically ~ 1 mm in diameter, that represent molten droplets formed in space, presumably by flash heating and rapid cooling during the T-Tauri phase of the Sun's formation. Chondrites also commonly contain millimeter-to-centimeter-sized calcium-aluminum-rich inclusions, one of the earliest materials formed within our solar system, olivine aggregates, and other small refractory objects formed by evaporation, condensation, and melting of nebular materials in the high-temperature proto-stellar environment (Hazen et al., 2008). The fine-grained matrix between chondrules and Ca-Al inclusions is largely silicate material, but generally contains a diversity of metals, sulfides, oxides, and phosphates as well. Achondrite meteorites are likely to be derived from chondrites that experienced partial melting and little differentiation in their parent planetesimals, while iron-nickel meteorites represent the core material of differentiated planetesimals.

Attending to their oxidation state chondritic meteorites are classified ranging from extremely reduced EH3.0 enstatite chondrites to oxidized CV3.0 carbonaceous chondrites, including CO3.0 and LL3.0 ordinary chondrites that formed under intermediate oxygen fugacities. Carbonaceous chondrites are among the most primitive objects of the solar system, characterized by significant carbon contents mostly occurring as insoluble organic matter. The macromolecular structure of this material resembles that of terrestrial kerogen with polycyclic aromatic and heterocyclic aromatic molecules with linked functional groups such as OH and COOH bridged by alkyl chains and ether linkages. In addition, carbonaceous meteorites contain numerous soluble organic materials related to familiar biochemical compounds such as amino acids, fatty acids, purines, pyrimidines, and sugars, along with alcohols, aldehydes, ethers, amides, amines, monocarboxylic, and dicarboxylic acids, aliphatic, and aromatic hydrocarbons, heterocyclic aromatics, ketones, phosphonic, and sulfonic acids, and sulfides (Tielens, 2013). Concentrations of the major representatives of these moieties vary widely from less than 10 ppm (amines) to tens of ppm (amino acids) to hundreds of ppm (carboxylic acids).

The elemental analysis of meteorites shows that phosphorus is a minor but ubiquitous element found in many types of meteorites, ranging within 0.1–0.4% in weight in all meteorite classes, with the exception of the so-called mesosiderites, where one finds samples containing up to 1.4% phosphorus in weight (Jarosewich, 1990). The average abundance of phosphorus in chondritic meteorites is close to 0.25%. Thus, phosphorus is the 13th most abundant element in meteoritic material, a figure slightly below that reported for IDPs containing this element.

Among the approximately 250 mineral species found in meteorites, one can find about twenty phosphorus-bearing compounds (Table 7.2).

TABLE 7.2 Phosphorus Bearing Minerals in Meteorites*

Family	Name	Formula	Meteorite Type
Phosphates	Apatite	$Ca_5(PO_4)_3(Cl,F,OH)$	CC (Orgueil), MM
	Whitlockite	$Ca_3(PO_4)_2$	EC, MM
	Farringtonite	$Mg_3(PO_4)_2$	EC, I
	Graftonite	$(Fe,Mn,Ca)_3(PO_4)_2$	I
	Sarcopside	$(Fe,Mn,Ca)_7(PO_4)_4F_2$	I
	Buchwaldite	$NaCa(PO_4)$	I
	Brianite	$Na_2CaMg(PO_4)_2$	I (Carlton)
	Chladniite	$Na_2CaMg_7(PO_4)_6$	I-IIICD
	Panethite	$(Na,Ca)_2(Mg,Fe)_2(PO_4)_2$	I (Dayton)
	Merrillite	$Ca_9NaMg(PO_4)_7$	EC, I, MM
	Johnsomervillite	$Na_{10}Ca_{16}Mg_{18}(Mn,Fe)_{25}(PO_4)_{36}$	I
Phosphides	Schreibersite	$(Fe,Ni)_3P$	I, EC, CC
	Barringerite	$(Fe,Ni)_2P$ (hexagonal)	I, LM, CC
	Allabogdanite	$(Fe,Ni)_2P$ (orthorrombic)	I (Onello)
	Melliniite	$(Fe,Ni)_4P$	Acapulcoite
	Florenskyite	Fe(Ti,Ni)P	Kaidun
	Andreyivanovite	FeCrP	Kaidun
	Monipite	MoNiP	CC
Sulfides	Pentlandite	$(Fe,Ni,Co)_8(PH_4)(S,O)_8$	CC
Silicides	Perryite	$(Ni,Fe)_8(Si,P)_3$	I, EC

*List of names and idealized chemical compositions of P-bearing mineral phases identified in different meteorite classes. Keys: CC = carbonaceous chondrites, EC = enstatite chondrites, I = iron, LM = lunar meteorites, MM = Martian meteorites.

Source: Fuchs, 1969; McCoy et al., 1993; Britvin et al., 2002; Hazen et al., 2008; McCoy, 2010; Pasek, 2015b; Jungck and Neiderer, 2017.

In stony meteorites the presence of phosphorus is mainly reported as phosphates (apatite, whitlockite, farringtonite), while in iron meteorites it is found predominantly in the phosphide schreibersite alloy form (see Section 4.3.2), and in smaller amounts in other minerals such as perryite, graftonite, sarcopside, panethite, or brianite (McCoy, 2010). As alloys, the oxidation state of phosphorus and of the metals can be nominally considered to be zero in schreibersite, although electron studies suggest an average oxidation state of phosphorus in these minerals of -1 (Pasek et al., 2017).

Iron-nickel meteorites are relatively simple in their major mineralogy, but they incorporate a host of exotic transition metal sulfide minerals, as well as more than a dozen novel phosphates including the graftonite and sarcopside forms, buchwaldite, farringtonite, and the Na-Ca-Mg phosphates brianite, chladniite, panethite, and johnsomervillite. Most of these rare phosphate phases probably formed during melting and recrystallization processes occurring in the cores of some asteroids (McCoy, 2010). Thus, it is thought that buchwaldite, brianite, and chladniite phosphates were formed by the oxidation of phosphides originally present in the parent bodies (Pasek et al., 2017). The rare phosphide barringerite was first reported in the Olleague pallasite (one lunar meteorite) and a deposit in China, and subsequently found in a chondrite as well. A phase that was first identified in pallasites is the so-called phosphorus olivine, in which P atoms replace about 10% of Si atoms in the equivalent tetrahedra of stoichiometric olivine (Buseck and Clark, 1984).

Mixed Fe-Ni phosphides are among the first phosphorus compounds to condensate from the solar nebula as part of a homogeneous accretion model and thus are the most ancient of P-bearing mineral phases in our solar system, though both phosphides and phosphates species are predicted by thermodynamic calculations (Figure 7.7, Pasek, 2008). Phosphorus bearing Fe and Ni sulfides, with an approximate general formula (Fe,Ni)(S,P), have been reported as characteristic accessory phases of the CM class carbonaceous chondrites, usually associated with schreibersite, barringerite, eskolaite (Cr_2O_3) and daubreelite ($FeCr_2S_4$). These sulfides represent a new type of natural phosphorus compounds, probably stemming from solid solutions among the constituting elements. They typically have relatively high P elemental abundances, within the range 1–17% in weight, exhibiting atomic proportions ranging from 3 to 25 for (Fe + Ni)/P and S/P ratios (Nazarov et al., 2009). The existence of a P-rich sulfide appears to be quite extraordinary from the geochemical standpoint because phosphorus is known to be either a siderophile or lithophile element with no chalcophilic tendency. The best chemical model for the composition of the P-rich sulfide is a solid solution with variable contents of Fe, Ni, S, and P, resulting from substituting Fe by Ni and S by P, so P is not present as PH_4 or as a thio-salt ($(Fe,Ni)_3(PS_4)_2$). Oxygen, if present, could be of secondary origin. Therefore, the P-rich phase is probably not a primary oxysulfide, but rather a solid solution of a sulfide and a phosphide close in composition to (Fe,Ni)S and (Ni,Fe) P, in which the phosphide has a higher Ni/Fe ratio than the sulfide. For instance, the atomic proportions of the main elements in the Erevan rich-sulfide are $(Fe,Ni,Co,Cr,Na,K)_{9.36}S_{7.20}P$. Alternatively, S can replace O in the $(PO_4)^{3-}$ to give phosphorothioates, so the sulfide could contain $P(S,O)_4$ tetrahedra; the sulfide could also accommodate $(PH_4)^+$ ions, and then its formula could be written as $(Fe,Ni,Co)_8(Fe,Ni,PH_4,Cr,Na,K)(S,O)_8$ (see pentlandite formula in Table 7.2). Thus, from the chemical standpoint, the P-rich sulfide is not a forbidden compound, and its existence only points to special formation conditions.

This suggests that the assemblage was formed by condensation from the solar nebula gas under reducing conditions after the refractory phases had condensed, so that P-bearing sulfides would be a primary phase rather than a secondary alteration product formed under the conditions of the CM chondrite parent body. This phase had to be stable in the preplanetary solar nebula after the formation of the Ca-Al inclusions and before the condensation of the Fe-Ni metal alloys. Indeed, during condensation schreibersite is replaced by barringerite, thus decreasing the Fe/Ni ratio of phosphides and increasing the S/P and Fe/Ni of the P-bearing sulfides, which could be formed in the solar nebula by a sulfidation of a precursor phase of extrasolar origin (Nazarov et al., 2009). P-bearing sulfides are unstable during thermal metamorphism under parent body conditions and probably transformed into phosphate-sulfide assemblages, which are widespread in chondrites.

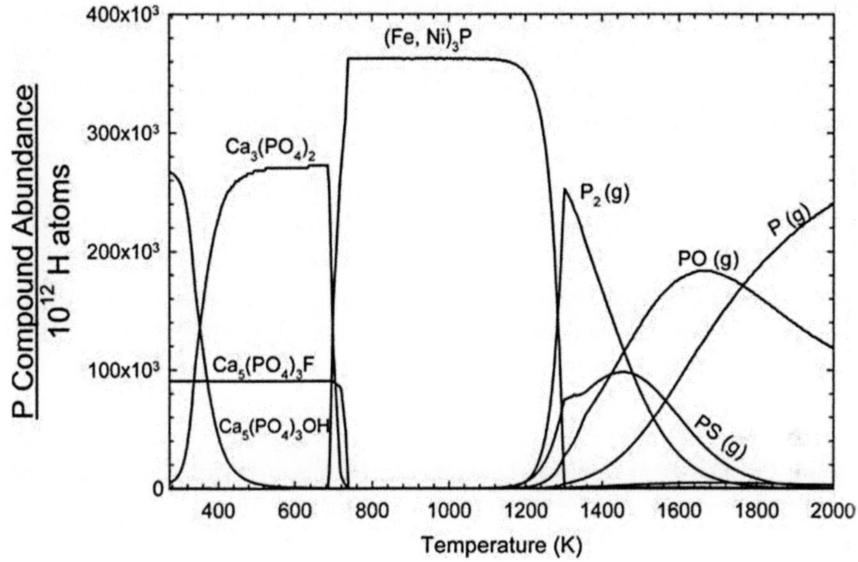

FIGURE 7.7 Condensation sequence for phosphorus-containing minerals obtained from thermodynamic calculations assuming 10^{-4} bar pressure and solar elemental abundance for the protosolar nebula (By courtesy of Matthew Pasek).

Although the possible presence of tributyl phosphate was earlier reported from ion chromatograms studies of the Murchison meteorite (Cronin and Pizzarello, 1990), no organophosphorus compounds were identified in meteorites until the analysis of Murchison meteorite by Cooper et al., (1992), where a homologous series of alkyl phosphonic acids (from methyl to butyl) was identified by mass spectroscopy, along with inorganic orthophosphate (~25 μmol/g). Each gram of meteorite contained 9 nmol methyl phosphonic acid and 6 nmol ethyl phosphonic acid. Enhanced isotopic D/H ratios suggested the reported phosphonic acids could be derived from ISM precursors (Cooper et al., 1997), starting from CP via hydrogenation and incorporation of water (Gorrell et al., 2006). A plausible prebiotic synthesis of these phosphonic acids by UV irradiation of orthophosphoric acid in the presence of formaldehyde, primary alcohols, or acetone was proposed by de Graaf and co-workers (1995). Compounds containing the C-P bond are extremely stable and survive conditions which would result in extensive hydrolysis of esters of phosphoric acids.

The presence of orthophosphates in both Murchison (18 ± 1 μmol/g) and NWA502 (2.0 ± 0.1 μmol/g) carbonaceous chondrites have been confirmed by quantitative NMR analysis (Pasek and Lauretta, 2005). In this regard, it has been demonstrated by Pasek and co-workers (2007) that schreibersite readily corrodes in aqueous solution to form a series of P-bearing compounds including orthophosphate (oxidation state +5), pyrophosphate (oxidation state +5), hypophosphite (oxidation state +4), phosphite (oxidation state +3), and diphosphate (oxidation state +3). In addition, in the presence of acetate/ethanol, several organophosphorus compounds are formed as well (Pasek and Lauretta, 2005). Thus, hypophosphorus, and orthophosphorus acids could be produced by the interaction of this mineral with water on a meteorite parent body (in the near-surface region) or in ablating comets or meteorites (Bryant and Kee, 2006; Gorrell et al., 2006). These could be a source for photochemical production of phosphate radicals, which in turn could serve to synthesize phosphonic acids. While previous studies looking into the formation of alkyl phosphonic acids have focused on the iron-nickel phosphide mineral schreibersite and phosphorus acid as a source of phosphorus, the possible synthesis of alkyl phosphonic acids from phosphine-mixed ices under stellar conditions has been recently studied (Turner et al., 2018), spurred by the observation of PH_3 in IRC +10216 circumstellar envelope.

7.3.4 Giant Planets

It has been long known that Jupiter and Saturn's upper atmospheres contain phosphine (PH_3), where it is the major phosphorus carrier, and that reduced P-bearing species are likely dominant in the crust of Jupiter and similar giant planets (Lewis, 1969). Direct spacecraft measurements have confirmed the presence of phosphine with a volume mixing ratio of 0.6 and 0.7 ppm in Jupiter and Saturn respectively, implying a P/H ratio in good agreement with the solar value. Indeed, the C/N, N/H, Ge/H, and O/H ratios measured on Jupiter suggest that the formation of this planet took place from the accretion of planetesimals rather than by condensation from a homogeneous nebula (Levine, 1985). In fact, according to Delsemme (1995), the giant planets may have developed their embryos by accumulating comets. Such a possibility traces back the question of the origin of phosphorus compounds actually observed in the reduced atmospheres of Jupiter and Saturn to its possible presence in such bodies.

In LTE conditions P_4O_6 would be expected to dominate the upper atmospheres of gaseous planets with approximate solar abundances. However, almost no P_4O_6 is found in these environments because of high disequilibrium abundances due to phosphine's quenching and rapid vertical mixing from regions where it is dominant (Visscher et al., 2006). In fact, in the deep atmosphere, under high pressure and temperature, Jupiter's thermochemical processes change the approximately solar elemental composition by converting H atoms into molecular form (H_2) and reactive atoms (e.g., C, N, O, and P) into saturated hydrides (methane, CH_4; ammonia, NH_3; water, H_2O). These molecules are transported upwards by convection into the cooler regions, where H_2O, NH_4SH, NH_3, and PH_3 condense to form clouds. Thus, the time scales for the conversion of PH_3 to P_4O_6 are much longer than the convective time scales, so the production of phosphorus oxide is kinetically inhibited.

Once exposed to sunlight phosphine suffers photolysis by photons with wavelengths in the range 1600 Å < λ < 2300 Å according to the reaction $PH_3 + h\nu \rightarrow PH_2 + H$, which can undergo a three-body reaction to form diphosphine, $PH_2 + PH_2 + M \rightarrow P_2H_4$, and through further photolysis yield red phosphorus P_4 particles that condense to form solid red aerosols (Sánchez-Lavega, 2011). These particles have been invoked as a possible chromophore agent to tint the reddish cloud features observed in Jupiter, particularly its Great Red Spot. At the same time, phosphine can be oxidized by water reaction at $300 < T < 800$ K, producing P_4O_6. Therefore phosphine detection occurs in an altitude range, which is limited at high altitudes by photolysis processes and at lower altitudes by its oxidization (Sánchez-Lavega, 2011). Potentially observable amounts of the organophosphorus compounds HCP and CH_3PH_2 were predicted by Kaye and Strobel (1983) on the basis of the C_2H_2 catalyzed photodissociation of CH_4 with PH_3, but have not been observed by spacecraft.

Phosphine has not been detected in the atmospheres of the gaseous planets Uranus and Neptune (Lunine, 1993). The presence of phosphine in the atmospheres of Jupiter and Saturn and its absence in that of both Uranus and Neptune, suggests that the gas-phase abundance of phosphorus is probably subsolar in the atmospheres of the later ones (Moreno et al., 2009), and raises a question about the origin of this compound in the giant planets.

Jupiter's chemical composition is also influenced by external sources of material. The most dramatic illustrations of this were the multiple impacts of comet Shoemaker-Levy 9 in July 1994 that injected large quantities of N-, O-, and, S-bearing molecules into Jupiter's stratosphere near 45°S. Substantial amounts of hydrogen cyanide (HCN), carbon monoxide (CO), and carbon monosulfide (CS), were produced in the ensuing shock chemistry and subsequent photochemistry. Carbon dioxide (CO_2), presumably a secondary product of the Shoemaker-Levy 9 collision formed from the photochemical evolution of CO and H_2O, and was detected by the Infrared Space Observatory Short-Wavelength Spectrometer observations (Harrington et al., 2004; Kunde et al., 2004). No phosphorus related compounds, which may be originally trapped inside the comet nucleus, where detected.

7.3.5 The Planet Mars

Petrologic investigations of Martian rocks have been accomplished by spectroscopy from orbiting spacecraft, by direct in situ analysis of rocks with instruments onboard Mars Pathfinder, Spirit, Opportunity, and Curiosity Mars rovers, and from indirect sources like those provided by the analysis of several Martian meteorites collected on earth. Meteorites from Mars are recognized by the presence of trapped Martian atmosphere gas into pockets of impact-melted glasses, along with a distinctive oxygen isotopic composition, and characteristic Fe/Mn ratios in certain minerals. Martian meteorites are classified into three main families referred to as shergottites, nakhlites, and chassignites after the reference meteorites Shergotty, Nakhla, and Chassigny, respectively. Although the locations on Mars from which they came are unknown, they sample many more locations than have been visited by rovers up to now, and elemental abundance analyses indicate a high phosphorus content, within the range 0.1–1.3 wt% (Banin et al., 1992), mainly related to the presence of relatively abundant phosphates in both shergottites and chassignites representatives, where apatite can amount up to 4% of the minerals identified (McSween Jr., 2015). In comparison, the celebrated ALH84001 Martian meteorite contains a much lesser amount (0.15%) of phosphate minerals. Among them, minor apatite crystals (~ 300 μm) occur as interstitial grains, along with larger (up to 800 μm) whitlockite and merrillite crystals.

Data from remote-sensing and Martian meteorites indicate that the planet's crust is dominated by igneous rocks of basaltic composition. On the other hand, ancient sedimentary rocks have become the focus of rover exploration spurred in part by the search of evidence of liquid water, and, by extension, possible fingerprints of life (McSween Jr., 2015). Mineralogical studies performed by Mars Pathfinder Lander have confirmed the presence of P-bearing compounds in five samples, giving large elemental phosphorus abundance values ranging from 0.9 to 1.5 wt% (Rider et al., 1997), in agreement with results previously obtained from the study of Martian meteorites. Apatite crystals have been reported as a minor but ubiquitous component in igneous volcanic rocks studied in the Gusev and Gale Craters, ranging from 2% to 12% of their mineral content. In the studied sedimentary rocks, the presence of apatite is scarce, and when present, it represents about 2% of the mineral inventory of the rock (McSween Jr., 2015). These results strongly suggest that phosphorus is more abundant on Mars surface than in the earth's crust, probably as a consequence of the high phosphorus abundance in the Martian mantle as compared to the terrestrial one.

7.3.6 The Planet Earth

Phosphorus is the eleventh most abundant element in the earth's crust. Due to the rocky nature of our planet, its abundance on the earth is higher than its cosmic ranking (17th). There are several hundred known phosphate minerals, which are formed from sedimentary processes, by metamorphic processes, by weathering of other minerals, and as igneous minerals. On earth, a large group of phosphate minerals forms by biological action, and most phosphates are found as ubiquitous yet accessory minerals of the apatite group in all important classes of igneous and metamorphic rocks, and also as well-separated crystalline masses formed by differentiation from cooling basic magma. In most rocks, phosphorus appears in small concentrations, usually about 0.15 to 0.2% in weight (Nriagu, 1984), a figure that compares well with chondritic compositions. This fact goes along with the consideration that the main reservoir of phosphorus in the planet earth most likely resides in the lower mantle or in the core (Nash, 1984), and suggests that the primary source of phosphorus was the in fall of planetesimals condensing from the cooling protoplanetary nebula. It was probably incorporated as solid phosphate, not only from early planetesimals, but also from comets and other minor bodies of the solar system during late accretion stages.

In the lithosphere phosphorus occurs in the pentavalent positive state, and it is found almost entirely as a part of the orthophosphate ion PO_4^{-3}, giving rise to minerals of the apatite group, by far the most ubiquitous: $Ca_5(PO_4,CO_3)_3(F,ClOH)$ (fluorapatite, chlorapatite, and hydroxylapatite), $Ca_3(PO_4)_2$ (whitlockite), $(Fe,Mn,Ca)_3(PO_4)_2$ (graftonite) and $(Fe,Mn,Ca)_7(PO_4)_4F_2$ (sarcopside). In igneous and metamorphic rocks, the most common species is fluorapatite ($Ca_5(PO_4)_3F$), while carbonate fluorapatite is the typical mineral in sedimentary phosphates. The second most important mode of occurrence of phosphorus in the lithosphere is in silicate minerals in igneous rocks, where it substitutes for silicon in SiO_4 tetrahedra.

For many years it was believed that phosphide minerals indicated an extraterrestrial origin for a rock. However, in the last few decades, this preconception has been suggested as erroneous due to the fact that there are indeed environments that can be highly reducing and reach high enough temperature to promote the reaction of metals with phosphorus to make metal phosphide minerals appear in terrestrial rocks. These environments are characterized by iron- and carbon-rich rocks being heated in the absence of air. For instance, the native iron deposits of Disko island Greenland, which bear schreibersite mineral, formed as basalt erupted through carbonaceous shales. Recently several new phosphide minerals were found in the Hatrurim formation, and it is thought that they formed in a route akin to the Disko native iron minerals (Pasek et al., 2017).

Phosphorus availability is thought to dictate the amount of primary biological productivity that can be sustained in the oceans on geologic time scales. Estimating phosphorus concentrations in the ocean along earth's history is thus critical for understanding the growth of the biosphere and the evolution of major biogeochemical cycles. Van Cappellen and Ingall (1996) have studied the biogeochemical cycles of C, P, O, and Fe, and have calculated the different reservoirs of phosphorus on the earth's surface. Thus, they have found two interesting facts. One is the similarity in the amount of phosphorus on land (4.3×10^{20} g) and on the ocean floor (3.9×10^{20} g), and the other is the low abundance of P in the oceans (6.3×10^{16} g) as compared to the former reservoirs. Quite remarkably, by assuming the average ratio P/C $\simeq 0.02$ in living beings and a total carbon mass of about 10^{18} in the biosphere, the amount of phosphorus trapped in the biosphere results in 2.0×10^{16} g, a figure which compares well with the oceanic phosphorus content.

In 1934 A. C. Redfield reported a suite of dissolved nitrate, phosphate, and oxygen measurements from various depths in the Atlantic, Indian, and Pacific oceans. These data showed a remarkable consistency, with nitrate to phosphate occurring at a ratio 16:1. Later the study was expanded to include a ratio of carbon to the phosphate of 106:1; this so-called Redfield ratio illustrates the power of elemental stoichiometry to describe the ocean biochemical cycling (Redfield, 1958). The averaged phosphate concentration on surface measurements in Atlantic, Indian, and Pacific oceans is usually low: approximately 7×10^{-5} g/l (about 7×10^{-7} M), due to the low solubility of phosphate salts. The vertical distribution of PO_4^{-3} in oceans reflects the biological consumption of phosphorus. Thus, the phosphate concentration is very low near the surface, and it tends to significantly larger values $[PO_4^{-3}] \approx 40$ g/l and $[PO_4^{-3}] \approx 80$ g/l for the Atlantic and Pacific oceans, respectively, at depths bigger than 1,000 m up to depths more than 4,000 m. In fact, a most remarkable observation is that the elementary composition of waters of the open ocean is almost equal to that of the plankton growing in that water, indicating that either the primordial ocean already had the above elementary composition, and the primitive organisms adapted themselves to these conditions, or alternatively, the present-day ocean water composition is a secondary one, due to the modifying and controlling influence of the biological systems (Halmann, 1974). This second scenario, namely, the control by biological metabolism, seems to be the most probable cause, especially if one takes into account the problem of phosphate concentration; that is, the fact that the conversion of orthophosphate into organic phosphates is very unfavorable thermodynamically in the presence of an excess of water.

Recent work suggests that during the Archean (and perhaps through the Proterozoic) the phosphorus concentration remained low (say less than about 20% of the modern concentration). The favored mechanism for phosphorus depletion in the Precambrian ocean is scavenging of P atoms from the water column by incorporation into ferrous minerals or by adsorption onto iron oxides. Alternatively, a limited recycling of P atoms in an oxidant-poor ocean may also account for low concentration values, for after the phosphorus in the surface ocean is exhausted due to biological activity, the required remineralization of sinking biomass releases P atoms back into the marine environment. Accordingly, this recycling increases the residence time of phosphorus in the ocean (Kipp and Stüeken, 2017).

On the other hand, under prebiotic conditions, phosphorylation may have occurred not in solution in the open sea but on the surface of minerals, or in localized environments under special conditions. Since apatite is less soluble than calcium carbonate and some other minerals with which it is associated in many occurrences, the problem of progressive phosphate concentration and phosphorylation under prebiotic conditions may have been accomplished, not only with the help of condensing agents but primarily by reactions on the surface of minerals, such as hydroxylapatite (Miller and Parris, 1964) and other phosphates available on the surface of the earth (see Section 8.3).

Atmospheric transport of phosphorus is significant. Phosphine gas has been unambiguously confirmed in the lower earth's atmosphere at trace levels in the troposphere (exhibiting concentrations of about 1ng m^{-3}) along with phosphate dust. In the atmosphere, phosphine reacts very rapidly with hydroxyl radical, with a half-life of about 28h to yield phosphate, which returns to earth in rainfall. Thus, it is estimated that atmospheric fallout of phosphorus is in the range of 6–13 $\times 10^{12}$ g per year, whereas the phosphorus flux from land to the atmosphere is estimated to be about 4 $\times 10^{12}$ g per year. Considering 10^{18} m^3 the rough volume of the troposphere (extending up to a height of about 8.5 km above sea level, Sánchez-Lavega, 2011), atmospheric PH_3 would amount 4 $\times 10^9$ g. Compared with the above figure for the atmospheric phosphorus transport, phosphine in the troposphere at any time is about 0.05% of the total global atmospheric phosphorus flux (Morton and Edwards, 2005).

In a study performed by Murphy et al., (1998) the chemical composition of aerosol particles at altitudes between 5 and 19 km was measured in situ. They did not analyze the data specifically for phosphorus, most of which was in the troposphere, often associated with aerosols from biomass burning. They did not obtain a clear signal of phosphorus in the collected meteoritic material. This may be due to the fact that the P atoms evaporated during the meteoroid entry throughout the atmosphere could recondense differently than more refractory elements like Fe, though it still might end up in the aerosol phase at some point. Alternatively, phosphorus may probably be below the detection limit. Other meteoritic elements such as Na and K, which have a similar abundance to P, are easily detected due to the fact that they ionize extremely easily, so they are detected with exceptionally high efficiency. In the meteoritic spectra, if present, phosphorus peaks would be at less than 10% of the height of the Na peak, and thus it is difficult to distinguish them unambiguously both from noise and from other ions with the same nominal masses (Murphy, 1999, personal communication).

7.3.7 The Moon

Phosphates are common accessory minerals in many lunar rocks (Papike et al., 1991), and returned lunar samples' analyses indicate that phosphorus is present as a minor element in each of the studied specimens. The major P-bearing phase is apatite, although whitlockite is also present, along with traces of schreibersite, probably as a consequence of meteoritic impacts in the production of the lunar regolith. In fact, although a majority of phosphorus is in phosphate minerals on the Moon, a non-negligible amount is present as the mineral schreibersite, which is common to all the examined Apollo samples, though the amount of schreibersite varies with the location of the collected rocks, ranging from 0.1% to 0.6%. Some of this material may have been delivered from exogenous meteoritic falls, and others may have been formed in situ by the reduction of phosphates via phosphorus volatilization after meteorite impact with the soil followed by phosphidation of suitable present metals (Pasek, 2015b).

For most lunar basalts, the phosphorus content is quite limited, ranging from 200 to 900 ppm. On the other hand, the Lunar Prospector gamma-ray data showed that the so-called KREEP glass (for potassium, rare earth elements, and phosphorus) in igneous rocks, which shows a relatively high phosphorus content of around 0.33 wt% (Gibson and Chang, 1992) is concentrated in the rim areas of Mare Imbrium, the nearside Maria and highlands of Mare Imbrium, and the Mare Ingenii South Pole-Aitken basin area on the farside; meanwhile the concentration of KREEP in highlands is relatively low and uniform. The data support models that the Imbrium impact excavated KREEP-rich material from the depth and distributed it over the Moon. Post-impact KREEP volcanism and KREEP injection into the upper crust are also responsible for the global distribution of KREEP-rich rocks on the Moon (Binder, 1998).

7.4 SUMMARY AND REVIEW QUESTIONS

In this chapter, we have further examined the chemical evolution of phosphorus compounds in their journey from ISM clouds to star-forming regions, protoplanetary disks, and planetary systems. The cosmochemical behavior of phosphorus includes both the lithophile phosphate phase and a high-temperature siderophile phosphide phase, which is rarely encountered on earth's surface. However, both lithophilic and siderophile phosphorus compounds are encountered in meteorites. Phosphates like apatite and whitlockite are the major carriers of phosphorus in lunar meteorites, basaltic achondrites, and Martian meteorites, whereas phosphides like schreibersite are the major carriers in iron meteorites, pallasites, and enstatite chondrites.

Phosphorus is a moderately volatile element, as it condenses after the formation of iron-nickel metal at about 1,300 K at 10^{-4} bar of pressure (see Figure 7.7). Phosphorus reacts with the metal to form phosphides at relatively high temperatures. The most common phosphide mineral is schreibersite, though several minerals with metals other than iron and nickel or metal to P ratios other than 3:1 are known from meteorites (see Table 7.2). Relative to other typical volatiles phosphorus is not appreciably volatile as there are no significant phosphorus gases, such as phosphine, at any rocky planet in the solar system. On the other hand, phosphorus is a lithophile element at the redox conditions on the surface of the earth, and hence orthophosphate is the dominant form of inorganic phosphorus on earth's surface today. The dominance of orthophosphate is predicted from thermodynamic calculations (see Figure 7.7), because no reduced oxidation state of P compounds are stable under terrestrial redox conditions. Other P-bearing phases may form in low concentrations by geologic processes, but these phases are not in thermodynamic equilibrium and slowly hydrolyze or oxidize to form orthophosphate on geological scale times (Pasek, 2008).

No phosphorus compounds have yet been detected in protoplanetary disks nor in exoplanets atmospheres. The P^+ ion has recently been identified in mass spectra of comet 67P/Churymov-Geramisenko (Altwegg et al., 2016), although the possible parent molecules for these phopshorus ions are still unknown. In most IDPs of possible cometary origin that have been studied in detail, phosphorus abundances consistent with carbonaceous chondritic ones are often measured. Sometimes higher abundances of P are seen, while in other IDPs it is undetected, although the general trend is that IDPs are enriched in P content as compared to meteoritic material.

The discovery of trace quantities of alkyl phosphonic acids in the Murchison meteorite, containing C - P bonds which are significantly more stable than alkyl phosphates, demands possible pathways for their formation and suggests alternative routes for prebiotic phosphorylation. Nevertheless, although the detection of organic phosphonates is intriguing, they make up a very small fraction (0.1%) of the total P content in Murchison meteorite, and the bulk of the meteoritic material that falls to earth today has no phosphonates (Pasek, 2008).

KEYWORDS

- carbonaceous chondrites
- exoplanets
- habitable zone
- hot Jupiters
- interplanetary dust particles
- Kuiper belt objects
- Martian meteorites
- meteorites
- Oort cloud
- planetary systems
- planetesimals
- planets
- prebiotic atmosphere
- presolar grains
- protoplanetary disk
- protostar
- rogue planets

CHAPTER 8

Prebiotic Chemical Evolution and the Phosphate Problem

"...we could conceive in some warm little pond, with all sorts of ammonia and phosphoric salts, light, heat, electricity, etc. present...that a protein compound was chemically formed, ready to undergo still more complex changes."

(Charles Darwin, 1871)

"An eventual understanding of life's origin will have to rely heavily on experimental exemplifications of the potential of chemical systems to undergo the transition from non-living to living."

(Eschenmoser, 2007)

8.1 THE PREBIOTIC SYNTHESIS NOTION

During the past century several possible scenarios have been considered regarding the origin of life, which mainly differ by the precise location where the key processes leading to the emergence of the first living beings are supposed to occur, that is, in the heavens or on the earth. On the one hand, the extraterrestrial view that bacteria-like microorganisms could be delivered to earth from outer space, usually referred to as panspermia, was earlier suggested by Arrhenius (1903) and later supported by Hoyle and Wikramasinghe (1981). On the other hand, a purely geocentric view was introduced by Oparin (1924) and Haldane (1929) by assuming that all the sort of biopolymers currently found in a living cell were originally assembled in ancient times starting from elemental parent molecules containing just a few atoms, through a progressive route of chemical processing in suitable geochemical places such as oceans, lakes or hydrothermal vents on the early earth. According to this view, the emergence of full-fledged microorganisms on our planet was preceded by a relatively short period of *prebiotic* synthesis during which fundamental basic molecules found in living beings were systematically synthesized under abiotic conditions, that is, in the absence of the living organisms which usually use them, in line with the earlier Wöhler's demonstration in 1828 that urea $[(NH_2)_2C=O]$ can be synthesized "without the aid of a kidney" from the inorganic compound ammonium cyanate $[(NH_4)OCN]$. Nowadays, however, we are aware that urea molecule has also been detected in the Murchison meteorite and in the dense molecular cloud Sagittarius B2 (Remijan et al., 2014), and it has been tentatively detected on grain mantles toward the protostellar source NGC 7538 IRS9 (Raunier et al., 2004; Belloche et al., 2019). Therefore, there is no fundamental reason to restrict the very notion of prebiotic synthesis to early earth places only, but the scenario should be properly extended to include suitable astrophysical objects as well (Lingam and Loeb, 2018).

It is generally assumed that prebiotic synthesis starts with the formation of an ensemble of relatively simple monomers, such as amino acids, fatty acids, nucleobases, and sugars, the latter two combining among them along with phosphate to yield nucleotides (Figure 8.1). Amino acids and nucleotides would subsequently lead to longer chain molecules, through condensation reactions joining monomers of the same kind to each other, eventually giving rise to macromolecular biopolymers such as proteins and nucleic acids, respectively (Figure 8.1).

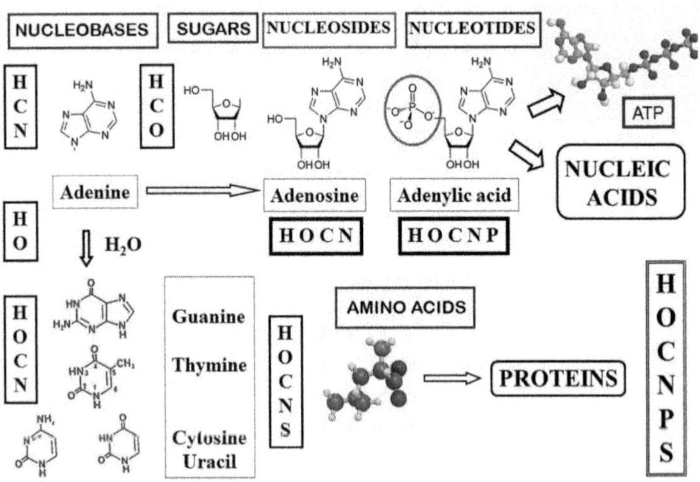

FIGURE 8.1 The basic building blocks of nucleic acids and protein biopolymers. The main bioelements composing them are indicated in the boxes.

After Berzelius' 1837 postulation of enzymes as peptide-related molecules, much of the thinking about the transition from inorganic to the organic world became focused on amino acids, the fundamental components of proteins. In this way, the prebiotic synthesis notion entered the domain of experimental science and public consciousness in 1953, following the celebrated Miller's organic synthesis experiments yielding amino acids that are found in proteins, such as glycine, alanine, aspartic acid and glutamic acid, along with non-proteinic amino acids, in a flask containing a gas mixture of hydrogen, methane, and ammonia molecules simulating a predominantly reducing primeval earth's atmosphere affected by high energy sources such as electric discharges (mimicking stormy lightning) acting for a week. The flask was connected by a tube to a reservoir of warm liquid water simulating primeval oceans (Miller, 1953, 1957). Amino acid formation in Miller's experiment proceeded via the so-called Strecker synthesis (see Table 8.1), hence indicating that hydrogen cyanide (HCN) and aldehyde (R-CHO) molecules, which are the actual precursor molecules in the Strecker synthesis, are formed during the sparking of the original gases (Miller and Orgel, 1974). Numerous experiments were conducted in the wake of this discovery testing the range of conditions under which such synthesis could take place, examining the use of different types of energy sources (electric discharges, UV radiation, high energy ionized particles) and different gas compositions, ranging from highly reducing mixtures to neutral (CO, N_2) or relatively oxidizing (CO_2, N_2) atmospheres. These latter experiments were spurred by some evidences indicating that early earth atmosphere, mainly composed of N_2, CO_2, and H_2O, with small quantities of H_2, CO, and CH_4, and negligible amounts of O_2, was not as reducing as it was originally thought (Zahnle et al., 2010). In this way, it was found that reducing gases generally produce not only a greater variety of organic compounds, but also a greater yield. Thus, when Miller's experiment is repeated with CO_2 and N_2 instead of CH_4 and NH_3, the organic yield went down by a factor of 100 or more (Miller and Schlesinger, 1984). In addition, several biological amino acids, such as arginine, methionine, and those containing aromatic residues, i.e., histidine, tryptophane, phenylalanine, and tyrosine, still remain difficult targets of prebiotic synthesis (Miller, 1998).

On the other hand, it was clear that the chemical synthesis of the full list of required biomonomers should include, not only amino acids, but sugars, fatty acids, and nucleotides as well, which could not be properly accomplished in the protein forming geochemical environments considered in Miller's experiments. Furthermore, on the basis of detailed geochemical and paleontological studies, the time window available for the transition from inanimate to living matter has been experiencing a progressive shortening, whose lower bond is currently placed at about 3.95 billion years ago (Tashiro et al., 2017), corresponding to the end of a period of intense meteorite bombardment, as testified by the analysis of the Moon's craters. For the upper bond, we can rely on geological records, such as stromatolites and microscopic fossils, which provide evidence that 3.8 billion years ago, life was already present on earth (Mojzsis, 1996). Therefore, the prebiotic synthesis stage should take place in just about 200 million years after the meteoritic bombardment episode.

Taken altogether these constraints open the door to an intermediate view on the origin of life on earth which assumes the extraterrestrial origin of certain biomonomers (or their precursors) via meteoritic and cometary delivery during the first stages of the planet formation. This scenario gained an increasing support during the 1960s and 1970s following: (1) the prebiotic synthesis of adenine by polymerization of HCN (Oró, 1960; Lowe et al., 1963), a well-known abundant molecule in all observed comets (Oró, 1961), and (2) the chemical analysis of Murchison meteorite, a freshly collected carbonaceous chondrite displaying a broad inventory of organic molecules of biological relevance, including amino acids, purines, pyrimidines, and organophosphorus compounds (Figure 8.2), thereby demonstrating that prebiotic synthesis of biochemical compounds is feasible in places of the solar system other than the earth.

In fact, carbonaceous chondrites contain a diverse suite of exogenous organic compounds, many of which are essential in contemporary biology on earth (Callahan et al., 2011). The extraterrestrial origin for most of the amino acids found in carbonaceous chondrites has been firmly established on the basis of the following results:

1. The detection of racemic amino acid mixtures. In this regard, it is noteworthy that the mixture of Miller's synthesized amino acids turned out to be similar to that of the Murchison meteorite in both composition and racemization, that is, they exhibit equal ratios of the optical isomers of both the amino acids and hydroxy acids formed.

2. A wide structural diversity, including many non-protein amino acids that are rare or non-existent in the biosphere. For instance, of the 74 amino acids found in samples of the Murchison meteorite, 8 are present in proteins, 11 have other biological roles, while the remaining 55 have only been found in extraterrestrial samples (Cronin, 1989).
3. Non-terrestrial values for deuterium, carbon, and nitrogen isotopic abundances. In particular, the analysis of the deuterium/hydrogen ratio of Murchison meteorite amino acids shows that they are highly enriched in deuterium as compared with the average solar system D/H value. These measurements suggest that these amino acids could not have formed from gases of the solar nebula, but from highly deuterium-enriched hydrogen and ammonia molecules present in presolar interstellar grains which became part of the condensed material in the meteorite parent planetesimal from the ISM gas. Further thermal evolution of these bodies may have led to transient hydrothermal phases during which the amino acids were synthesized following the Strecker mechanism, catalyzed by the magnetite mineral (Fe_3O_4) usually present in many meteorites.

Determining the origin of the meteorites' nucleobases is more challenging due to their low abundance relative to most other organics (see Figure 8.2), meteorites heterogeneity, experimental artifacts, and possible terrestrial contamination. In fact, at variance with the richer structural diversity reported for amino acids, most purines (adenine, guanine, hypoxanthine, and xanthine) and the one pyrimidine (uracil) reported in meteorites to date are biologically common, and could be attributed to terrestrial contamination. In a recent analysis three unusual and terrestrially rare nucleobase analogs, namely, purine, 2,6-diaminopurine, and 6,8-diaminopurine were detected in Murchison and Lonewolf Nunataks 94102 meteorites, hence suggesting these purines are indigenous and not terrestrial contaminants. Interestingly, these compounds can be synthesized as products of ammonium cyanide (NH_4CN) chemistry under prebiotic conditions, thereby providing a plausible mechanism for their origin in the corresponding parent bodies (Callahan et al., 2011).

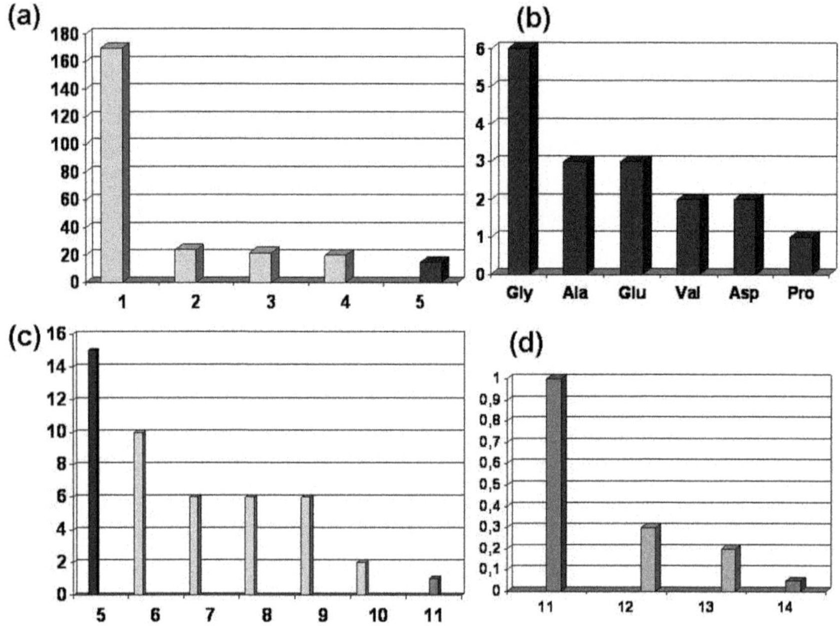

FIGURE 8.2 Main organic compounds found in the Murchison chondrite meteorite (Source: Wood and Chang, 1985; Cooper et al., 1992). Panels (a), (c) and (d) are measured in ppm concentration values, whereas panel (b), giving the concentration of main amino acids, is measured in μg per gram of meteorite (Source: Miller and Orgel, 1974). Keys: 1: Monocarboxylic acids (C2 – C8), 2: Aliphatic hydrocarbons, 3: Aromatic hydrocarbons, 4: Urea, 5: Amino acids, 6: Ketones (C3 – C5), 7: Alcohols (C1 – C4), 8: Aldehydes (C2 – C4), 9: Hydroxy acids (C2 – C5), 10: Amines (C1 – C4), 11: Purines, 12: Methyl phosphonic acid, 13: Ethyl phosphonic acid, 14: Pyrimidines.

In this way, a line of thought, connecting abiotic organic synthesis occurring in circumstellar shells and ISM clouds, with the possible emergence of life on earth and similar Earth-like planets discovered in other planetary systems (see Section 7.2.2), has grabbed the limelight during the last decades, spurred by evidences clearly showing that organic molecules of biological interest not only formed before the appearance of life on earth, but even before the origin of the earth itself. Thus, the ever-increasing molecular inventory reported by spectroscopic observations of the ISM, has disclosed the presence of typical precursor molecules of biomonomers (see Table 8.1), such as H_2, PO, H_2O, CO_2, HCN, H_2S, NH_3, H_2CO, HCOOH, CH_3CHO, HC_3N, NH_2CN, and $(CH_3)_2CO$, in dense molecular clouds. These molecules include most of the functional groups usually present in biochemical compounds. Furthermore, the simplest sugar glycolaldehyde (CH_2OHCHO) was detected, along with its related derivative ethylene glycol ($HOCH_2CH_2OH$), toward the galactic center in the dense molecular cloud Sagittarius B2(N), a star formation site (Hollis et al., 2000, 2004). Glycolaldehyde can react with propenal (CH_2CHCHO) to form ribose, a building block of the RNA backbone. Interestingly enough, recent observations indicate that both prebiotic molecules are widespread around Sgr B2, covering a sightline size of approximately 36 pc, and their abundance seems to decrease from the cold outer region to the warmer central region associated with star formation activity, hence suggesting that these molecules are likely to form through a low-temperature process (Li et al., 2017). Besides Sgr B2, glycolaldehyde molecule has been detected in a few protosolar-type stars, the protostellar shock region L1157-B1 (where the PN molecule has also been detected, see Figure 1.5), and the comet Lovejoy (see Figure 7.6).

Quite remarkably the simplest amino acid glycine (NH_2CH_2COOH), although not yet confirmed in the ISM (Kuan et al., 2003; Cunningham et al., 2007), has been detected in the comet P67/Churyumov-Gerasimenko (Altwegg, 2016). Nevertheless, as of August 2019 no interstellar fatty acid, nucleotide derivative, or other metabolite has been reported in the ISM. In particular, several nitrogen heterocycles, including pyrimidine (c-$C_4H_4N_2$), have recently been sought without success in the Taurus molecular cloud named TMC-1 (Cordiner et al., 2017). The non-detection of these compounds in a cold dense cloud complements previous non-detections of similar molecules in star-forming regions and in the envelopes of evolved stars. These negative results contrast with the detection of benzonitrile (c-C_6H_5CN), one of the simplest nitrogen-bearing aromatic molecules, by McGuire and co-workers (2018) in the ISM, as well as the observation of its relative compound toluene (c-$C_6H_5CH_3$) in comet 67P/Churyumov-Gerasimenko (Altwegg et al., 2017), suggesting that interstellar chemistry may favor the presence of heteroatoms as functional side-groups, rather than within the ring structure. Anyway, the presence in the ISM of a large and diverse set of molecules of prebiotic importance, combined with the fact that their synthesis must differ from the chemical pathways that were possible in the primitive terrestrial environment, clearly indicates that many prebiotic molecules are particularly easily synthesized all over the universe (Oró et al., 1990).

These molecules are formed in both gas-phase reactions in the ISM and on the ice mantles deposited on the surface of interstellar grains, which can be subsequently incorporated in the planetesimal bodies of protoplanetary systems. On this basis, icy mixtures of water, methane, carbon dioxide, ammonia, formaldehyde, and methanol, simulating interstellar grains mantles analogs, have been irradiated by ionizing radiations and/or UV light in vacuum conditions to ascertain interstellar ices photochemistry in several laboratories. In doing so, significant amounts of amines, amides, ketones, alcohols, quinones, and ethers, along with relatively complex organic polymers (polyoxymethylene, a polymer of H_2CO, and related derivatives) were obtained during the 1990s (Schutte et al., 1993a, 1993b; Bernstein et al., 1995, 1999). Subsequently, irradiating a mixture of CO, CO_2, H_2O, NH_3, and CH_3OH ices at a temperature of about 12 K, Muñoz-Caro and co-workers (2002) produced 16 different amino acids (including glycine, alanine, valine, serine, proline, and aspartic acid), carboxylic acid salts, nitrogen-heterocyclic species, pyrroles, and furans. Some of these amino acids have been identified in carbonaceous chondrites, hence supporting the possible connection of certain amino acids all the way from ISM to comets. In this vein, the inclusion of PH_3 ice in ISM grain model analogs experiments has recently been considered in order to account for the alkyl phosphonic acids observed in Murchison meteorite (Turner et al., 2018).

Thus, the key aspects sustaining a full-fledged astrobiological approach to the origin of life are those related to the cosmic connection linking astrophysical scenarios down to earth in a straightforward manner. By all indications, the incorporation of interstellar matter in meteorites and comets in the presolar nebula provides the basis of a "cosmic dust connection" proposing that some molecules of biological interest originated in interstellar grains and were subsequently transported to earth (Ehrenfreund, 1999). For instance, from ^{15}N and deuterium isotopic ratios measurements of meteorites and IDPs, it is now clear that at least half of the carbonaceous chondrites insoluble organic matter originated from sources that existed before the solar system was formed, so that a significant amount of material of interstellar origin could have been delivered to earth. These results provide evidence of a direct relationship between organic-rich interstellar grains, comets, asteroids, and carbonaceous chondrites. Certainly, some sort of chemical link between comets and different ISM sources is strongly suggested by inspecting the list of shared common molecules given in Figure 7.6.

Additional evidence is provided by the recent observation of the first known interstellar object, dubbed Oumuamua: a highly elongated object with a size of about 200 × 20 × 20 m, discovered on October 2017. In fact, it is generally accepted that during the formation of the solar system, significant numbers of cometary and asteroidal bodies were ejected into the interstellar space. It can reasonably be expected that the same happened for other planetary systems. Detection of such interstellar objects would allow us to probe the planetesimal formation processes around other stars, along with the effects of long-term exposure to the ISM conditions. Indeed, spectroscopic observations of Oumuamua indicate the presence of an insulating mantle produced by long-term cosmic ray exposure preventing this object to exhibit a cometary-like coma formation, even though it passed within 0.25 AU of the Sun (Fitzsimmons et al., 2017).

In summary, the findings described in this Section clearly indicate that a relatively large inventory of precursor organic materials for prebiotic synthesis is ubiquitously distributed throughout several astrophysical environments. In this way, the relatively easy formation of these basic building blocks of life, along with their broad distribution all over the universe constitute the empirical foundations supporting our current astrobiological concept of life's origin on earth or elsewhere.

8.1.1 The Basic Building Blocks

"It remains a paradox that hydrocyanic acid, one of the most toxic molecules to living systems, had apparently been essential to the prebiotic production of one of the most important molecules for life, adenine."

(Oró, 1995).

All types of basic biomonomers shown in Figure 8.1 can be formed starting from a relatively small set of simple precursor molecules such as H_2, CO, HCN, H_2O, NCO^-, NH_3, H_2CO, HC_3N, or NH_4CN, and energy coming from a wide variety of possible sources, including electric discharges and/ or radiation processes in the gas or liquid phases, usually in the presence of some appropriate catalyst. In Table 8.1, we list some proposed prebiotic reactions which have been extensively discussed in the literature during the last decades. For a detailed review, the interested reader is referred to the recent work by Kitadai and Maruyama (2018).

TABLE 8.1 Possible Prebiotic Reactions Leading to Several Basic Biomonomers*

Monomer	Precursors	Catalysts	References
Sugars	H_2CO	$Ca(OH)_2$ $CaCO_3$	Butlerow (1861) (formose)
Adenine	(aq.) $NH_4CN \rightarrow HCN + NH_3$		Oró (1960)
Guanine	$HCN + NH_3 + H_2O$		Orgel-Lahrmann (1974)
	$H_2 + CO + NH_3$	Yes	Fischer-Tropsch
Amino acids	$HCN + H_2O$		Oró-Kamat (1961)
	$HCN + NH_3 +$ R-CHO (aq.)	Fe_3O_4	Strecker
	$H_2 + CO + NH_3$		Fischer-Tropsch
Pyrimidines	$HC_3N + H_2O + NCO^-$ (aq.)		Ferris et al., (1968)
Fatty acids	$H_2 + CO + NH_3$	Yes	Fischer-Tropsch

*The formation of sugars from the reaction of H_2CO was discovered by Butlerow in 1861 and is known as the formose reaction. Adenine can be synthesized in the presence of water by the polymerization of five HCN molecules in a truly remarkable pathway, which does not require complex condensing agents. Amino acids can be obtained from HCN and aldehydes via the Strecker reaction, where the C = O group in the aldehyde is replaced by ammonia to form NH, and this undergoes a reaction with HCN to form a cyano amino compound that finally hydrates to the amino acid. Pyrimidine bases are formed in aqueous solution under mild conditions from cyanoacetylene (HC_3N) and either urea ($((NH_2)_2C = O)$) or the cyanate ion (NCO^-) to give considerable amounts of cytosine, and then uracil by hydrolytic deamination. Thymine can be produced in small yields from uracil from a methylation reaction. Fatty acids can be obtained by means of Fischer-Tropsh type reactions, which also produce purines and amino acids. We note the important role played by the HCN molecule, which appears as a precursor in the synthesis of both purines and amino acids, being itself a precursor of cyanoacetylene molecule (HC_3N) used in the pyrimidines synthesis.

Attending to the number of different precursor molecules required for their prebiotic synthesis we see that sugars and adenine can be regarded as compounds more ready to obtain than guanine, amino acids and the pyrimidines cytosine, thymine, and uracil. Sugars are made up of the three more abundant elements in the universe: H, O, and C, and these compounds share the empirical formulae $(CH_2O)_n$, formally making them oligomers of formaldehyde (H_2CO). Ribose, the sugar used in RNA, is but one of the isomeric pentamers where n = 5. For a long time, the formose reaction was considered the only abiotic pathway to carbohydrates. However, the complex, coupled reactions involved in the process result in a large number of compounds, which subsequently react to yield an intractable tar of insoluble products. For instance, ribose is obtained in very low concentration via formose reaction and, once formed, it would have a half-life of less than 73 min at 100°C or 44 years at 0°C at neutral pH (Larralde et al., 1995). These results suggested that this pathway was not an appropriate source of prebiotic carbohydrates. Indeed, the prebiotic synthesis of deoxyribose was accomplished by Oró (1965) from the condensation of glyceraldehyde ($CH_2OHCH(OH)CHO$) with acetaldehyde (CH_3CHO). Recent results suggest that the generation of numerous sugar molecules, including ribose, may be possible from photochemical and thermal treatment of H_2O, CH_3OH and NH_3 ices, probably occurring in the late stages of the solar nebula formation (Meinert et al., 2016, 2017; Kawai, 2017), or in the mantles of interstellar grains. At the same time, there has been a resurgence of interest in formose related reactions as several new mechanisms, involving the presence of phosphorus compounds, have been discovered that produce a less diverse mixture of sugars, give rise to a higher yield of ribose and make the ultimate phosphorylated sugar derivatives considerably more stable than they are in the free form, due to the presence of electric charges in the phosphate group (Lambert et al., 2010).

Adenine is a nitrogen heterocycle which, like sugars, contains three types of atoms only: H, C, and N, at variance with the remaining nucleobases, guanine, cytosine, and thymine, which contain oxygen atoms as well (see Figure 8.1), thereby requiring the contribution of O-bearing precursor molecules, such as CO or H_2O (two of the most abundant molecules in the universe, by the way). Being a pentamer of HCN, adenine obeys the simple stoichiometric relationship $(HCN)_5$. The intermediates identified in the prebiotic synthesis of adenine include the HCN dimer, trimer, and tetramer, and the overall yield of the reaction is relatively low: when solutions of ammonium cyanide are refluxed for a few days the yield of adenine is about 0.5%. On the other hand, guanine can be formed by the condensation of 4-aminoimidazole-5-carbonitrile with urea.

Amino acids, the primary building block of proteins, are molecules that contain an amine group (NH_2), and a carboxylic group (COOH), along with a side chain. In the Strecker reaction, HCN, and aldehydes condense with each other in the presence of ammonia forming amino nitriles that, upon hydrolysis, yield amino acids. This reaction does not provide stereo control over the C center and results in racemic mixtures of amino acids. Depending on the starting conditions, about 12 of the 20 biological amino acids have convincing prebiotic synthesis mechanisms (Miller, 1998). An additional route involves the condensation of several HCN molecules, which upon hydrolysis yield glycine, alanine, and aspartic acid.

Finally, the Fischer-Tropsch type reactions are very versatile, giving fatty acids, amines, amino acids, and nitrogen-heterocycles, including pyrimidine and purine bases and porphyrin-like polypyrrole pigments. The range of amino acids afforded by these reactions is not large, but it includes the aromatic amino acids phenylalanine and tyrosine, which are not produced by the main alternative prebiotic routes.

What monomer/s was/were formed first? Nucleic acids contain phosphate, ribose (RNA) or deoxyribose (DNA), and five possible nucleobases: the purines adenine and guanine, and the pyrimidines cytosine, thymine (DNA) or uracil (RNA). Quite interestingly, the prebiotic synthesis of pyrimidines from cyanocetylene and cyanate (see Table 8.1) produces cytosine and uracil, but it lacks thymine, suggesting that RNA preceded DNA as a genetic molecule. This idea was spurred by the discovery that certain ribonucleic acids (RNA) are capable of acting as catalysts: the so-called ribozymes. The finding of catalytic organic molecules that are not proteins significantly contributed to undermining the previous paradigm concerning the fundamental role of proteins over nucleic acids during the first stages of life origins, since one can now envision a primitive cell in which genetic information is encoded by RNA and metabolic functions are carried out by ribozymes, a view referred to as the *RNA world* (Gilbert, 1986). According to the RNA world hypothesis, all reactions in protocells were catalyzed by these nucleic acid enzymes before the origin of protein-based catalysis currently prevalent in most living beings. It is unclear, however, how the very first ribozymes came into existence in the absence of pre-existing molecular catalysts capable of synthesizing long enough, fully functional RNA oligomers.

8.1.2 Joining the Blocks?

"*Exploitation of phosphates to orchestrate selective, robust prebiotic chemistry, coupled with the central and universally conserved roles of phosphates in biochemistry, provide an increasingly clear message that understanding phosphate chemistry will be a key element in elucidating the origins of life on earth."*

(Fernández-García, Coggins and Powner, 2017).

Once the monomers were available, the major next step was their concentration and assembling. Nucleotides, the building blocks of nucleic acids, consist of three kinds of biomonomers: a sugar (ribose or deoxyribose), a nucleobase and a phosphate group (Figure 8.1). Reaction of a sugar with a nucleobase provides a nucleoside. This reaction has been found to be notoriously difficult under prebiotic conditions. This is one of the reasons to doubt that such nucleosidation reactions have played a role in a primordial environment (Eschenmoser, 2007). Phosphorylation of a nucleoside yields a nucleotide. The use of phosphorylated moieties from the very beginning proved to be very useful in order to obtain stable sugar phosphates from glycolaldehyde phosphate ($CHOCH_2$-O-$PO_3^{=}$) precursors. Nevertheless, phosphorylation of prebiotic molecules in an aqueous environment faces two problems: (i) the water molecule itself can and will react with the phosphorylating agent, thus rendering it ineffective, and (ii) in most cases the phosphorylation process involves the formation and removal of a water molecule, which becomes thermodynamically unfavorable in an aqueous environment (Karki et al., 2017).

A similar problem appears in order to obtain polypeptide chains of sufficient length to promote catalysis, since amino acids link together via a condensation polymerization reaction which requires the release of water as a condensation step. The required condensation processes could have taken place in open systems, allowing the water molecules to escape, such as near the ocean-atmosphere interface, or in a drying pool, or in mineral pores. Interestingly enough, mineral surfaces have been identified that select and organize molecules in the path from geochemistry to biochemistry. In fact, adsorption on solid surfaces or internal sorption in minerals was proposed as an escape from the dilemma of dilution and activation barriers in the emergence of life by Goldschmidt (1945) and Bernal (1949), who suggested that clay minerals might have been particularly effective on the primitive earth (Cairns-Smith, 1985).

Notwithstanding this, although there have been some successful demonstrations that properly activated ribonucleotides can polymerize to form RNA, it is far from obvious how such ribonucleotides could have formed from their constituent parts (ribose, phosphate, and nucleobases). As previously mentioned, ribose is difficult to form selectively, and the addition of nucleobases to ribose is inefficient in the case of purines and does not occur at all in the case of the canonical pyrimidines. Thus, instead of starting from very simple (say containing up to five atoms) precursor molecules that join to each other following a relatively large number of steps to yield a biomonomer generally composed of up to 15–20 atoms, one may consider the very possibility of starting from more complex molecules instead (say, containing 8–10 atoms each) which then assemble together in just a few number of reaction steps (Figure 8.3). It is interesting to note that the starting molecules in pyrimidine ribonucleotide synthesis shown in Figure 8.3, namely, glycolaldehyde, and cyanamide, are relatively abundant species in the ISM. However, to the best of my knowledge, the next reactants 2-aminooxoazole and glyceraldehyde have not been detected in the ISM (Cordiner et al., 2017) or in comets (see Figure 7.6).

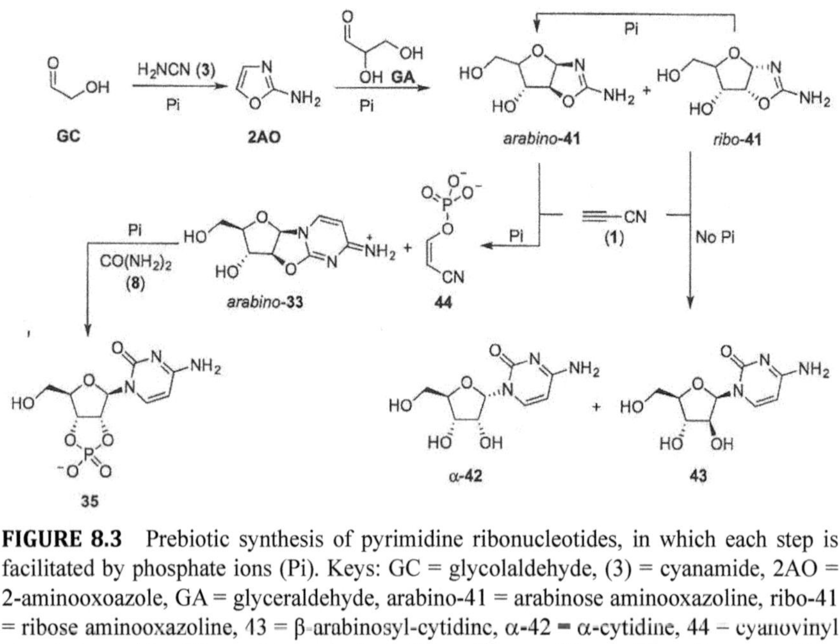

FIGURE 8.3 Prebiotic synthesis of pyrimidine ribonucleotides, in which each step is facilitated by phosphate ions (Pi). Keys: GC = glycolaldehyde, (3) = cyanamide, 2AO = 2-aminooxoazole, GA = glyceraldehyde, arabino-41 = arabinose aminooxazoline, ribo-41 = ribose aminooxazoline, 43 = β-arabinosyl-cytidine, α-42 = α-cytidine, 44 – cyanovinyl phosphate, arabino-33 = arabino-anhydrocytidine, (8) urea, 35 = cytidine ribonucleotide.
Source: Fernández-García, Coggins, and Powner, (2017); Work licensed under a Creative Commons Attribution 4.0 International License.

In this way, the very possibility of making prebiotically plausible nucleosides and nucleotides in a sequence of reactions, where nucleobases are allowed to form in the presence of inorganic phosphate and sugar precursors, instead of separately synthesizing prebiotic nucleobases and prebiotic sugars, has been explored in detail (Fiore and Strazewski, 2016). In particular, activated pyrimidine ribonucleotides can be formed in a short sequence that bypasses free ribose and the nucleobases independent monomers. The starting materials for the synthesis (cyanamide, cyano-acetylene, glycolaldehyde, glyceraldehyde, and phosphate) are plausible prebiotic molecules, and analogously the conditions of the synthesis are consistent with potential early-earth geochemical models (Powner et al., 2009). Although phosphate is only incorporated into the nucleotides at the late stage of the reaction sequence, its presence from the start is essential since it controls three reactions in the earlier stages by acting as a general acid/base catalyst, a nucleophilic catalyst, a pH buffer, and a chemical buffer. This result highlights the importance of considering blended chemical systems in which reactants for a particular reaction step can also control other steps in an intertwined way.

8.1.3 The Appearance of Membranes

"The formation of membranes must be taken into account in all the comprehensive pictures of the origin of life."

(John Desmond Bernal, 1949).

The origin of life did not depend on the appearance by chance of a few single monomers, but on the gradual evolution of molecular systems formed by groups of biochemical molecules exhibiting different specific properties (reproductive, energetic, catalytic) which are mutually inter-twined. It is, therefore, necessary to hold them together within a reduced enough volume space, a task currently performed by cellular membrane structures separating intracellular from the extracellular medium. The coupling of membrane compartmentation with the possibility of RNA activity provides a scenario for the possible origin of the interaction between the RNA world and protocell compartments, leading to the formation of the first evolving protocell. Thus, encapsulation may have strongly affected the dynamics of prebiotic chemistry by segregating molecules and reactions, increasing local concentrations, promoting coevolution of biochemicals, and allowing selective permittivity, which was necessary for the origin of the living cell homeostasis. Thus, molecular crowding within a confined volume is crucial to the functioning of contemporary biology (Dass et al., 2018).

Indeed, all current living cells are encapsulated by membranes that play essential biological functions, including transport of organic and inorganic compounds, molecular communication, and energy conservation by establishing a chemiosmotic gradient across their membranes. They also provide a docking site for many enzymes. Modern cell membranes mainly consist of phospholipids, built by phosphate, glycerol, and two long fatty acids or isoprene units (see Figure 4.18c in Section 4.6). These hydrophobic molecules are able to assemble themselves spontaneously into single spherical-like lipid bilayers surrounding an aqueous core, forming microstructures about 1–10 μm in diameter, which are called liposomes. Though it has been shown possible to synthesize two representative phospholipids (namely, phosphatidylcholine, and phosphatidylethanolamine) under prebiotic conditions (Oró et al., 1978), phospholipid glycerol esters are too complex, so it is considered highly improbable that the required building blocks were available in sufficient quantities under abiotic conditions. A possible alternative is that the initial membrane structures were assembled by single chain fatty acids, fatty alcohols and monoglycerides, which can be either formed in volcanic hydrothermal systems via Fischer-Tropsch type synthesis, or synthesized in reduced gas clouds in the early solar system and delivered to earth in carbonaceous meteorites.

8.2 FIRST METABOLIC PATHWAYS

"At some time in the history of life on earth, prebiological chemistry became biological chemistry."

(McDonald, 2015).

"Microbial evolution is one of metabolic diversity rather than cell complexity."

(McGuinness, 2010).

Metaphorically speaking the problem of the origin of life may be envisioned as similar to assembling a polyhedron composed of many faces. In the previous section, we have worked out the, so to speak, "structural face," dealing with the material pieces out of which the living cells are assembled. To this end, we have described the search of possible routes for the prebiotic synthesis of macromolecules of biological relevance, such as nucleic acids, proteins, and phospholipids. Nevertheless, although the appearance of these molecules constitutes a very fundamental aspect of the problem, probably it is not the most essential one. Indeed, by exclusively following a purely structural route, we could possibly arrive to the origin of the first virus-like particles, but not likely to the origin of the first bacterium cell. For bacteria, at variance with the nicely geometrically encapsulated viruses, are characterized by possessing a full-fledged metabolism, which rules the intertwined flows of matter and energy necessary to sustain self-contained biochemical machinery within the cell, running the required rhythms of life in a completely *autonomous* way.

Life on earth is divided into three domains: Archaea, Bacteria, and Eukarya. Single-celled organisms are common in all three domains, and all organisms in the domains Archaea and Bacteria are single-celled. On the other hand, all multicellular organisms fall within the domain Eukarya. This classification was proposed by Woese and colleagues (Woese et al., 1990) based on the nucleotide sequence of the gene for a small ribosomal unit (termed "16S" rRNA for Bacteria and Archaea, and "18S" rRNA for Eukarya), since all living organisms contain ribosomal RNA, which plays an essential role in protein synthesis. Given its ubiquitous presence, researchers usually utilize rRNA gene sequences to determine the evolutionary relationships among organisms. To this end, they derive a phylogenetic tree which provides information about the progress of biological evolution by showing the relative order in which a given gene sequence has changed over time.

One of the most fundamental observations in biology is that all living beings share the same core metabolic pathways and biochemical building blocks. This profound fact, known as the unity of biochemistry, implies that such a core metabolism must be ancient and date back to the so-called last universal common ancestor (LUCA). In this view, present-day metabolism is seen as a direct consequence of prebiotic chemistry, rather than an accidental by-product, and the origin of nucleic acids is intimately linked to proto-metabolic reactions (Copley et al., 2007).

Then, the question naturally arises concerning the origin of a primordial "metabolism." Certainly, this quest constitutes another fundamental facet in the origin of life polyhedron, so that, along with the progressively increasing structural complexity route, one must also consider the origin of the main energy sources and energy carriers in the first metabolic pathways. In general, metabolic pathways function primarily in either catabolism or anabolism modes. Catabolism is the breakdown of certain molecules into a relatively small number of intermediate products, in order to convert chemical energy into a suitable form appropriate for biological processes. Anabolism is the use of catabolic products for the synthesis of biomolecules such as amino acids, nucleic acid bases, and lipids. Generally, catabolic pathways are exergonic, yielding a net release of energy, while anabolic ones are endergonic, requiring a net consumption of energy. The chemical energy produced in catabolic processes is transferred to a set of energy carriers that make possible the energy-consuming anabolic biosynthesis processes. The most common energy carrier is adenosine triphosphate (ATP, see Figures 8.1 and 4.15, 4.18b in Chapter 4), since the phosphodiester bond of the terminal phosphate group in ATP can be hydrolyzed to yield an amount of free energy over 30 kJ mol^{-1} (0.31 eV).

All modern organisms drive metabolic synthesis reactions by using electrochemical energy from redox reactions. In these processes, reduction entails becoming more electron-rich (i.e., gaining electrons), whereas oxidation refers to becoming more electron-deficient (i.e., losing electrons). In a similar way to the flow of electrical current in a conducting material, where positive and negative leads are required for a functioning circuit, electrons in metabolic chemical reactions flow by hopping from a chemical species which becomes oxidized to another one which becomes reduced. Accordingly, we may expect that early metabolism developed in environmental settings where redox gradients occurred naturally (Domagal-Goldman et al., 2016). It seems to be a consensus that what eventually arose were chemical systems capable of autocatalytic molecular replication via reaction cycles, and that had the potential to evolve. Metabolic cycles are supposed to replicate chemical products, and some proposals of primeval metabolism include the reductive citric acid cycle, supposed to be capable of operating without assistance by enzymes (Eschenmoser, 2007). One of the most ubiquitous coenzymes in current metabolic pathways is nicotinamide-adenine dinucleotide, which functions as an electron and proton carrier. The functional part of this coenzyme is a nicotinamide ring (vitamin B3), which can be obtained from cyanoacetylene, propionaldehyde ($HC \equiv C - CHO$), and ammonia (Miller and Orgel, 1974). Remarkably, all these molecules have been detected in the ISM.

Organisms can be described in terms of both how they obtain their metabolic energy and how they obtain bioelements to build the required biomolecules. Organisms that use light as an energy source are called phototrophs. Light energy (photons) is utilized by these living beings through photosynthesis, and the underlying mechanisms used to harness this energy are redox reactions. In oxygenic photosynthesis, photons are employed to split H_2O into O_2 and protons and electrons. The protons and electrons are then spent to create energy storage molecules, which ultimately serve to reduce CO_2 into organic compounds. Thus, the net redox moieties involved in oxygenic photosynthesis are H_2O being oxidized to O_2, and CO_2 being reduced to organic compounds. There are also photosynthetic mechanisms that utilize photons for energy but do so with electron sources other than water and create by-products other than oxygen.

Within the oxygen-rich environment of present-day earth, essentially all phosphorus sources are based on the fully oxidized orthophosphate. However, the prebiotic earth was certainly oxygen-poor, an environment in which lower oxidation state phosphorus species, such as phosphonic and phosphinic acids and their derivatives, would certainly be more stable. There is circumstantial evidence that certain anaerobic bacteria can reduce phosphate directly. Earlier on Tsubota (1955) determined that 100 mg/l of hypophosphite and significant phosphite were produced in an anaerobic soil culture initially containing 2g/l orthophosphate. More recently, Schink and Friedrich (2000) have identified sulfate-reducing bacteria (isolated from sediments of the Canale Grande in Venice) which exploit the reverse process, oxidation of phosphite ion to orthophosphoric acid, as a key step in microbial energy metabolism through the reaction $4HPO_3^{2-} + SO_4^{2-} + H^+ = 4HPO_4^{2-} + HS^-$. Genetic analysis of this anaerobic strain revealed strong sequence similarities with a sub-class of sulfate-reducing proteobacteria. Microorganisms of this genre could have thrived on a litho(auto)trophic metabolism involving the oxidation of reduced phosphorus compounds, in which case, phosphate oxidation by sulfate-reducing bacteria may represent an ancient evolutionary trait (McGuinness, 2010).

The question then is how could reduced oxidation state phosphorus compounds have been produced on and/or delivered to the early earth. Gulick (1955) pointed out that both phosphoric (H_3PO_3) and hypophosphoric (H_3PO_2) acids could have been accessible on the early earth if the redox potential of the environment was appropriately reducing.

Alternatively, suitable reduced phosphorus compounds could arise from corrosion of the meteoritic mineral schreibersite (see Section 4.3.2) to yield phosphite radicals in aqueous solution. These radicals can form activated polyphosphates, which can phosphorylate organic compounds such as acetate to give phosphonates and organophosphates (Pasek and Lauretta, 2005; Pasek et al., 2007).

Questions regarding as to whether the first organisms appeared in hot or cold environments, or they were autotrophic or heterotrophic ones have progressively spurred a growing interest. The Oparin-Haldane heterotrophic theory of the origin of life was originally accepted on the basis that a heterotrophic organism is simpler than an autotrophic one. There are, however, some recent examples of autotrophic proposals based on the reduction of atmospheric CO_2. In fact, the deepest branches of the universal life tree are occupied by anaerobic sulfur-dependent hyperthermophiles that fix CO_2 by a reductive Krebs cycle. However, it is important to distinguish between ancient and primitive. Hypertermophiles may be ancient, but they are hardly primitive, since they contain the same elaborate protein biosynthesis and most of the modern organisms' enzymes. Truly primitive organisms would be representative of the RNA world or some of their immediate descendants, but no such organisms have been found to date. Accordingly, the study of hyperthermophiles is an invaluable source of information on the nature of the LUCA of all extant life forms, but an extrapolation into prebiotic times should not be taken for granted (Lazcano and Miller, 1996).

The ranges of temperatures available for prebiotic reactions vary greatly. In principle, the building blocks of life could be synthesized in interstellar clouds and on comets at very low temperatures approaching the absolute zero, or near hydrothermal vents where temperatures may be over 100°C. Also, the temperatures postulated for the actual origin of life widely vary, some hypothesis favoring a hot environment (Miller and Lazcano, 1995), other favoring a cold one (Bada and Lazcano, 2002). Those advocating a hot origin of life support their view on the observation that many genetically primitive microorganisms are thermophiles found in hot places such as hydrothermal vents which could fulfill both the chemical diversity and available energy sources requirements. The hypothesis that all life forms evolved from hyperthermophilic organisms was suggested because the oldest extant life forms (e.g., Crenarchaeota, Nanoarcheota) are chemoautotrophic hyperthermophiles found near the underwater hydrothermal vents at midoceanic ridges rich in dissolved hydrogen sulfide, a possible habitat of pioneer organisms (Rothschild and Mancinelli, 2001). However, organic chemist's intuition sees some problems with such a hot origin stemming from liability of most involved organic compounds exposed to too high temperatures (Eschenmoser, 2007).

8.3 THE PHOSPHATE PROBLEM

"The when- and whereabouts of phosphate entering the scene is one of the lasting unknowns, which any scenario-building must eventually come to grip with."

(Eschenmoser, 2007)

The synthesis of the first organophosphates has been a major question in origin of life studies due to the indispensable role of phosphate in modern biochemistry, which indicates a strong evolutionary pressure for the early incorporation of phosphate in nucleic acids, membranes, and metabolic energy carriers. Inorganic phosphate is present within a typical cell cytoplasm at a concentration of a few mM, and organophosphate compounds are present at twice this concentration. Taking these concentrations as a reference value, the amount of dissolved phosphorus in prebiotic environments should be 1–10 mM, and organophosphates should be about 70% of this figure. However, due to the strong affinity of phosphate towards divalent cations, it is generally assumed that, on a prebiotic earth, most phosphate was sequestered in low-solubility minerals, such as those belonging to apatite group, rendering it unavailable for use in prebiotic chemical reactions, with phosphate concentrations typically around 10^{-6} M. This obstacle would present a significant challenge to the possible emergence of an RNA world, as well as to other tentative models for the origin of phosphate-based life on earth. Additionally, the most plausible routes to phosphorylation of organic molecules in metabolic pathways are through water releasing condensation reactions, which are thermodynamically unfavored in aqueous solutions, where the phosphate ion is inherently unstable. Thus, albeit the formation of organophosphate compounds is an essential step for the emergence of life, the synthesis of these compounds is difficult and does not occur spontaneously. For instance, the phosphorylation of glucose given by Eq. (1) has a Gibbs free energy of $\Delta G = +13.8$ $kJmol^{-1}$ (0.14 eV) at 298 K if the phosphate attaches to the 6' carbon in the glucose molecule.

$$C_6H_{12}O_6 + H_2PO_4^- \rightarrow C_6H_{12}O_6PO_3^- + H_2O. \tag{1}$$

The positive value indicates that the above reaction moves toward the reactants, rather than toward the products, so that the ratio of the concentration of glucose-6-phosphate to the product of the concentration of glucose and phosphate is quite small, namely:

$$\frac{[C_6H_{12}O_6PO_3^-]}{[C_6H_{12}O_6][H_2PO_4^-]} = e^{-\frac{\Delta G}{RT}} \cong 0.0038, \tag{2}$$

where, R = 8.314 Jmol^{-1}K^{-1} is the perfect gas constant. Notwithstanding this, glucose is actually one of the easier compounds to phosphorylate, so that most phosphorylation reactions of biological interest exhibit much lower yields.

The most popular condensation agents (which are compounds that promote dehydration reactions) include cyanate, cyanamide, imidazoles, or sulfur-containing organics, such as thioesters (-S-), although these compounds frequently require lower pH values than those expected on the early earth surface (Burcar et al., 2016). In current living beings, the required reactions take place through involved reaction pathways during metabolic cycles. But then a fundamental question naturally arises: how were the phosphates first incorporated during the earlier stages of the development of life on earth? These difficulties constitute the basis leading to the so-called *phosphate problem* for the origin of life, as it was originally termed by Orgel (1989). Indeed, prebiotic phosphorylation attempts typically produce remarkably low yields, even when employing high concentrations of reactants (Keefe and Miller, 1995). So, what were the appropriate geochemical settings and primitive phosphate sources that could lead to the origin of the first organophosphates of prebiotic relevance? (Maciá et al., 1997; Pasek et al., 2017).

Historically, attempts to form phosphorylated biomolecules have proceeded via a dehydration reaction between an orthophosphate and an organic molecule containing a hydroxyl (-OH) or amino (-NH_2) groups. In this way, a number of nucleosides and sugars can be phosphorylated, albeit at low to moderate rates, and requiring temperatures within the range 80–160°C to this end. In fact, phosphorylation reactions are endergonic, with a ΔG^0 at 298 K of +10 to +20 kJ mol^{-1} (0.10–0.21 eV) for forming orthophosphate esters, or +40 to +60 kJ mol^{-1} (0.41–0.62 eV) to yield reactive organophosphates (Pasek and Kee, 2011). Furthermore, the reactions must be heated to dryness, since dehydration reactions are thermodynamically unfavorable in water, so that heating eliminates H_2O molecules, thus providing an entropic driving force in order to maintain non-equilibrium conditions. In the same vein, acidic solutions generate more organophosphates, as there are more protons attached to phosphate groups at $pH < 7$, which are subsequently available for condensation.

Despite these relative successes, many putative attempts are plagued by geochemical problems, since phosphorus is assumed to be comparatively scarce in aqueous environments in primitive earth. In addition, several of the proposed condensing agents were unlikely to have been abundant in the prebiotic stage, and thus it has been suggested that condensed phosphates, such as pyrophosphate or polyphosphates (see Section 4.5.2) were a more likely source of chemical energy for prebiotic phosphorylation (Exercise 16). We note that phosphorylation by a polyphosphate has a leaving group of a phosphate or smaller polyphosphate. In this way, the "water problem" generally associated with phosphorylation is removed, since releasing phosphate does not compete with already present water solvent. However, only four polyphosphate minerals have been recognized by the International Mineralogical Association to date (see Section 4.5.2), and few of them were likely abundant on the early earth, so that this route is also difficult to justify. Alternatively, cyclic trimetaphosphate has been proposed as an excellent phosphorylating agent at concentrations as low as 10^{-6} M (Pasek and Kee, 2011).

Interestingly enough, it has been recently shown that hydroxyapatite in a urea/ammonium formate/water solution containing magnesium sulfate can undergo a significant alteration, giving rise to a more soluble phosphate mineral: struvite ($NH_4MgPO_4.6H_2O$), which is about 10^3 times more soluble than apatite. In addition, this solvent favors nucleoside phosphorylation from this solubilized phosphate. This scenario, which reminds the warm little pond mentioned in Darwin's celebrated account opening this Chapter, could have played a potential role in urea-rich aqueous pools experiencing daily evaporation episodes on the early earth (Burcar et al., 2016). In this regard, it has been recently proposed that this approach could be extended to survey a wider spectrum of phosphorus derivatives, such as compounds containing P-N bonds, namely, amidophosphates (Gibard et al., 2019) as plausible prebiotic reagents (Figure 8.4).

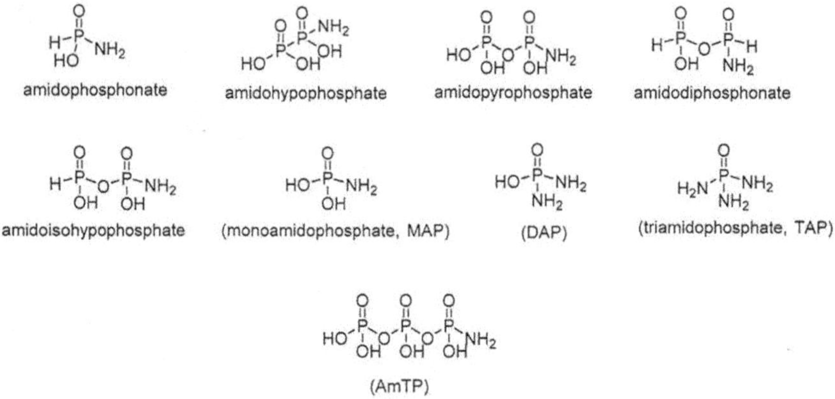

FIGURE 8.4 Plausible prebiotic nitrogenous analogs of inorganic phosphates.
Source: Karki, Gibard, Bhowmik, and Krishamurthy, (2017); Work licensed under a Creative Commons Attribution 4.0 International License.

An alternative way to circumvent the phosphate problem is by considering from the very beginning more reduced phosphorus compounds, such as hypophosphate, phosphite or hypophosphite, whose oxidation states are +4, +3, and +1, respectively (Figure 8.5). The suggestion that these compounds may have played a role in the development of life due to the increased solubility of phosphite and hypophosphite (about a factor 10^3 and 10^6 the solubility of phosphate in the presence of divalent cations, respectively), was first put forward by Gulick (1955). This idea, however, was dismissed at that time since no source of such reduced phosphorus compounds was provided. Interestingly, the mineral schreibersite (see Section 4.3.2), common in iron meteorites, has been recently proposed as a possible source. Specifically, it has been shown that schreibersite releases phosphite, hypophosphite, and pyrophosphate molecules on reaction with water, in addition to phosphate (Bryant and Kee, 2006; Pasek et al., 2007). The reaction proceeds by oxidation of phosphorus by water to first generate phosphite radical species (PO_3^{2-}), which then combine together to produce hypophosphite, and with H^+ and OH^- radicals to generate the other observed products (Figure 8.5). We note that the phosphite and phosphate moieties are the most abundant species formed by this reaction (~ 80%), followed by pyrophosphate and hypophosphate.

As we see, this proposed phosphate source invokes a significant contribution of extraterrestrial schreibersite mineral, which should be delivered to earth during the Hadean age. What about such a possibility?

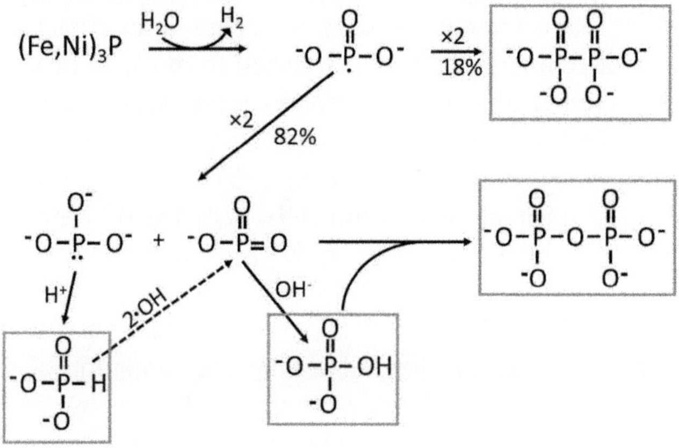

FIGURE 8.5 Schematic of the corrosion pathway of schreibersite. The first step is the oxidation of phosphorus in the mineral to yield PO_3^{2-} radicals which combine together to produce hypophosphate (top right box). Alternatively, PO_3^{2-} radicals can combine with H^+ and OH^- radicals to give phosphite and phosphate, respectively (bottom boxes). The resonant metaphosphate ion can then react with phosphate to yield pyrophosphate (middle right box). Reprinted from Pasek, Gull, and Herschy, (2017). *Chemical Geology,* 475; Copyright (2017), with permission from Elsevier.

8.4 THE ROLE OF COMETS AND ASTEROIDS IN PREBIOTIC SYNTHESIS

"The synthesis of adenine from hydrogenic acid, the presence of this compound in comets, and the atmospheric explosion which occurred in Tunguska, Siberia, in 1908, caused most likely by a comet, were the three factors that intuitively led me to propose the hypothesis of the role of comets in the formation of biochemical molecules on primitive earth."

(Oró, 1995)

The occurrences of late accretion processes are supported by independent lines of research which suggest that the surface of our planet was covered with a veneer of solid and volatile-rich material. From an observational and experimental point of view, the evidence of such past collisions is made clear particularly by the presence of heavy noble gases (Ar, Kr,

Xe) in our atmosphere and that of other inner planets. From laboratory data, Owen, and his co-workers (1992) have found that ice deposited at approximately 50 K, the temperature at which comets presumably formed in the Uranus-Neptune region of the solar nebula, is able to trap heavy noble gases in characteristic proportions. These proportions have been observed in ancient micrometeorites collected from Antarctica.

The late accretion of volatiles becomes a necessity if one takes into account the formation of the earth-Moon system by the massive collision of a Mars-sized body (Cameron and Benz, 1991), since both the earth and the Moon would have become completely depleted of water and other volatiles. Indeed, isotopic studies of noble gases in earth's mantle relative to its current atmosphere suggest that our planet lost most of its original carbon, linked to the molecules CH_4, CO, and CO_2, when earth shed its original atmosphere (see Section 7.1). After the formation of the earth-Moon system, our planet continued to receive collisions by smaller planetesimals and comets, as it is recorded by the lunar cratered surface, which can be used as a calibrator of some of these past events. A body of the size that produced the impact known as Mare Imbrium on the Moon would have probably caused a significant erosion of the earth's hydrosphere and atmosphere, although it would probably not have caused the total ejection of the water of the seas. There may have been larger collisions on the earth, but this is uncertain. At any rate, we need to be concerned only with the formation of biochemical compounds and the emergence of life after the last major collision to which our planet was subjected. From then on, the matter falling on the earth from comets and other minor bodies of the solar system was captured, and it is generally assumed that both prebiotic synthesis and life began in this environment.

In short, independent lines of evidence indicate that our planet was subjected to cometary collisions during the late accretion period, and that impacts from comets provided a variety of organics to early earth, though only a very small fraction of earth's water budget, whose most likely primary source was the delivery by certain type of asteroids (Morbidelli et al., 2000). Therefore, most organic matter that was incorporated into living systems probably originated somewhere else than early earth, the more probable source being meteorites, comets, and IDPs which bombarded our planet 3.8–4.0 billion years ago.

The amount of cometary matter captured by the early earth was estimated some time ago to be up to 10^{18} grams (Oró, 1961), which is approximate of the same order of magnitude as the mass of the current biosphere. More accurate estimates were subsequently obtained, so that we can consider an influx of about 10^{23} g of cometary matter during the first 2 x 10^9 yr as a reasonable estimate (Oró et al., 1990, 1992, 2006). Taking the value 0.1% in weight as a confident lower amount of the phosphorus contained in this cometary material (see Sections 7.3.1 and 7.3.2) then the influx of extraterrestrial phosphorus to the earth during the time period when prebiotic synthesis presumably took place would be around 10^{20} g, which is about 12% of the total phosphorus in the earth's crust (i.e., on land and ocean floor, see Section 7.3.6). If, in addition to the contribution due to comets, we also include the mass fraction delivered by other extraterrestrial sources such as asteroids, meteorites, and IDPs, the above figure would be certainly higher (Chyba and Sagan, 1992). In particular, meteorites may have been an important source of reactive phosphorus in pyrophosphate and triphosphate form (Pasek and Lauretta, 2005). Detailed calculations are unnecessary to our purposes, since the main point we wish to emphasize here is that the contribution of phosphorus delivered only by comets would suffice to cover the requirements for this element in the current biosphere (~ 4 x 10^{16} g) by a large margin.

A remarkable aspect of the interaction between IDPs and the atmosphere of the earth is that, notwithstanding IDPs enter the atmosphere at high velocities, in excess of 11 km/s, they decelerate so gently that they are slowed without severe heating and mechanical stress. As a result, frictional heat flows slowly, and dust particles are able to radiate the heat without melting. Once decelerated, IDPs reside in the upper atmosphere, quietly settling down and eventually mixing with the terrestrial particulate material commonly found at lower altitudes (Llorca, 2005). Quite interestingly, on the basis of detailed experimental work, it has been reported that had nucleotides and nucleosides been originally present in IDPs having sizes of 10^{-6}–10^{-5} m, these compounds could be delivered onto the surface of earth-like planets, surviving temperatures up to 500°C generated during atmospheric entry (Marcano et al., 2004).

The substance classes we believe are present in the cometary dust are highly reactive in warm water. Thus, the unsaturated carbohydrates may react to give sugars, the N-containing species may react to give nucleobases, and the mineral core may serve as the phase boundary necessary to maintain a local concentration gradient and also as a source for dissolved phosphoric acid (Kiessel and Krueger, 1987). Thus, the simultaneous presence of precursor amino acid molecules methylamine (CH_3NH_2) and ethylamine ($CH_3CH_2NH_2$) together with glycine in both P67/ Churymov-Gerasimenko and Wild 2 comets suggests that the pathways for glycine formation on interstellar dust grain ices, in a temperature range from 40 to 120 K, could account for the synthesis of glycine in cometary material (Altwegg et al., 2016). Accordingly, the presence of glycine, phosphorus, and many organic molecules of biological importance in the coma of comet P67/Churymov-Gerasimenko (see Section 7.3.1) supports the idea that comets delivered key molecules for prebiotic chemistry to the solar system and the early earth (Oró, 1961), significantly increasing the concentration of life-related chemicals.

Although comets remain most of their time far away from the Sun, they occasionally approach it due to small gravitational perturbations. During their approach to the star, their temperature progressively increases, promoting the sublimation of ices along with an intense episode of photochemical processing of their originally pristine materials. In the case of periodic orbit comets, this processing assumes a cyclic nature, undergoing successive episodes of condensation due to the recurrent loss of most volatile materials forming their comae and tails. In this sense, periodic comets somehow mimic usual handling in chemical laboratories, leading to the synthesis of complex organics by means of cyclic sublimation/condensation processes along their orbits. Similar alternating evaporating processes probably occurred on the primitive earth as a result of the daily cycles of night and day. The episodes taking place in comets would differ significantly from those on the earth both in their duration (extending for many days or weeks depending on the orbital speed around the Sun over the entire perihelion passage interval) and the temperature variation range, spanning wider amplitude in comets than on earth. In any event, it is conceivable that due to the continued repetition of the evaporation/condensation processes, relatively high degrees of polymerization could be reached in both scenarios.

As we have previously mentioned, the insolubility of calcium phosphate in water is a significant stumbling block in the chemistry required for the origin of life on earth. The discovery of alkyl phosphonic acids in the Murchison meteorite suggests the possibility of delivery of these water-soluble, phosphorus-containing molecules by meteorites to the early earth. This contribution could have provided a supply of organic phosphorus for the earliest stages of chemical evolution. Thus, although phosphonic acids were probably not components of early genetic systems, these compounds may have been precursors to the first nucleic acids. If so, the first RNA molecules may have been preceded in evolution by the emergence of a chemically more "robust" system, a primitive precursor of RNA in which phosphates were replaced by phosphonic acids, so that they might have played a role in chemical evolution, analogous to, but before the emergence of the first biophosphates. Indeed, methylene phosphonic acids are structurally similar to orthophosphate esters, and have been synthesized as analogs of nucleotides. In this sense, also a number of different alternative backbones for potential early genetic systems have already been examined (Miller, 1997).

8.5 SUMMARY AND REVIEW QUESTIONS

Prebiotic chemistry is the study of chemical reactions which result in the formation of organic molecules, without the involvement of biological processes. This includes reactions taking place under conditions such as those of the primordial earth environment or the ISM, where organic compounds can be synthesized from inorganic building blocks. Indeed, the fact that many of the known interstellar organic molecules are also present in protoplanetary disks and in comets (see Figure 7.6) supports the view that ISM environments could possibly be the first formation sites for the simpler prebiotic molecules, which subsequently may have been delivered to the early Earth by comets, asteroids, and meteorites during the Hadean age.

There are a number of serious problems that have not yet been resolved in the field of the origin of life. Primarily, we must consider the chemical reactions necessary for the emergence of biochemistry itself. For instance, what key molecules are necessary for the presence of self-assembled molecular systems capable to reproduce and undergo some type of molecular evolution towards the creation of higher-order structures? And, how are these elementary building blocks being synthesized in the absence of any biological machinery? In this regard, finding efficient routes leading to the formation of the ribose and deoxyribose sugars to make RNA and DNA nucleic acids still remains an open question.

The incorporation of phosphorus into biomolecules was likely one of the key steps of prebiotic chemistry that led to the origin of life. What was the first phosphorus-containing biomolecule?

Did it fill a role in metabolic, replication, or structural processes? It seems probable that condensed phosphate forms, such as polyphosphates, played a substantial role here. Related to this question is: what was the first use of ATP? In modern life, ATP is circulated much more frequently through metabolic processes than in RNA synthesis, but is this necessarily true for the origin of life? Some have speculated that ATP arose first as a metabolic molecule, and only later was used to construct RNA. It seems more likely that ATP was co-opted by metabolism from its role as an RNA building-block component during the RNA world stage.

The phosphorus used by early life had to come from a mineral source, probably including calcium phosphates (see Section 4.5.4). But these minerals are pretty stable and poorly soluble in water. For instance, one would expect to see less than 10^{-6} M of phosphate in the water resulting from apatite dissolution, the other possible primordial phosphorus bearing minerals providing similarly low phosphate amounts. Prebiotic chemists have attempted to form phosphorylated organic compounds by using four major routes. Two of them aim to remove water, either by heating the reactants above the boiling point of water, or by replacing water by an alternative solvent, such as formamide. The other routes proceed by heating solutions to evaporate water, hence decreasing water activity, adding condensing agents, or introducing polyphosphates in the solutions (Pasek and Kee, 2011). Unfortunately, all these methods share a common shortcoming: very few of them (if any) accurately represent phosphorylation processes under plausible geologic conditions, a conundrum which was termed the phosphate problem by Orgel. The phosphate problem stems from the fact that the geochemistry of phosphorus is different from all of the other main biogenic elements. Indeed, phosphorus is almost exclusively in phosphate minerals on the earth's surface, lacking significant volatile phases such as CO_2, H_2O, CH_4, NH_3 or H_2S for the remaining biogenic elements. As a result, the cycling of P atoms on earth is slow, since it is tied to the slow rising of mountains, weathering of these mountains to sediment, burying of sediments and their eventual uplifting to form new mountains by plate tectonics.

Where are possible locations that might have provided suitable settings for the onset of the first metabolic cycles? What were the likely routes to get phosphorus into metabolism? Earlier approaches considered that prebiotic phosphorylation reactions might have proceeded by chemically simple, dehydration, and condensation reactions within early earth geological environments, but all these assumptions face with the drawbacks we mentioned before (for a recent review, see Liu et al., 2019). Thus, recent work has suggested that reduced phosphorus compounds, released by meteorites minerals corrosion, may provide a suitable route to prebiotic organophosphorus compounds (Figure 8.5). Accordingly, it has been suggested that phosphorus may not have been ecosystem-limiting on the early earth because phosphate could be reduced to phosphite by ferrous ion, even at the relatively low expected temperatures (<200°C), before the presence of significant O_2 levels in the atmosphere, an event which took place about 2.4 billion years after the earth formation (Herschy et al., 2018). To this end, the possible existence of confined environments, either in interstellar grains, planetesimals, meteorites, comets or in different geochemical scenarios on the earth, would have had a great influence in phosphorylation chemistry in particular, and prebiotic chemistry in general (Dass et al., 2018).

In the light of the possible prebiotic scenarios described through this chapter a number of key general questions regarding the emergence of life, either on our planet or elsewhere in the universe, naturally appear, namely:

1. What is the origin of the observed homochirality of biomacromolecules? In fact, chemical synthesis normally leads to a racemic mixture, as reported for amino acids detected in meteorites. The solution to this problem will probably be found in the selective formation of homochiral oligomers, because of their increased stability or efficiency in their catalytic activities, or improved structurally dependent functions (Oró, 1995). In this regard, it has been recently suggested that when polymerization is constrained to proceed in only one direction along with the template, as it occurs in DNA synthesis, evolution favors chiral monomers and homochiral polymers. This evolutionary advantage stems from the ability of a chiral monomer to bond with the template in only one orientation relative to the template monomer, along the direction of polymerization (Subramanian and Gatenby, 2018).
2. Why are there only twenty amino acids used in proteins in living organisms on earth? A wide range of experiments has shown that "non-protein" amino acids can be readily incorporated into proteins (Hohsaka and Sisido, 2002). An answer to this question will be probably related to the translation mechanisms of the coded RNA information into peptides.

3. Why does genetics use only the sugars ribose and deoxyribose in RNA and DNA nucleic acids, respectively? Studies by Eschenmoser and co-workers (1999) have demonstrated that other sugars have been found to work equally well in several DNA analogs.
4. Why are there only four nucleobase pairs in DNA? Why does uracil replace thymine in RNA? There are other base pairs that appear to function as satisfactorily, and which have been successfully incorporated in synthetic nucleic acids (Yang et al., 2007).

Similarly, a number of questions regarding the origin of complex organic molecules in chondrite meteorites still remain: did they form within the meteoritic body or outside in the ISM and were subsequently incorporated? Are there any interactions between organic molecules and minerals present in meteorites?

Since we have direct knowledge of life only on earth and, by all indications, known life on earth has evolved from a single ancestor; we have just one source of information about what life is. Accordingly, most answers to the questions listed above probably will reveal in the years to come that our biological understanding is nowadays strongly biased by our current knowledge regarding what starting organic materials happened to be present on earth when life began, and by the fact that life evolved too rapidly to sample all chemical possibilities (Benner et al., 2004).

KEYWORDS

- **abiotic processes**
- **anabolism modes**
- **astrobiology**
- **autotrophic/heterotrophic organisms**
- **biomonomers**
- **catabolism**
- **condensation reactions**
- **early phosphorylation problem**
- **last universal common ancestor**
- **late accretion processes**
- **metabolic pathways**
- **Miller's experiments**
- **panspermia**
- **phospholipid membranes**
- **phosphorylation**
- **photosynthesis**
- **polymerization**
- **prebiotic synthesis**
- **ribozymes**
- **RNA world**

CHAPTER 9

Phosphorus Technology: Man-Made Compounds

"Questions about our sources of phosphorus and the applications for which we deploy it rise the provocative issue of the human role in the ongoing depletion of phosphorus deposits, as well as the transfer of phosphorus from the land into the seas."

(Christopher C. Cummins, 2014)

In the previous chapters, we have followed the journey of phosphorus atoms and their molecular compounds, starting from the formation of phosphorus nuclei inside the cores of massive stars, followed by the release of phosphorus atoms into the envelopes of aged stars to form simple molecules there, and then considering the chemical evolution of these P-bearing species, leading to the synthesis of more complex compounds in different places of the ISM. Eventually, some of these moieties were incorporated onto planets of our solar system (and possibly in bodies belonging to other planetary systems too) to play a relevant role in prebiotic chemical synthesis, ultimately leading to the emergence of life on earth. In this way, we have disclosed subtle links between chemical evolution steps occurring in diverse astrophysical sites, and those taking place in our homeland blue pale planet. It is timely now to narrow down our perspective from the workings of Nature all over the universe to focus on the workshops of men laboring on the earth, and its close planetary neighborhood.

9.1 INDUSTRIAL APPLICATIONS OF PHOSPHORUS COMPOUNDS

Although unnoticed to the ancient metallurgists forging plows and swords alike, the presence of little traces of phosphorus in the melted alloys they worked with played an important role in the resulting hardness and ductility properties of the tools and weapons used by ancient civilizations. These small amounts of phosphorus compounds were naturally present in the employed iron ores, along with other impurities. In a similar way, farmers were not originally aware of the necessary contribution of this vital element for the proper growth and health of their crops. However, they were well acquainted of the convenience of burning the stubble in the fields in order to use the resulting ashes to enrich their lands, thereby increasing the harvesting productivity. As we will see in this chapter, all these old uses of phosphorus are still in use today, and new ones have progressively been discovered as the growth of the industry, and its related technological improvements have spurred the search for novel applications and innovative materials with enhanced properties.

In Figure 9.1 we illustrate the main phosphorus compounds of current industrial interest as well as the routes followed for their synthesis from their common natural source: the so-called *phosphate rock*, which is mainly fluorapatite, usually containing some impurities including calcium and magnesium carbonates, Fe_2O_3, Al_2O_3, and silica (SiO_2). This mineral occurs widely, but most important reserves are in Morocco and Western Sahara, China, Algeria, Syria, South Africa, Russia, Jordan, Egypt, the US (Florida, North Carolina), and Tunisia.

World phosphate rock production during 2017 was lead by China (140 million tons), the US (27.7 million tons), Morocco, and Western Sahara (27 million tons), and Russia (12.5 million tons), and it is expected to increase this production from 263 million tons worldwide in 2017 to about 320 million tons in 2021. The leading areas of growth are planned in Africa and the Middle East. In Morocco, mine production capacity is expected to double owing to the expansion of existing mines and development of a new mining complex (Jasinski, 2018). World resources of phosphate rock are estimated to be more than 300 billion tons, so that no imminent shortages of phosphate rock are expected in the near future.

About two-thirds of the phosphorus consumed worldwide is used in the form of phosphate fertilizers, which are obtained from relatively low-quality phosphoric acid made directly from rock phosphate ores. Phosphoric acid of higher purity is necessary for some food processing and for etching semi-conductors. It is known as *thermal phosphoric acid* and is made by burning previously obtained elemental white phosphorus in moist air. The balance of the phosphate rock mined is for the manufacture of elemental white phosphorus, which is used to produce phosphorus compounds for a variety of industrial applications (Table 9.1).

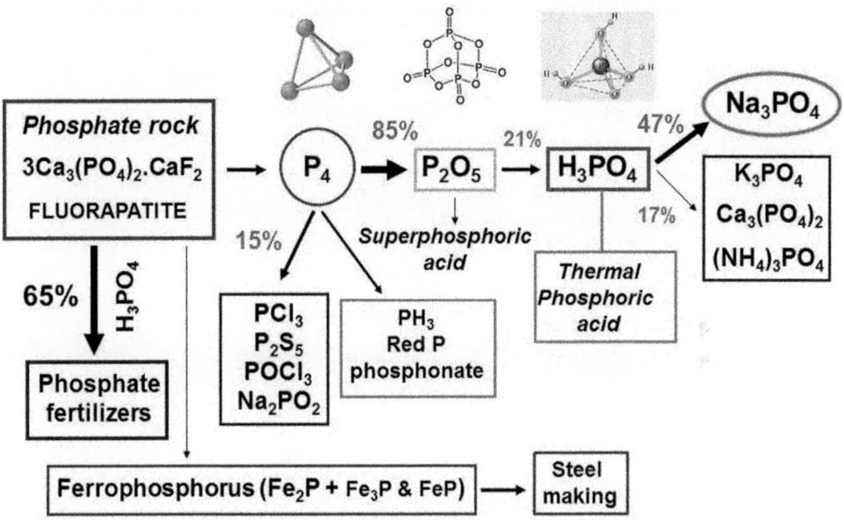

FIGURE 9.1 Different phosphorus compounds are derived from phosphate rock ore in the current phosphorus industry. The more relevant ones are phosphate fertilizers and white phosphorus, most of which is subsequently transformed in phosphorus pentoxide as well as in several chemical intermediates. In turn, about a quarter of the produced P_2O_5 is transformed into high purity phosphoric acid to obtain phosphates. Superphosphoric acid is a blend of orthophosphoric acid and polyphosphoric acid. Polyphosphoric acid is composed of linear polyphosphate species, including pyrophosphate, tripolyphosphate, tetrapolyphosphate, and longer chains.

TABLE 9.1 Some Major Phosphorus Compounds for the Industry*

Compound	Chemical Formulae	Application
White phosphorus	P_4	Used to produce P_2O_5, PCl_3, P_2S_5, phosphoric acid, red phosphorus, etc.
Phosphorus pentoxide	P_2O_5	Condensation reagent for organic synthesis in agrochemical and pharmaceutical industries.
Phosphoric acid	H_3PO_4	Used to produce phosphates and phosphonates.
Phosphites	HPO_3^{-2}	Fungicide, fertilizer.
Hypophosphites	HPO_2^{-3}	Plating processes.

TABLE 9.1 *(Continued)*

Compound	Chemical Formulae	Application
Phosphorus trichloride	PCl_3	Pesticide intermediates, flame retardant, antioxidants, and stabilizers, paint additives, lubrication oils, pharmaceuticals.
Phosphorus pentasulfide	P_2S_5	Lubricating oil and grease additives, insecticides, ore flotation.
Red phosphorus		Flame retardants, pyrotechnics, matches, phosphites manufacture.
Phosphides	AlP, Zn_3P_2 Fe_2P	Rodenticides, fumigants. Used to produce phosphine. Nanoelectronics.
Phosphine	PH_3	Fumigant, dopant for semiconductors, catalyst to improve fuel efficiency (aircrafts)..
Phosphonates	CH_3PO_3 $CH_3CH_2PO_3$ $C_3H_8NO_5P$	Herbicides, insecticides, antibiotics, flame extinguishers, corrosion inhibitors, enzymes, chemical additives.

*Main uses of phosphorus compounds.

Adapted from Morton and Edwards (2005).

White phosphorus is also required for the production of a limited number of chemicals intermediates (products used in the manufacturing processes for other chemical products), amounting to about 15% of the P_4 compound usage. The obtained phosphorus derivatives are released to consumers through various products including fertilizers, fungicide, insecticide, herbicides, rodenticides, fumigants, or flame retardants (Table 9.1)

More than 95% of the phosphate rock mined in the United States was used to manufacture phosphoric and superphosphoric acids, which were used as intermediate feedstocks in the manufacture of granular and liquid ammonium phosphate fertilizers and animal feed supplements. Thus, concentrated H_3PO_4 is one of the major acids of the chemical industry and is manufactured on the multi-million tons scale for the production of fertilizers, detergents, pharmaceutical products, or food industry applications. Approximately 50% of the wet-process phosphoric acid produced was exported in the form of upgraded granular diammonium and monoammonium phosphate fertilizer, and merchant-grade phosphoric acid (Jasinski,

2018). For years phosphites (HPO_3^{-2}) have been widely marketed as fungicides. More recently the foliar application of phosphites as more efficient than phosphates as plant nutrients has been promoted. Hypophosphite (HPO_2^{-3}) has been used in electrodeless plating processes such as those used in compact-disk manufacture (Morton and Edwards, 2005). Metallic phosphides are obtained from calcium phosphate (which has been previously dissociated from phosphate rock) by reducing with carbon according to the chemical reaction: $Ca_3(PO_4)_2 + 8C \rightarrow P_2Ca_3 + 8CO$.

9.1.1 *White Phosphorus Manufacture*

There are various allotropic forms of phosphorus (see Section 4.2) but only two forms have commercial significance - white and red phosphorus - and white phosphorus is the most commercially important, accounting for 99% of demand worldwide. According to the *Chemical Economic Handbook, Phosphorus, and Phosphorus Chemicals* (IHS February, 2017), in 2016 elemental phosphorus was mainly manufactured in China, Kazakhstan, the US, Vietnam, and India, with China having by far the largest share at about 87%.

Elemental phosphorus production is by an electrothermal process, and consequently, the energy demand is very high: each ton of white phosphorus produced requires about 14 MWh. Accordingly, its manufacture is carried out mainly where comparatively cheap energy, such as hydroelectric power, is available. As we previously mentioned, the employed raw material is fluorapatite, $3Ca_3(PO_4)_2.CaF_2$, commonly known as 'phosphate rock.' To optimize the production efficiency, phosphate rock is heated strongly in a rotating electrical oven, which results in the fusion of small particles required for satisfactory furnace operation. A modern furnace has a diameter of approximately 12 m and is 8 m high and can produce 30,000 tons of phosphorus a year. The furnace, containing three vertical carbon electrodes, is fed with a mixture of coke, sand, and phosphate rock in a mass ratio of 16:30:100, typically. After decomposition of fluorapatite into CaF_2 and $Ca_3(PO_4)_2$, the calcium phosphate overall reaction may be written as:

$$2\,Ca_3(PO_4)_2\,(s) + 6\,SiO_2\,(s) + 10\,C\,(s) \rightarrow 6\,CaSiO_3\,(s) + P_4\,(g) + 10\,CO\,(g)$$

And it proceeds by a first reaction: 2 $Ca_3(PO_4)_2$ + 6 SiO_2 → 6 $CaSiO_3$ + 2 P_2O_5, yielding phosphorus pentoxide, which is subsequently reduced by the coke: 2 P_2O_5 + 10 C → P_4 + 10 CO. Gaseous phosphorus and carbon monoxide (CO), escaping from the top of the furnace, are passed into a spray of water at 343 K. The CO is either burnt off or recycled as a source of fuel. Most of the phosphorus P_4 molecules condense into white phosphorus (melting point 317 K) using cold water. White phosphorus ignites spontaneously in air, and therefore is stored under warm water.

9.1.2 Compounds Derived From White Phosphorus

9.1.2.1 Red Phosphorus

Red phosphorus, unlike white phosphorus, is not spontaneously flammable, although it is easily ignited. Accordingly, red phosphorus is used in pyrotechnics and matches. It is also used as a flame retardant in plastics (particularly polyamides) where its rapid oxidation consumes all the oxygen present, thereby stopping the fire. It is manufactured from white phosphorus, which is run from its storage tank into a steel pot where it is kept under a layer of water. A lid, fitted with a safety pipe, is securely fastened and the pot is heated to 550 K for 3 to 4 days. The water escapes as steam through the safety pipe, and phosphorus vapor loss is prevented by a reflux condensing system. After 48 hours, the temperature is raised to 673 K when most of the unchanged white phosphorus (boiling point 553 K) distills off. The pot contents are kept wet and ground to give a slurry of red phosphorus. After decanting most of the water, sodium carbonate is added. On boiling, the remaining white phosphorus is destroyed. Finally, the red phosphorus is removed, and vacuum dried.

Red phosphorus is heated with powdered metals to give phosphides (see Figure 9.2) such as aluminum phosphide (AlP), zinc phosphide (Zn_3P_2), and calcium phosphide, (Ca_3P_2). Their principal use is to kill rodents such as rats and as a fumigant for stored cereal grain. The phosphides react with moisture, hydrolyzing to form phosphine, and thus, they have to be handled very carefully.

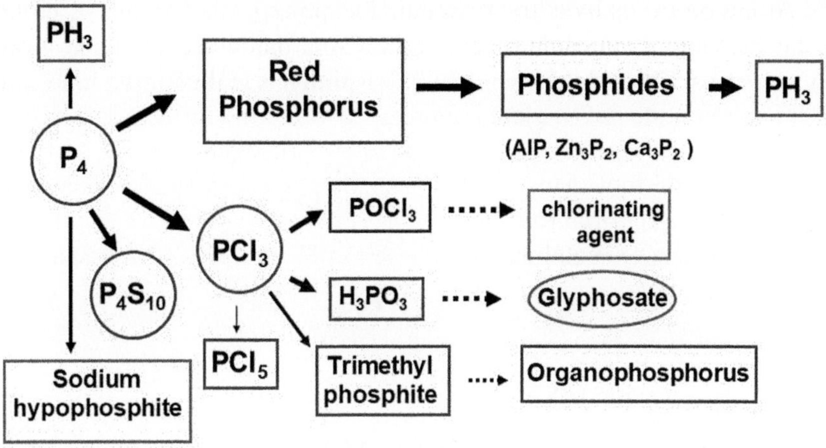

FIGURE 9.2 White phosphorus and its chemical intermediates related industry.

9.1.2.2 Phosphorus Trichloride

Phosphorus trichloride is obtained by the direct reaction between white phosphorus and chlorine: P_4 (s) + 6 Cl_2 (g) → 4 PCl_3 (g). The resulting compound is a colorless liquid (melting point –118.8°C) which boils at 74.22°C. It is highly reactive to atmospheric moisture and is transported in lead-lined, glass-lined or nickel vessels. The total volume of phosphorus trichloride manufactured each year is about 700,000 tons, and it has a number of important applications (see Figure 9.2), which include the production of phosphorus oxychloride ($POCl_3$), used as a chlorinating agent, by direct reaction with oxygen: 2 PCl_3 + O_2 → 2 $POCl_3$. In recent years the most important application of PCl_3 has been the production of the herbicide glyphosate, which requires that it is first converted to phosphorus acid (H_3PO_3). This is made by hydrolyzing phosphorus trichloride according to the reaction: PCl_3 + 3 H_2O → H_3PO_3 + 3 HCl. The phosphorus acid is concentrated in evaporators and the molten product cooled. It solidifies into flakes. Glyphosate is produced by heating a mixture of phosphorus acid with glycine and then adding methanal. Its empirical formula is $C_3H_8NO_5P$, though there are dozens of glyphosate-formulated products in the world herbicide market. It is perhaps the most prominent phosphonate, and it is the world's most widely used and commercially valuable herbicide for crops since 1971.

Other important examples of organophosphorus compounds used to make herbicides, pesticides, and flame retardants are obtained from phosphorus trichloride via trimethyl phosphite, as is illustrated in Figure 9.3.

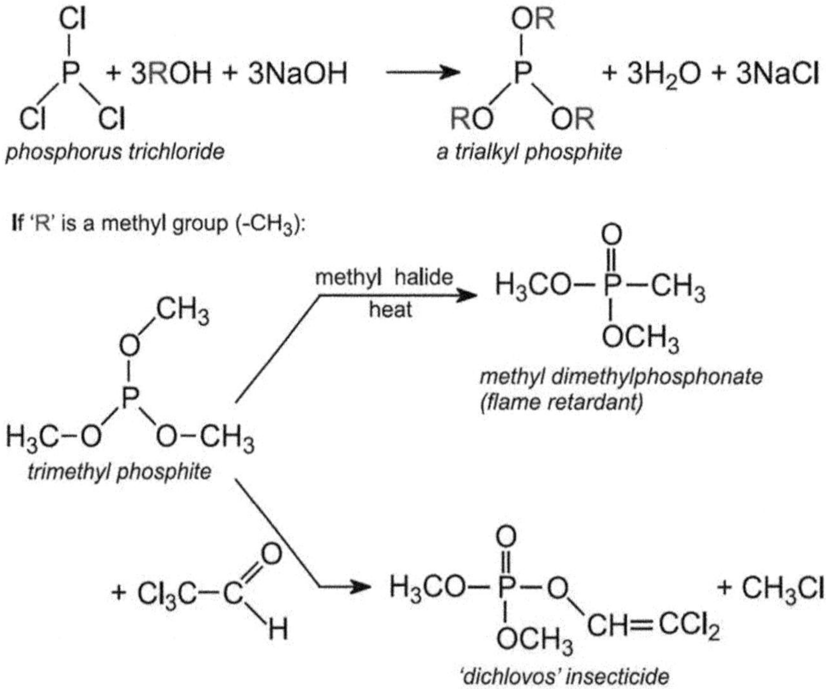

FIGURE 9.3 Synthesis routes of organophosphorus compounds related to phosphorus trichloride.

9.1.2.3 *Phosphorus Sulfides*

White phosphorus reacts with sulfur to form a variety of sulfides, P_4S_{10}, P_4S_7, P_4S_5, or P_4S_3. By far the most important of these is P_4S_{10}, which is normally referred to as phosphorus pentasulfide, P_2S_5. It is used to make lubricating oil additives which act as antioxidants, corrosion inhibitors, and antiwear additives, and as an intermediate in the manufacture of insecticides such as sulfur-based organophosphorus pesticides. There are over 52 registered pesticides for which P_2S_5 is an intermediate. It is reacted with either methanol, ethanol, or isopropanol to produce respectively dimethylphosphorodithioic acid (DMPA), diethylphosphorodithioic acid (DEPA), or diisopropylphosphorodithioic acid (DIPA). These are key intermediates for the organophosphate pesticide industry. Small amounts of phosphorus sesquisulfide, P_4S_3, are used in the match industry.

9.1.2.4 *Sodium Hypophosphite*

Sodium hypophosphite ($NaPO_2H_2$) is the sodium salt of hypophosphorus acid and is often encountered as the monohydrate, $NaPO_2H_2 \cdot H_2O$. Sodium hypophosphite is obtained from white phosphorus and sodium phosphate. To this end, the calcium salt is first made by reacting white phosphorus with a slurry of lime at the boil. Subsequently, the calcium salt reacts with sodium sulfate to form $NaPO_2H_2$. It is a solid at room temperature, appearing as odorless white crystals. It is soluble in water, and easily absorbs moisture from the air. This compound should be kept in a cool, dry place, isolated from oxidizing materials. It decomposes into phosphine, which is irritating to the respiratory tract. Sodium hypophosphite has a number of commercial applications, but the most important (accounting for around 90% of demand) is for electrodeless nickel-plating, in which it acts as a reducing agent. With this method, a durable nickel-phosphorus film can coat objects with irregular surfaces, such as in avionics, aviation, and the petroleum field. A major growth area for this technique has been the manufacturing of hard disc drives for computers.

9.1.2.5 *Phosphine*

About 1,500 tons of phosphine are made each year and are used to make flame proofing derivatives and for pest control. For example, the gas is used in large grain silos, to protect the grain. This compound can be obtained directly from white phosphorus (see Figure 9.2), which is heated under pressure in water, and acidified by phosphoric acid, at about 550 K to yield phosphine through the reaction: $2\ P_4\ (s) + 12\ H_2O\ (l) \rightarrow 5\ PH_3\ (g) + 3\ H_3PO_4\ (aq)$.

9.1.3 Steel Making

Reduced phosphorus compounds play a significant role in steelmaking at various stages of production (Figure 9.4). The process starts with iron ore, which can contain from 0.02 to 1% phosphorus by weight. The ore is heated up to 1,300–1,600°C in a blast furnace to produce the so-called pig iron, and massive quantities of iron slag are produced to rid the ore of excess phosphate. Pig iron end products can contain between 0.05 and 0.5% phosphorus by weight, and it is then used as a feedstock for steel making. In some cases, it is desirable to add ferrophosphorus (a combination of phosphorus with iron in different proportions, see Figure 9.1) in order to:

1. Enhance the mechanical properties of the final steel;
2. Reduce energy costs by lowering the casting temperature;
3. Enhance abrasion resistance, and
4. Improve corrosion resistance (Morton and Edwards, 2005).

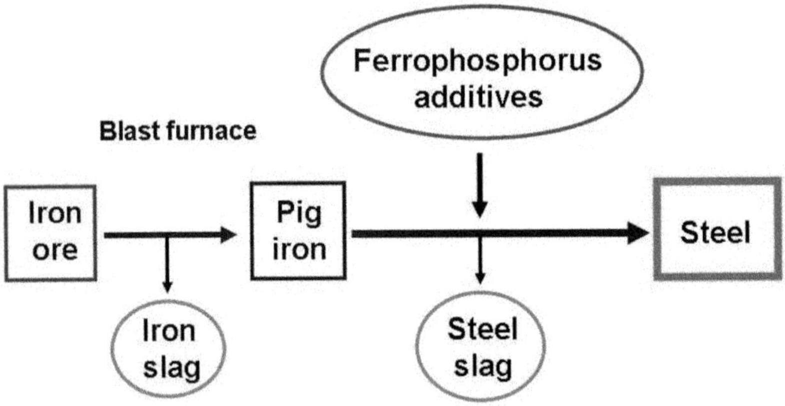

FIGURE 9.4 The role of phosphorus compounds in steel making.

Thus, in low-carbon steels, phosphorus is considered as a useful alloying agent; it increases the tensile and yield strengths, the creep strength, and the corrosion resistance of the steel on moist atmospheres. It also causes the formation of an adherent scale which prevents steel sheets from sticking together when rolled. For the sake of illustration, about 5,070 metric tons of ferrophosphorus was added to steel as an additive in the U.S. in 2001 (Fenton, 2001).

However, a too high phosphorus content (e.g., > 0.5 wt%) often leads to the detrimental formation of a brittle phosphide (the mineral steatite) network (Schipper et al., 2001). Assuming that the representative average phosphorus content for all steel types is 0.04%, about 360,000 metric tons of phosphorus is used annually within the steel. This figure is about 5% of that mobilized for use in fertilizers, and is therefore quantitatively significant.

In addition of its use as a steelmaking additive, Fe_2P crystals have attracted a great technological importance as they show unique magnetic properties, such as large uniaxial anisotropy, magnetocaloric, and magnetoelastic properties, due to which this compound finds wide use in industrial applications. In recent years, it has been also identified as potential electrocatalyst for hydrogen evolution reaction.

9.1.4 Materials for Energy Conversion

Environmental concerns regarding refrigerant fluids as well as the convenience of using nontoxic and nonexpensive materials, have significantly spurred the interest in looking for novel, high-performance thermoelectric materials for energy conversion in small scale power generation and refrigeration devices, including cooling electronic devices, or flat-panel solar thermoelectric generators. This search has been mainly fueled by the introduction of new designs and the synthesis of new materials. In fact, the quest for good thermoelectric materials entails the search for solids simultaneously exhibiting extreme properties. On the one hand, they must have very low thermal conductivity values. On the other hand, they must have both electrical conductivity and Seebeck coefficient high values as well. Since these transport coefficients are not independent among them, but are interrelated, the required task of optimization is a formidable one. (Maciá-Barber, 2015).

Materials for thermoelectric applications are evaluated in terms of their figure-of-merit value, given by the expression $ZT = \frac{\sigma S^2}{\kappa}$, where $\sigma(T)$ is the electrical conductivity, $S(T)$ is the Seebeck coefficient, $\kappa(T)$ is the thermal conductivity, and T is the temperature. As we see, ZT is determined by the thermal conductivity (appearing in the figure-of-merit denominator) and the so-called power factor σS^2, which appears in the figure-of-merit numerator. Thus, large figure-of-merit values require both small thermal conductivity values and large Seebeck coefficient and electrical conductivity values. Therefore, in searching for promising thermoelectric materials, one must focus on materials exhibiting strong couplings between the electrical and thermal currents. Certainly, to discover more efficient thermoelectric materials, having high ZT values, would significantly increase the economical competitiveness of thermoelectric devices, considerably expanding their range of possible applications. Nevertheless, according to the recorded experimental data, a practical barrier appeared in the way to achieve this goal. In the first research period, spanning from the 1940s to the 1990s, the upper limit was set at about $ZT \approx 1$. Subsequently, during the last two decades, the upper limit has been shifted up to $ZT \approx 2$–3 by considering the study of bulk materials with very complex crystalline structures, as well as nanostructured materials.

Due to the toxicity of Pb and Te elements present in currently used thermoelectric materials, the fabrication of these compounds is greatly restricted. This generates a strong motivation for finding suitable "lead-free" alternative materials for next-generation thermoelectric devices (Maciá, 2014). To this end, the potential of black phosphorus (see Section 4.2.3), a layered semiconducting material with a direct bandgap of ~0.3 eV, for potential thermoelectric applications has been recently studied experimentally. Relatively large Seebeck coefficient values, ranging from $S = +335\ \mu VK^{-1}$ at room temperature to $+415\ \mu VK^{-1}$ at $T = 385$ K, have been reported (Flores et al., 2015). In addition, the electrical conductivity of this material steadily increases as the temperature increases, so that the power factor attains 2.7 times the room temperature value at moderately high working temperatures. Black phosphorus is also emerging as a promising semiconductor with a moderate band gap for nanoelectronics and nanophotonics applications, as well as gas sensing devices (Ling et al., 2015).

On the other hand, a prospective study of palladium phosphide sulfide (PdPS), a layered material with an orthorhombic crystal structure, has been recently performed by considering band-structure based numerical calculations (Kaur et al., 2017). The obtained results indicate that this material could exhibit a Seebeck coefficient of $S = 300\ \mu VK^{-1}$ at room temperature, and could reach values as high as $S = 800\ \mu VK^{-1}$ at 800 K, for both p-type and n-type electrical conduction. These high Seebeck coefficient values are accompanied by very low thermal conductivity values close to $0.1\ Wm^{-1}K^{-1}$ at room temperature, hence leading to a remarkably high thermoelectric figure of merit value.

The concept of lithium-ion batteries was first introduced into the market by Sony in 1992. Most of the research and commercial cathode materials for rechargeable lithium-ion batteries are compounds with layered, spinel or olivine type structures, where Li^+ ions are inserted. Advancement in the electrochemical energy storage technology has seen the development of many important lithium-ion battery electrode materials, exemplified by the graphite anode and lithium iron phosphate $LiFePO_4$ cathode (Figure 9.5), which has recently attracted much attention as promising cathode material for lithium-ion batteries, due to its environmental benigness and potentially low cost, instead of expensive $LiCoO_2$ generally used compound. The overwhelming advantage of iron-based compounds is that, in addition to be inexpensive and naturally abundant, they are less toxic than Co, Ni, and Mn. Its commercial use has already started, and there are several companies that base their business on lithium phosphate technology. Still, there is a need for a manufacturing process that produces electrochemically active $LiFePO_4$ at low enough cost (Jugovic and Uskokovic, 2009).

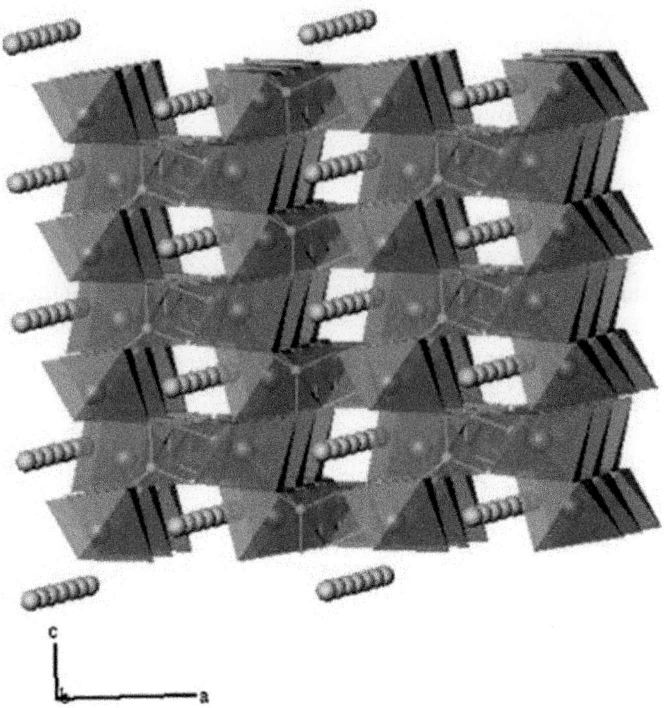

FIGURE 9.5 The olivine structure of $LiFePO_4$ showing the FeO_6 octahedra, PO_4 tetrahedra, and the one-dimensional tunnels in which the lithium ions reside.
Reprinted from: Jugovic and Uskokovic, (2009); Copyright (2009), with permission from Elsevier.

The practical application of Cs-containing phosphates in optics, electronics, catalysis, and other modern technologies is quite significant. Thus, double phosphates with the general formula $CsMPO_4$ (where M stands for a divalent metal in tetrahedral coordination geometry) have been investigated as ferroelectric materials. A number of Cs-bearing phosphates with high thermal stability are considered as suitable matrices for nuclear waste storage. Moreover, these materials can provide further commercial and medical applications as ^{137}Cs γ-radiation sources. In addition, Cs-containing phosphates have been intensively studied in recent decades as prospective hosts of luminescent ions for LEDs, for example, $CsMgPO_4$ doped with Eu and the pyrophosphate $Cs_2MP_2O_7$ doped with La (Zatovsky et al., 2018).

9.2 ADVANCES IN NANOTECHNOLOGY

9.2.1 Phosphorene

Triggered by the discovery of graphene, a single-layer material composed of carbon atoms arranged according to a honeycomb structure, the family of two-dimensional crystals has grown considerably, today encompassing a rich variety offering almost all desirable electronic properties that are required for nanoelectronics, ranging from insulators, such as hexagonal boron nitride, to direct bandgap semiconductors belonging to the family of layered transition metal dichalcogenides, such as MoS_2, $MoSe_2$, WS_2, and WSe_2, and ending with the so-called Dirac semimetals: hexagonal monolayers of group 14 elements C, Si, Ge, and Sn, referred to as graphene, silicene, germanene, and stanene, respectively. Of particular interest to us is the so-called black phosphorene, namely, a single-layer material obtained by exfoliation of bulk black phosphorus (see Section 4.2.3), which can bridge the energy gap between that of graphene and that measured for transition metal dichalcogenides two-dimensional materials. Phosphorene is considered as a novel candidate for high performance electronic and photonic devices, owing to its remarkable properties such as high carrier mobility (up to 50,000 $cm^2V^{-1}s^{-1}$) and a tunable bandgap. Indeed, the electronic structure of black phosphorus is characterized by the presence of a semiconducting gap whose width is tunable with the number of layers, ranging from 0.3 eV (4140 nm, mid-IR) for bulk material to 1.3 eV

(955 nm, near-IR) for a phosphorene monolayer. Therefore, efforts in extending phosphorene gap to the high energy end, especially to the visible regime, are an attracting direction to pursue. Thus, red photoluminescence centered at 605 nm has recently been reported from few-layer (~20 nm thick) black phosphorus samples produced by a thermal annealing process, a commonly used method to produce thin layers of 2D materials from thick flakes.

The black phosphorus band gap can also be controlled by in-plane strain. Indeed, the puckered honeycomb lattice structure of phosphorene (see Figure 4.8) leads to in-plane anisotropic properties, including thermal conductivity and mechanical properties. Thus, tensile strain enhances electron transport along the zigzag direction, while a biaxial strain is able to tune the optical bandgap from 0.38 eV (at 0.8% strain) to 2.07 eV (at 5.5% strain). These properties allow for a new degree of freedom in designing novel black phosphorus-based devices with appealing features to be considered for nanoelectronic applications such as field-effect transistors, photodetectors with a very broad spectral range from the visible to infrared region, gas sensors, high-performance lithium-ion batteries and thermoelectric devices (Cho et al., 2017).

Black phosphorus crystals have strong in-plane bonds, and the weak Van der Waals interlayer interaction enables the exfoliation into few-layer black phosphorus sheets or phosphorene (single-layer black phosphorus). Scotch tape-based mechanical cleavage has been demonstrated to reliably produce pristine, ultrathin sheets of both graphene and phosphorene, or other graphene analogs. However, it is not a scalable process. To avoid this problem, few-layer black phosphorus is typically fabricated via liquid exfoliation from bulk black phosphorus, which depends on the energy match between the solvent and the surface of two-dimensional material for exfoliation. In liquid exfoliation, the layered bulk solid is immersed into a liquid and then ultrasonically exfoliated. In this way, ultrathin sheets with a diameter of about several hundred nanometers and a height of about 2.0 nm have been prepared. In doing so, the saturated NaOH N-methyl-2-pyrrolidone solution has presented efficient black phosphorus exfoliation rates. Phytic acid (see Section 9.3.1) with abundant polar phosphorus and hydroxyl functional groups can also assist bulk black phosphorus in exfoliating it into ultrathin nanosheets. The exfoliation effect of different solvent types on bulk black phosphorus has been systematically investigated, and it is proved that polar solvents are more favorable for the fabrication of ultrathin nanosheets. However, considering the possible medical uses (see Section 9.3.2), oxygen-free water is also used as the solvent for the exfoliation of bulk black phosphorus (Yang et al., 2018).

Following these experimental realizations, other possible phosphorus-based layered structures have been investigated. Phosphorene is predicted to have many allotropes that are either semiconducting or metallic depending on their 2D atomic structure. Among them, the so-called *blue-phosphorene*, arranged in a more flat configuration than black phosphorene, stands out as the most stable one, although having only a few meV higher cohesive energy than that of black-phosphorene. Blue phosphorene has recently been grown on a gold substrate by molecular-beam epitaxy (Xu et al., 2017). It also has a semiconducting direct bandgap whose width varies with their shape. According to numerical calculations, blue phosphorene is a potential candidate for thermoelectric applications with a ZT close to 1.2 at room temperature (Sevik and Sevincli, 2016).

The most commonly used solar cells are based on silicon and the so-called III-V semiconductors, such as GaAs, InAs, and InP, all of them having bandgaps below 1.5 eV. Combining a group 14 element (C, Si, Ge, Sn) with P can be expected to result in a material with a convenient band gap width, and recently it has been theoretically proposed a new 2D structure, phosphorus carbide, to this end (Guan et al., 2016). However, as no corresponding layered bulk phase of CP is known in Nature, the experimental realization of this material by direct exfoliation is not available today. In this regard, it is interesting to recall that CP radical is relatively abundant in circumstellar regions around evolved stars (see Section 6.2.2), so that it could be appealing to search for the possible existence of extended networks of hexagonal monolayers (analog to PAHs) made up of C and P atoms in the ISM.

On the other hand, a layered material composed of P and Ge, with stoichiometry GeP_3, was already reported in the 1970s, and its synthesis is well-known (Figure 9.6). By comparing with the phosphorene structure shown in Figure 4.8, we see that the replacement of every fourth P atom by Ge introduces an additional valency, which reinforces interlayer interactions in the bulk phase leading to an arsenic-type honeycomb structure, and is also responsible for its metallic character.

Studies regarding the cleavage energy indicate that exfoliation to a monolayer from the bulk material could be a reachable goal (Jing et al., 2017). The resulting 2D crystalline layer would exhibit an indirect bandgap of 0.55 eV and high carrier mobilities similar to those of phosphorene.

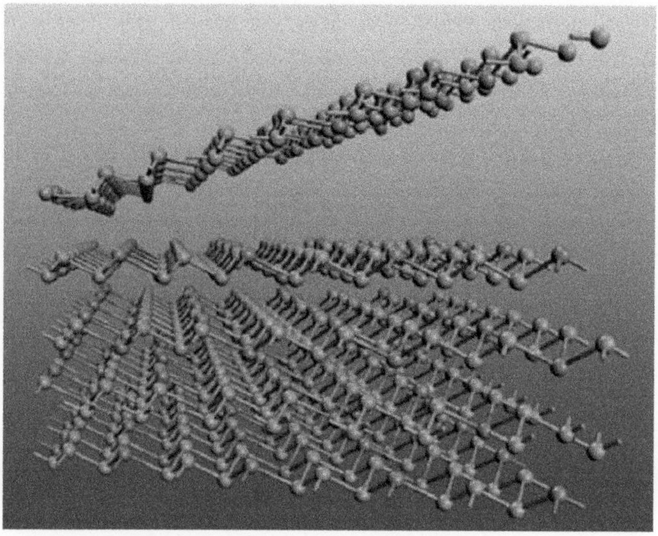

FIGURE 9.6 (See color insert.) Atomic structure of GeP_3.
Reprinted with permission from Jing, Ma, Li, and Heine, (2017); Copyright (2017) American Chemical Society.

In addition to 2D arrangements of phosphorus, one may also explore one-dimensional ones. In this way, it has been found that three different polymers, $(CuI)_2P_{14}$, $(CuI)_3P_{12}$ and $(CuI)_8P_{12}$, can be obtained by polymerization of elemental phosphorus in a CuI matrix (Figure 9.7). The former two were isolated after removal of CuI by an aqueous solution, yielding the nano-rod like polymer P_4–P_8 shown in Figure 9.7a, and the helical strand P_2–P_{10} shown in Figure 9.7b, which can properly be regarded as new allotropes of phosphorus (Scheer et al., 2010).

9.2.2 Computers at the Nanoscale

It has been demonstrated in recent years that digital information can be stored in biological and synthetic macromolecules. In these so-called digital polymers, the monomer units that constitute the chains are used as molecular bits and assembled through controlled synthesis into readable digital sequences. For example, it has been reported that suitably ordered oligonucleotide sequences enable storage of several kilobytes of data in DNA chains. Alternatively, it has been demonstrated that binary messages can also efficiently be stored in different types of synthetic macromolecules. In this way, polymers might be used to store data on a scale that is around 100 times smaller than that of current hard drives.

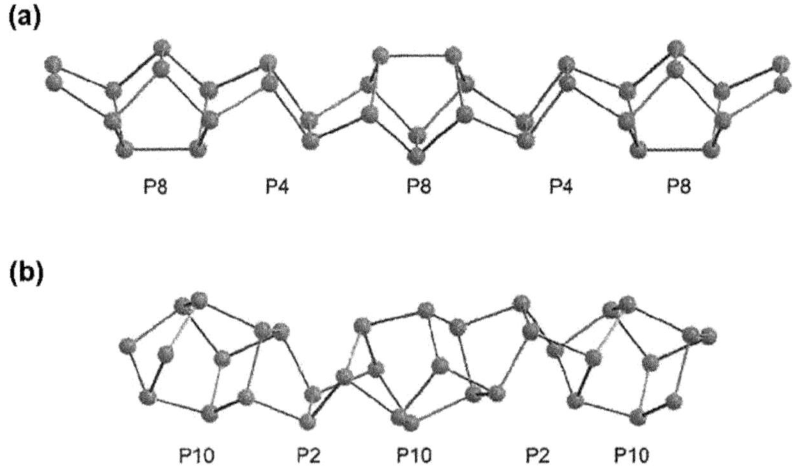

FIGURE 9.7 New phosphorus allotropes based on $(CuI)_nP_m$ compounds. They consist of regular sequences of phosphorus atoms cages arranged along linear tubes.
Adapted with permission from Scheer, Balázs, and Seitz, (2010); Copyright (2010) American Chemical Society.

Along this vein, it has been recently reported that mass spectroscopy sequencing of long-coded polymer chains can be achieved through careful macromolecular design (Al Ouahabi et al., 2017). To this end, poly-phosphodiester molecules made up of two kinds of monomers containing phosphate groups, which correspond to 0s and 1s, respectively, were synthesized and sequenced (Figure 9.8). In particular, a monomer containing a propyl phosphate group represents a binary 0 and a 2,2-dimethyl phosphate a binary 1 synthetic building block (or synthon). After every eight of these monomer "bits" (which themselves make up a byte), a molecular separator in the form of a weak alkoxyamine group was added, to obtain poly-phosphodiester chains containing several bytes of information. This technique has succeeded in reading a 78-element polymer chain containing 64 bits, seven tags, and seven spacers, hence illustrating the use of biological molecules such as DNA as a data-storage medium.

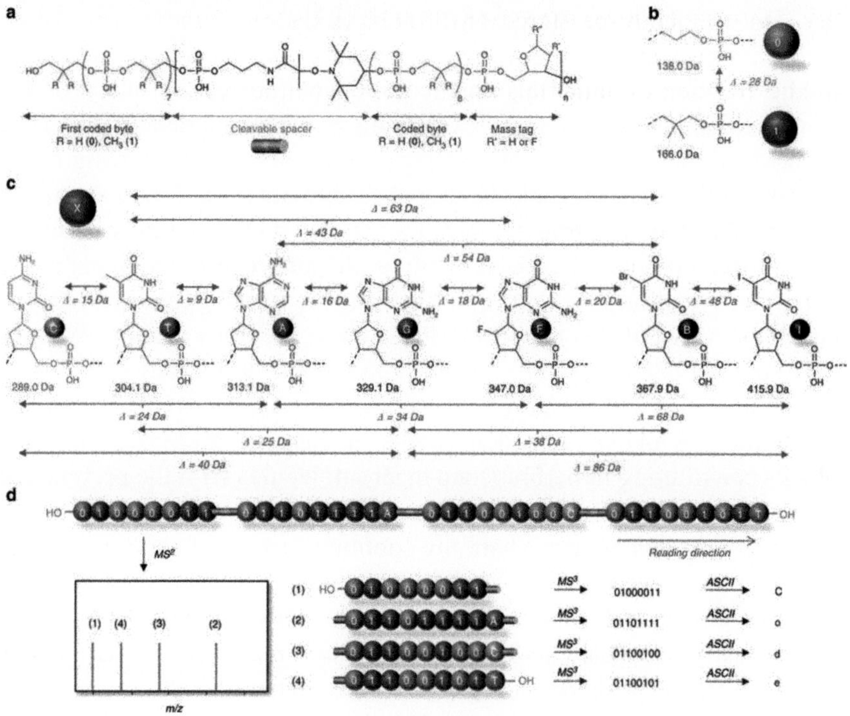

FIGURE 9.8 (See color insert.) General concept for the sequencing of long digital polymer chains. (a) Molecular structure of a digital polymer contains n+1 coded bytes (in red; a byte is a sequence of eight coded monomers that represent 8 bits). Two consecutive bytes are separated by a linker noted in black, which contains an N-O-C bond that can be preferentially cleaved during mass spectroscopy analysis. In order to sort out the bytes after cleavage, n bytes of the sequence are labeled with a mass tag noted in blue. (b) Molecular structure and mass of the two coded synthons that define the binary code in the polymers. (c) Molecular structure and mass of the tags that are used as bytes labels. In order to induce identifiable mass shifts after cleavage, the mass of a byte tag (noted in blue) shall not be a multiple of 28, which is the mass difference between a 0 and a 1 coded unit. In addition, the mass difference between two tags (noted in grey) shall not be a multiple of 28. (d) Schematic representation of the mass spectrometry sequencing of a digital polymer containing 4 bytes of information.
Source: Al Ouahabi, Amalian, Charles and Lutz, (2017); Creative Commons CC BY 4.0 license.

In addition to DNA other phosphorus bearing molecules of biological interest have been proposed to be used in nanoscale computers technology. Thus, ATP, the molecule that provides energy to all living beings in our biosphere (see Section 8.2), may also be able to power the next generation of supercomputers (Nicolau et al., 2016).

9.3 MEDICAL AND PHARMACEUTICAL USES

In the frontier of materials science and healthcare, the study of the so-called nanomaterials, that is, materials exhibiting a nanometric scale in one of their dimensions at least, has been receiving a growing attention during the last decade. Within the nanomaterials broad family in this section, we will briefly consider several representatives based on phosphorus compounds. These include phosphorene monolayers along with relatively thin black phosphorus sheets containing just a few monolayers each, which we have described in the previous section. We will also comment on relatively simple organophosphorus molecules, such as phytic acid or larger microsized bulk samples, such as hydroxyapatite crystallites or phosphorus dendrimers, a class of molecular building blocks constituting hyperbranched macromolecules with the presence of phosphorus element at the core (Caminade et al., 2018). All materials considered in this Section share the common feature of exhibiting some useful property for medical applications. Indeed, since phosphorus is a main biolement in the human body, the metabolism of phosphorus atoms can be easily controlled, and phosphorus-based materials are particularly prone to interact with the biological systems, generally exhibiting a good biodegradability, free of undesirable immune responses and toxicity (Bakshi, 2017).

9.3.1 Phytic Acid

Myo-inositol hexaphosphate (IP6), also called phytic acid, or phytate in its salt form, is the major phosphate store in plant seeds, and it is also present in all animal organs and tissues (Figure 9.9a). Different biological functions have been described for this compound. It acts as a potent crystallization inhibitor of calcium salts (i.e., prevents calcification), but it also has been described as an antioxidant (inhibitor of hydroxyl radical formation) and even as an anti-cancer agent. Quite remarkably, phytic acid is readily adsorbed on the surface of hydroxyapatite, the mineral constituent of bones, where it inhibits its dissolution, thereby decreasing the progressive loss of bone mass leading to osteoporosis. In addition, a positive action of phytic acid inhibiting osteoclastogenesis (the bone remodeling process) at the cellular level, without impairing the differentiation of cells to osteoblasts (the bone-forming cells) has also been described.

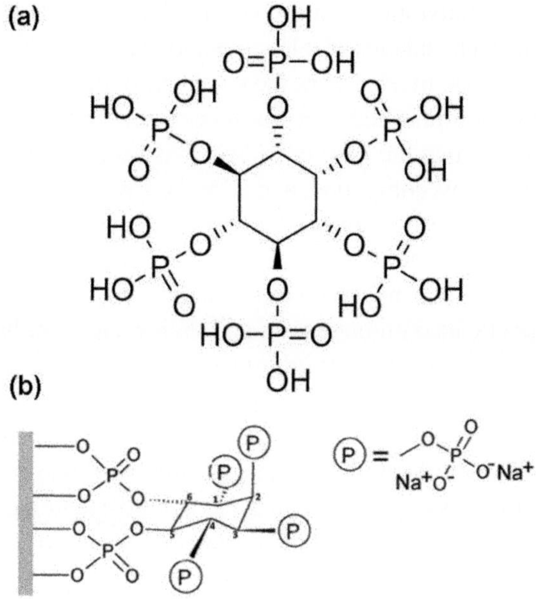

FIGURE 9.9 (a) Phytic acid molecular structure, exhibiting a remarkable hexagonal symmetry. (b) Illustration of a method to directly functionalize Ti surfaces covalently with IP6, without using a cross-linker molecule, through the reaction of the phosphate groups of IP6 with the TiO_2 layer usually present on Ti substrates.

Besides its described effects on bone, polyphosphates have shown interesting antimicrobial effects, inhibiting the growth of several Gram-positive and Gram-negative bacteria, including oral bacteria. Their antibacterial effects have been related to the ability of polyphosphates to chelate divalent cations, contributing to cell division inhibition and loss of cell wall integrity. In fact, phytic acid has been reported to have synergetic antibacterial effects in the presence of sodium chloride in *Escherichia coli*.

It is known that phosphates and phosphonates can covalently link to metal oxide surfaces, like TiO_2. In fact, alkyl phosphonic acids are commonly used as cross-linker agents to functionalize surfaces with other molecules to tune the surface properties to those of interest. Phosphonate linkers present the advantage of being more stable than other commonly used coupling agents like silanes, which suffer from hydrolytic instability in aqueous environments at physiological pHs. The use of robust and stable coatings under physiological conditions in biomedical applications is of high interest, and phosphates or phosphonates bound to metal oxides are stable in these conditions. Recently, the synthesis of biomaterials containing phytic acid has been reported. For instance, Chen and co-workers (2014) developed a biocompatible phytic acid-magnesium coating, based on the chelating ability of this compound, and Li and co-workers (2015) optimized this coating with a composite of phytic acid-magnesium and bioglass, and found that the coating decreased the degradation rate of a magnesium alloy and could be used for temporary biodegradable implants. Given the positive effect of phytic acid on bone cells and their related antibacterial effect, Cordoba, and co-workers (2016) introduced a method to directly functionalize Ti surfaces covalently with phytic acid, through the reaction of the phosphate groups of phytic acid with the TiO_2 layer usually present on Ti substrates (Figure 9.9b). In this way, since this phosphonate possesses six phosphate groups, this compound can act as a multifunctional bone implant. The grafting reaction consisted of an immersion in a phytic acid solution to allow the physisorption of the molecules onto the substrate, followed by a heating step to obtain its chemisorption. The reaction was highly dependent on the solution pH, only achieving a covalent Ti–O–P bond at pH = 0.

9.3.2 Nano-Black Phosphorus

Nanomaterials based on black phosphorus, including both phosphorene monolayers and very thin black phosphorus sheets containing just a few monolayers each, have attracted enormous attention in biomedical applications owing to a number of beneficial properties. For instance, the corrugated plane configuration of black phosphorus nanosheets (see Figure 4.8) endows them with relatively large surface area, which is beneficial for drug loading. They can also dissociate oxygen molecules, preventing oxidation stress in cells and tissues. On the other hand, black phosphorus nanoparticles exhibit a high thermal conversion efficiency upon proper illumination, and they are utilized as photothermal therapy agents for cancer treatment, showing excellent therapeutic effects and no observable toxicity at a cellular level.

Indeed, nanomaterial-based phototherapy has emerged as a research hotspot due to its unique advantages, such as high therapeutic efficiency, minimal invasiveness, good tumor targeting, few side effects, and low systemic toxicity. Photothermal and photodynamic therapies are two kinds of phototherapy with different therapy mechanisms. The former one is a tumor-localized hyperthermia treatment, which utilizes the photo-absorbing agents in order to efficiently convert light energy into heat, leading to cell death. The later one relies on black phosphorus nanoparticles to generate reactive oxygen chemical species by the absorbed light to induce malignant cells death.

9.3.3 Hydroxyapatite Nanocrystals

Understanding how certain living organisms form the extremely specialized mineralized structures they contain, and identifying the organic molecules controlling the final crystal size, shape, and spatial arrangements determining their unique mechanical properties, is highly relevant, not only on the fundamental knowledge side, but also as a source of inspiration for the design of advanced biomaterials. In fact, bio-inspiration is currently among the most exciting concepts ruling advancements in materials science, medicine, and other technological applications. Synthetic calcium phosphates are among the most interesting and versatile biomimetic materials, since they chemically resemble the inorganic phase present in hard tissues (e.g., bone, dentine, fish scales, or horns of different animals) as well as pathological calcifications (e.g., dental, and urinary calculi, tendon mineralization, calcification of blood vessels). Most of the excellent features of these synthetic phosphates, including biocompatibility and bioactivity, can be significantly enhanced by improving their biomimetism, that is, by mimicking the size, morphology, nanostructure, and chemical composition of their biological counterparts. Therefore, the synthesis of calcium phosphates under physiological conditions, the so-called *biomimetic growth*, mediated by organic additives resembling the small molecules or large macromolecules of the organic matrix of bones and teeth, is a matter of intensive research nowadays (Iafisco and Delgado-López, 2018).

Along with biomimetic materials the development of new *bioactive* materials, able to bond to living tissues, is a promising alternative for the production of implants and scaffolds for tissue engineering, as we mentioned in the previous section. Many bioactive materials are prepared by a sol-gel method, including glasses and organic-inorganic hybrids. When in contact with physiological fluids, they form carbonated hydroxyapatite nanocrystals similar in composition and structure to biological apatites. This is an essential stage in the formation of a bond between the bioactive glasses and the living tissues. Furthermore, in past years, bioactive organic-inorganic hybrids with mechanical behavior closer to that of bones have been synthesized in order to expand the clinical applications of phosphorus-bearing glasses. To ascertain the role of calcium in the in vitro bioactivity of glasses and hybrids, and why the presence of phosphorus in sol-gel glasses first slows down their initial reactivity, but then speeds up the carbonated hydroxyapatite formation rate, Vallet-Regí and co-workers (2005) studied the nanostructure of three bioactive materials: two sol-gel glasses in the SiO_2–CaO and SiO_2–CaO–P_2O_5 systems and a SiO_2–CaO–poly(dimethylsiloxane) organic-inorganic hybrid. The nanostructural characterization indicates that the addition of P_2O_5 to the glass leads to the crystallization of a silicon-doped calcium phosphate, while in the materials without any phosphorus content binary glass and hybrid calcium is located in an amorphous silica network. Therefore, common features of the bioactive behavior of glasses and organic-inorganic hybrids are that inclusion of calcium into the SiO_2 network increases the carbonated hydroxyapatite formation rate and addition of a third component, namely, phosphorus pentoxide in glasses and SiO2–CaO–poly(dimethylsiloxane) in hybrids, hinders incorporation of calcium in the material and consequently its bioactivity.

Calcium pyrophosphate hydrates ($Ca_2P_2O_7{\cdot}nH_2O$, see Figure 4.14) are, together with calcium orthophosphates (see Section 4.5), the most common calcium salt crystals found in pathological osteoarticular cartilage and menisci (MacMullan et al., 2011). Calcium pyrophosphate crystals deposition disease occurs when these crystals form deposits in the joint and surrounding tissues. The crystal deposits provoke inflammation in the joints, which can cause the joint cartilage to break down. The disease may take a few different arthritis-related forms: osteoarthritis, a chronic rheumatoid arthritis, inflammatory arthritis, or an acutely painful inflammatory condition called pseudo-gout. In most cases, the cause of calcium pyrophosphate crystals formation is unknown, although deposits generally increase as people get older. Thus, almost half of people over 85 have the crystals, though fortunately enough many of them do not have symptoms.

9.4 SUMMARY AND REVIEW QUESTIONS

By inspecting Figures 9.1–9.4, we see that the main phosphorus compounds used in industrial activities are phosphate salts and the related phosphoric acid moiety, followed by elemental white and red phosphorus, phosphorus pentoxide, ferrophosphorus additives, and phosphine. The use of these phosphorus compounds covers four main industrial sectors, namely, agriculture (fertilizers, fungicides, pesticides, fumigants, and herbicides), siderurgy (steel making process) and chemical and pharmaceuticals industries, where a relatively broad number of P-bearing chemical intermediates are used. Along with these classical sectors, a number of emerging technological niches have been identified in the last two decades, especially in the areas of energy conversion (thermoelectric materials, ion batteries, solar cells, ferroelectrics, LEDs technology), and the rapidly evolving arena of nanotechnology. In this research area, new phosphorus allotropes, such as linear cage arrangements (see Figure 9.7), or phosphorene and GeP_3 flat layers (see Figure 9.6) appear as promising novel materials (Han et al., 2018). Particularly remarkable new applications have recently been reported in computing technology using digital polymers based on tagged DNA-like nucleotides, in biomaterials research by exploiting the design of properly functionalized hydroxyapatite nanocrystals, or in the development of phosphorus nanoparticles for cancer phototherapy strategies.

I would like to conclude this chapter by considering the possible influence of this growing phosphorus-based industrial activity on the circulation of phosphorus molecules through the atmosphere, hydrosphere, and biosphere of our planet in the years to come. In fact, it is currently understood that human activities have drastically accelerated earth's major biogeological cycles, altering the balance of nitrogen and phosphorus cycles in particular. Recent studies indicate that enhanced nitrogen deposition increases the natural limitation of phosphorus or other nutrients in many ecosystems, leading to an anthropogenic eutrophication which has a large impact on specific ecosystems, particularly when freshwater ecosystems receive phosphorus leached from land along with anthropogenic phosphorus coming from fertilizers, pesticides, and detergents domestic use, as well as industrial sewage. Accordingly, a faster accumulation of P than N atoms takes place in human-impacted freshwater ecosystems, which could have large effects in the trophic webs and biogeochemical cycles of estuaries and coastal areas (Yan et al., 2016). Some reflection on this state of affairs must be encouraged when thinking about the sustainable growth plans to be adopted by world leaders in order to preserve a healthy earth for the next generations of human beings.

KEYWORDS

- ATP powered computers
- bio-inspired materials
- biomaterials
- blue phosphorene
- cesium-containing phosphates
- ferrophosphorus additives
- glyphosate
- lithium-iron batteries
- mass spectroscopy sequencing
- nanophotonics
- organophosphorus compounds
- phosphate fertilizers
- phosphate rock
- phosphides
- phosphine
- phosphorene
- phosphorus sulfides
- phosphorus trichloride
- photodynamic therapies
- sodium hypophosphite
- steel making
- thermoelectric materials

CHAPTER 10

Outlook and Perspectives

"The probability of formation of a highly complex structure from its elements is increased, or the number of possible ways of doing it diminished, if the structure in question can be broken down into a finite series of successively inclusive sub-structures."

(John D. Bernal, 1960)

10.1 THE THREE MAIN QUERIES

In the initial chapters of this book, we introduced the phosphorus enigma in the following terms: the relative abundance of phosphorus measured in diverse astrophysical objects indicates that this element is, at the same time, scarce but ubiquitous throughout the universe. The scarcity of phosphorus at a cosmic scale is illustrated in Figure 1.1, where we see that this element ranks the 18th position attending to its elemental abundance value. In fact, the relative abundance of phosphorus as compared to other reference compounds takes on systematically low values in the different astrophysical objects where this element has been detected. For instance, the average phosphorus content in IDPs is of the order of 0.3% in weight, and it ranges within 0.1–0.4% in different meteorite classes, with the exception of mesosiderites, where we find samples containing up to 1.4% phosphorus in weight. The presence of phosphine in the external atmospheres of Jupiter and Saturn exhibits volume mixing ratios of about 0.6 and 0.7 ppm, respectively, whereas this molecule has been detected with an abundance of 10^{-8} relative to H_2 in the circumstellar shell around the carbon star IRC+10216. Similar figures have been reported for other P-bearing molecules, such as PO and PN around several evolved giant stars.

At the same time, we realize that albeit its paucity phosphorus is widespread all over the universe, as it is summarized in Figure 1.6. Thus, phosphorus ions have been observed in the photospheres of stars belonging to different spectral classes and evolutive stages, including main sequence, giants, supergiants, Wolf-Rayet, and white dwarf stars. Also, phosphorus bearing compounds have been reported in different astronomical sites, where they exhibit a wide variety of oxidation states, ranging from the highly oxidized phosphate minerals found on earth, Moon, and Mars crusts, stony meteorites, and IDPs of probable cometary origin, going through the moderately reduced schreibersite mineral $(Fe,Ni)_3P$, found in iron meteorites and lunar regolith, to the highly reduced molecule PH_3, which has been detected in the atmospheres of Jupiter and Saturn as well as in a protoplanetary nebula beyond the solar system (see Table 1.2). Furthermore, compounds where phosphorus exhibits intermediate oxidation states have also been observed, such as PO and PN molecules detected in a number of star-forming regions, or organophosphorus compounds containing the C-P bond, as it occurs in alkyl phosphonic acids detected in carbonaceous meteorites or the CP, CCP, HCP, and NCCP molecules present in several circumstellar regions.

In contrast to the apparent scarcity of phosphorus, and the wide range of oxidation states among P-bearing compounds throughout the universe, as indicated by astrophysical observations, phosphorus moieties exist in high abundances in living systems, mainly as phosphates, where they perform many fundamental biochemical tasks (see Table 1.1). Accordingly, the elemental abundance of phosphorus ranks at the 5th (6th) position in the chemical inventory of unicellular (pluricellular) organisms, respectively. This fact naturally raises the question regarding the way such a relatively scanty element in the chemical inventory of the universe as a whole has become so relevant for all living beings on earth during the entire history of life evolution in our planet. In particular, how can we explain the crucial role of phosphates in both structural (sugar-phosphate backbone of nucleic acids, phospholipids in membranes) and energy transfer molecules (ATP) present in most metabolic routes in all present-day living forms? A question which is closely related to that concerning the primary source of phosphorus, and specially phosphates, during the initial stages of prebiotic evolution.

Therefore, the phosphorus enigma includes three main aspects concerning:

1. The very formation of phosphorus nuclei in stellar cores;
2. The peculiar chemistry of P atoms in their way from stellar atmospheres and outer envelopes around stars, through the gas and solid phases of the ISM, ending with their final incorporation in small planetesimal bodies in newly forming planetary systems; and
3. The phosphorylation problem is related to the extremely low solubility of phosphate minerals and the low efficiency of condensation reactions in liquid water during the early prebiotic stages on earth.

For the sake of conciseness, hereafter I will refer to these three main queries as the phosphorus nucleosynthesis problem, the phosphorus chemistry puzzle, and the prebiotic phosphate conundrum, respectively.

10.2 SOME PRELIMINARY ANSWERS

10.2.1 The Phosphorus Nucleosynthesis Problem

As we learned in Chapter 5, the phosphorus nuclei are too heavy ($Z = 15$) to have been formed during the primordial nucleosynthesis stage, which stopped after the synthesis of lithium ($Z = 3$) and beryllium ($Z = 4$) nuclei took place. Indeed, by all indications the first phosphorus atoms were synthesized through thermonuclear reactions occurring inside the cores of very massive Population III stars when the universe was about 2–4 × 10^8 years old. Thereafter phosphorus has mainly been formed in massive enough stars, say $M \geq 12\ M_{\odot}$. This requirement is directly related to the fact that the ^{31}P nucleus is not only relatively heavy, but it also contains an odd number of nucleons. Thus, when considering the nucleosynthesis of ^{31}P nuclei we realize that these nuclei do not belong to the so-called α-series elements, which are successively formed by incorporating an α particle (^{4}He) at each reaction step, starting from carbon nuclei, namely, $^{12}C \rightarrow {}^{16}O \rightarrow {}^{20}Ne \rightarrow {}^{24}Mg \rightarrow {}^{28}Si \rightarrow {}^{32}S$ (see Figure 5.2). This route naturally leads to the synthesis of $Z = 4n$ nuclei, including the quite abundant main biogenic elements C, O, and S. On the other hand, since the nuclear reactants involved in the formation of ^{31}P nuclei have relatively large electric charges, intense repulsive Coulomb barriers must be overcome for the required nuclear reactions to proceed. To this end, extreme pressure and temperature values are to be attained in the stellar cores. These conditions are only reached in a minor subset of stars, just accounting for about 5–10% of a typical galaxy population, which are massive enough to explosively ignite the C and Ne fuels, leading to the synthesis of ^{31}P nuclei through a relatively involved nuclear reactions network (sketched in Figure 5.3), preferentially by proton capture on ^{30}Si nuclei or ^{34}S nuclei (see Table 5.1), and secondarily during hydrostatic Ne-burning shells in the pre-supernova (SN) stage, likely involving α capture on ^{27}Al nuclei. Other types of explosive nucleosynthesis, such as that occurring in classical novae and in type-Ia SN, have also significantly contributed to the current inventory of phosphorus in the Galaxy (see Figures 5.5 and 5.6). In this regard, spectroscopic observations of the young SN remnant Cassiopeia A, indicating abundance ratios of P to Fe (the major nucleosynthetic product in the SN material) up to 100 times the average P/Fe value in the Galaxy, strongly supports the view that phosphorus nuclei are produced in type-II SN and ejected to the ISM. Notwithstanding this, some recent results derived from spectroscopic studies of the Crab nebula SN remnant indicate that there seems to be much less phosphorus in this source than in Cassiopeia A. Thus, the two explosions seem to differ from each other, perhaps because Cassiopeia A resulted from the explosion of a rare supermassive star (Greaves and Cigan, unpublished). These preliminary results then suggest that material blown out into space could vary dramatically in phosphorus abundances from site to site in the Galaxy.

In summary, the scarcity of phosphorus in the universe can be explained as stemming from two main factors, namely:

1. The relatively small number of stars massive enough to attain the physical conditions necessary to ignite the required nuclear fuels; and
2. The relatively involved nature of nuclear reaction networks leading to the nucleosynthesis of ^{31}P nuclei, which results in a very low phosphorus yield in a natural way.

On the other hand, the explosive conditions necessary for phosphorus formation guarantee that this element will be efficiently spread out in rapidly expanding SN remnants and novae outflows progressively diluting into the ISM. In this way, the reported ubiquity of phosphorus atoms throughout the universe can properly be accounted for as well.

10.2.2 *The Phosphorus Chemistry Puzzle*

To the best of my knowledge, no phosphorus bearing molecule has been directly observed in stellar photospheres as yet, though local thermal equilibrium model calculations indicate that diatomic PO and PH molecules, as well as triatomic PH_2 (HCP) compounds, may be formed in the extended atmospheres of O-rich (C-rich) stars, respectively. Once formed, as these species flow away from the star, they can undergo photochemical processes, leading to their partial destruction and the formation of other compounds. Indeed, a number of P-bearing molecules, including PN, PO, CP, HCP, CCP, PH_3 and NCCP, have been detected around C-rich and/or O-rich giant evolved stars. However, despite recent progress on both the observational and theoretical fronts, the chemistry of phosphorus in circumstellar envelopes remains uncertain, and there is still some debate regarding which molecules are the dominant reservoirs of phosphorus in these environments. Models assuming thermochemical equilibrium typically predict that HCP will be the dominant P-bearing species in C-rich envelopes, while PH_3 or PO will dominate in O-rich environments. However, current models cannot satisfactorily explain the high abundance of PN observed in O-rich circumstellar shells. In Figure 10.1, some plausible gas-phase routes leading to the formation of different P-bearing molecules are shown for the sake of illustration. As we see, the presence of several intertwined networks indicates a rich and complex chemistry.

As the gas temperature progressively decreases with increasing distance from the star, a number of P-bearing compounds, such as ammonium monophosphate, $NH_4H_2PO_4$, phosphorus oxide, P_4O_6 and schreibersite, $(Fe,Ni)_3P$, can condensate, thereby being incorporated in solid-phase grains. In this way, the next stage of chemical evolution takes place by means of physical and chemical processes coupling gas-phase molecules and different ices mantles deposited onto the solid surfaces of grains interspersed throughout the medium among the stars.

By comparing the elemental abundances of the main bioelements in different astrophysical objects, we observe a systematic enhancement of the C, N, O, and S relative abundances in the condensed matter phases out of which planetesimals are made in protoplanetary disks. This enrichment is due to the processing of different chemical compounds during their transition from the gas phase to the condensed phase in both ISM and protoplanetary stages. This point can be properly illustrated by obtaining the ratios between the elemental abundance of bioelements in the gas phase of interstellar diffuse clouds, and that measured in comet Halley's dust (see Tables 3.1 and 3.2). In doing so, we realize that O, C, and S atoms are more abundant in Halley's dust grains composition than they are in the ISM gas phase by a factor of about 1.4, whereas the enrichment factor of N atoms is close to 1.25. The enhancement factor value of P atoms is still uncertain, although recent Rosetta spacecraft measurements suggest it may be close to 2. From these data, we conclude that biogenic elements are depleted from the gas phase by different amounts and concentrated in interstellar grains composing the ISM solid phase. A similar conclusion is reached by considering the relative abundances (with respect to H_2) of the P-bearing molecules detected in both circumstellar envelopes and dense molecular clouds (Exercise 13).

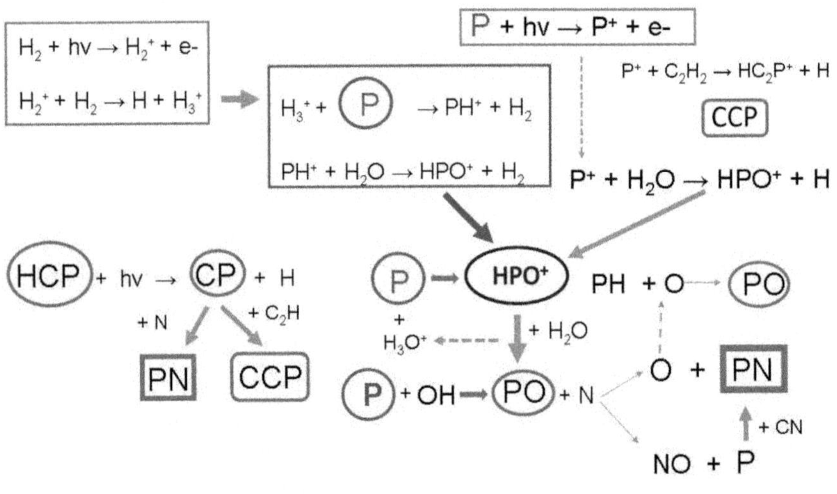

FIGURE 10.1 Gas-phase chemical reactions leading to the formation of different phosphorus bearing molecules in the circumstellar envelopes and the ISM.

At their initial stages these grains, formed in the circumstellar envelopes of aged stars and SN remnants, are mainly composed by a core of highly refractory materials, and have undergone little processing. As chemical evolution proceeds, the grains are progressively covered by different ices (CO_2, NH_3, CH_4, H_2CO, H_2S) resulting from their interaction with molecules present in the gas phase which, in turn, are experiencing a substantial chemical processing, via ion-neutral and neutral-neutral chemistry molecular networks. These grains may survive passage through the diffuse ISM, particularly if they are coated with thick enough mantles before escaping from the stellar neighborhood. In this regard, the phosphonic acids observed in meteorites (see Figure 8.2) can be related to the compounds containing the extremely stable C-P bond in the ISM. In fact, the CP radical has been observed around evolved stars, where it may be indicating the presence of the HCP molecule adsorbed onto dust particles. Eventually, some of these grains would aggregate to form cometary nuclei, though most of them will be incorporated in planetesimals out of which bigger planets are grown in the protoplanetary disks around newborn stars.

Quite remarkably, phosphate apatite mineral has been identified in presolar grains extracted from four meteorites, including two carbonaceous chondrites (Orgueil and Murchison) and two ordinary chondrites (Dimmit and Cold Bokkeveld) representatives (Jungck and Niederer, 2017). This finding provides evidence suggesting the possible formation of highly oxidized phosphorus compounds during SN aftermath events. In fact, the presence of apatite grains containing anomalous $^{38}Ar/^{36}Ar$ isotopic ratios in Orgueil carbonaceous chondrite meteorite has been interpreted as further evidence indicating that the formation of our solar system could have been triggered by a SN explosion. According to the scenario proposed by Jungck and Niederer (2017), the observed apatite crystals stem from phosphorus produced in a SN explosion, which condensed forming apatite layers grown around a graphite mineral core as the wave of explosion gases pushed outwards interstellar dust originally located around the SN progenitor star. Along with this mineral formed in the ISM, most iron meteorites contain mixed Fe-Ni phosphides, which were among the first phosphorus chemicals to condensate from the solar nebula (see Figure 7.7), and thus they are among the most ancient of P-bearing mineral phases in our solar system.

Of the seven P-bearing species detected in the ISM, the only ones identified in molecular clouds to date are PN (Ziurys, 1987; Turner and Bally, 1987) and PO (Rivilla et al., 2016; Lefloch et al., 2016). These molecules have been observed in several star-forming regions such as Sgr(B2), W51 e1/e2, W3(OH) and L1157 (see Section 2.3.2), both with similar abundances relative to molecular hydrogen ($\sim 10^{-10}$), the PO abundance being higher by a factor ~ 2. This ratio [PO]/[PN] is similar to those found in the O-rich circumstellar envelopes of the evolved stars VY CMa and IK

Tau, suggesting that phosphorus seems to be equally distributed in the form of PO and PN in both circumstellar and interstellar material. In fact, chemical modeling indicated that the two molecules are chemically related and are formed via gas-phase ion-molecular and neutral-neutral reactions during the cold collapse of dense molecular clouds (see Figure 10.1). The molecules freeze out onto grain at the end of the collapse, and evaporate during the warm-up phase once the temperature reaches ~ 35 K (Rivilla et al., 2016). In sources experiencing shock waves, the key player for the formation and destruction of PO and PN molecules is atomic nitrogen, which in turn is mainly released from ammonia. Thus, the reaction N + PO → PN + O contributes to ~ 85% to the total destruction of the PO (Aota and Aikawa, 2012; Lefloch et al., 2016).

Since earlier detections of PN were reported in hot and turbulent high-mass star forming cores, the very possibility that this molecule was produced by processes related to either high temperature gas phase reactions or grain disruption was originally considered, via thermal desorption of phosphine (PH_3) from grain mantles at temperatures above ~ 100 K, followed by rapid gas phase reaction which transform it into PN, PO, or atomic P in a time scale of about 10^4 years (Turner and Bally 1987; Charnley and Millar 1994). However, recent observations indicate the presence of PN molecules with column densities ~ 10^{11}–10^{12} cm^{-2} in relatively cold and quiescent gas clouds, hence challenging the chemical models that explain the formation of PN (Fontani et al., 2016). Thus, the possible formation routes for PN and PO molecules are currently strongly debated (see Section 6.3.1), albeit all the proposed scenarios explicitly consider the presence of grains playing a significant role in the ISM phosphorus chemistry (Jiménez-Serra et al., 2018). On the other hand, the lack of P-containing molecules in molecular clouds is surprising, and suggests that most P is condensed on dust grains or trapped in more complex molecules, such as PH_2CN, CH_3PH_2, PH_2CHO or $CH_3CH_2PH_2$ which are very difficult to be spectroscopically detected (Halfen et al., 2014; Ziurys et al., 2015; Turner et al., 2016).

The spectroscopic observations of Hale-Bopp in the near-infrared showed that particles emitted from the nucleus contained silicates, particularly magnesium-rich crystalline olivine (forsterite) and crystalline magnesium-rich pyroxenes, along with a large proportion of amorphous silicates. The detection of crystalline materials provided the first hint that some minerals in cometary nuclei had encountered high temperatures during their history before being stored in the comet nucleus, implying that they did not only consist of amorphous grains directly inherited from the presolar cloud.

Additional processing would occur in comets, which are formed from the low-temperature aggregation of these interstellar grains. Comets remain most of the time far away from stars, but occasionally approach them due to small gravitational perturbations. During their approach to the star, their temperature increases, and their originally pristine materials undergo an intense episode of photochemical processing. In the case of periodic orbit comets, this processing is cyclic, and the successive episodes of condensation due to the recurrent loss of most volatile materials form their comae and tails. In this way, periodic comets mimic usual chemical handling at an astronomical scale. Significant progress in our understanding of cometary dust has been achieved by a series of space missions performing a flyby of comets nuclei during the last three decades. These studies suggest that cometary dust is not only rich in organic compounds, but also diverse, dark, and quite porous. In particular, results showed that comet 1P/Halley has a chondritic composition within a factor of two for most of the elements detected. However, dust particles are enriched in three elements: carbon, nitrogen, and hydrogen by factors 11, 8, and 4, respectively. The relevance of cometary chemistry was illustrated in Figure 3.4, where we compared the elementary abundances of bioelements in the living matter as referred to the solar system and the mean compositions of cometary ices, respectively. We can see that the H content of living matter is very similar to its abundance in comets. In the case of cometary ices, a similar result is obtained for C, O, and to a lesser extent, for S and N as well. Conversely, the bioelements are overabundant in the living matter by several orders of magnitude as compared to their solar system abundance. Such an enhancement is particularly significant in the case of P. Accordingly, we conclude that during their transition from stars to protoplanets the bioelements are efficiently incorporated to the raw materials out of which new planetary systems will be formed, and this differential enrichment can be clearly appreciated in the pristine materials composing most comets.

10.2.3 The Prebiotic Phosphate Conundrum

As we mentioned previously, the main form of phosphorus in living beings is phosphate. This raises the question regarding the primary origin of such phosphates and the way they were originally incorporated in many organic molecules of biological interest, that is, the phosphorylation processes occurring during the earlier stages of prebiotic evolution. For one thing, the route to carrying highly oxidized phosphorus into newborn planets probably involves meteorites and IDPs, which implies that phosphate-containing minerals may be randomly distributed throughout the Galaxy. Indeed, if oxidized phosphorus compounds are mainly sourced from SN and nova outflows, and then travel across space in meteoritic rocks, it is possible that a young planet could find itself lacking this bio-essential compound because of the place where it was born. In that case, life might really struggle to get started out of phosphorus-poor chemistry on another world otherwise similar to our own. Alternatively, one could consider the possibility of getting relatively reduced phosphorus compounds from space sources and then oxidizing them in-situ on the planet or planetesimal surface.

Most studies along this line of thought performed to date have focused on chemical processes taking place in an aqueous environment. At this point, it is worth mentioning an alternative (and likely complementary) scenario based on condensed-phase chemistry instead. To this end, I will consider a possible set of interstellar grains modeled as multi-layered structures. Their inner core is assumed to contain a suitable phosphorus bearing compound, say schreibersite, phosphorus oxide (P_4O_6) or ammonium monophosphide ($NH_4H_2PO_4$), resulting from condensation in circumstellar envelopes. This core is covered with a first layer mainly containing formaldehyde ice, followed by another layer, mainly composed of hydrogen cyanide (HCN) ice (Figure 10.2). Under exposition to prolonged UV and cosmic ray irradiation, the outer layers may polymerize in the solid phase, yielding relatively high concentrations of sugars (including ribose H_2CO pentamer and deoxyribose) and adenine (the pentamer of HCN molecule) in their respective layers. In this way, these important biomonomers could be abiotically synthesized within a small volume, which favors subsequent polymerization with the previously obtained sugars and nucleobases by *in-situ* condensation reactions. Further thermal treatment during the protostar-protoplanetary stage could provide the necessary energy to transform the P_4O_6 core into phosphate, taking advantage to this end of the water stemming from the H_2O molecules previously released during the formation of sugar-bearing nucleosides. Accordingly, the phosphorylation of the nucleosides present in the processed cover layer of these grains will follow in a quite natural way, following an outer mantle – inward core sequence. The final result would be the synthesis of fully self-assembled AMP, ADP, and ATP moieties, which could be later liberated when the carrier grains were incorporated in planetesimals and/or cometary nuclei to experience further chemical processing upon grain fragmentation. Although I understand this proposal is highly speculative, I think it well deserves a closer experimental scrutiny in order to check its feasibility. Indeed, I deem the proposed scenario may help to understand the ubiquitous presence of ATP molecules as energy carriers in current biochemistry, as well as the possible emergence of RNA based proto-metabolism in a rather natural way.

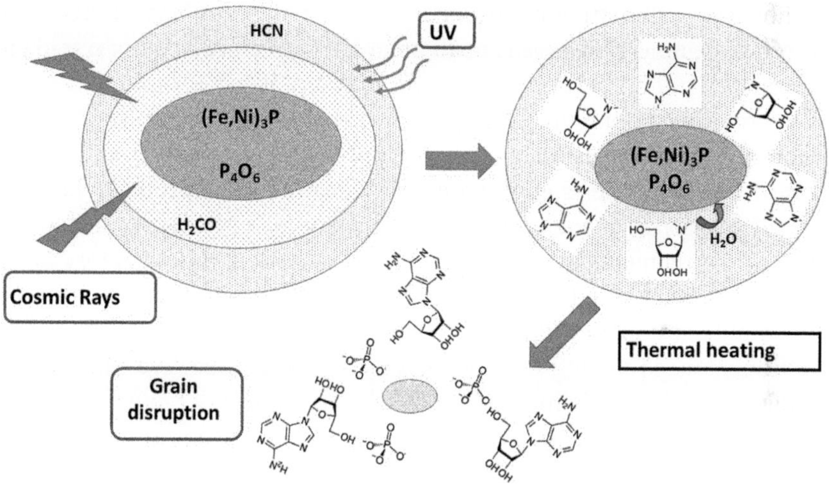

FIGURE 10.2 Schematics showing a possible route to the prebiotic synthesis of adenylic acid in a multi-layered interstellar grain subjected to UV and cosmic ray irradiation during its ISM stage and subsequently undergoing thermal heating during the protoplanetary disk stage.

For instance, it has been suggested that nucleobase arrays adsorbed on mineral surfaces (i.e., adenine physisorbed to graphite) might act as templates for the biomolecular assembly of amino acids (Heckel, 2002). On the other hand, the possible radiation-induced polymerization of formaldehyde molecules in the solid-state under extremely cold, typically interstellar temperatures was discussed some time ago by Goldanskii and co-workers (1973). It was found that the time of addition of one new link to a polymer chain increases exponentially in accordance with the Arrhenius law from 140 to 80 K, but approached a constant value (approximately 10^{-2} s) at temperatures below 10 K. This asymptotic behavior was explained in terms of quantum tunneling effects, which establish a low-temperature limit to this chemical reaction rate.

The possible formation of carbon-phosphorus bonds at low-temperatures has recently been probed in interstellar ice analogs containing phosphine-methane mixtures, in order to account for the presence of phosphonic acids in Murchison meteorite (Turner et al., 2016, 2018). Indeed, the discovery of trace quantities of alkyl phosphonic acids in this meteorite suggests the possibility of delivery of water-soluble phosphorus-containing molecules by meteorites to the early earth, hence providing an alternative to phosphorylation by orthophosphate and its polyphosphates derivatives. In fact, since alkyl phosphonic acids are related to orthophosphoric acid via substitution of the P-OH grouping by a highly stable P-C bond, this could have provided a supply of organic phosphorus for the earliest stages of prebiotic evolution. However, the bulk of the meteoritic material that falls to earth today has no phosphonates, and these compounds make up only a small fraction (~ 0.1%) of the total phosphorus in Murchison meteorite. Most of the remaining phosphorus may be trapped in the refractory high-temperature condensate form Fe_3P, which along with silicates, diverse oxides and carbon condensates constitute the cores of interstellar grains. This opens a promising alternative way to collect reactive phosphorus products capable of being incorporated into primitive biomolecules by means of the corrosion of schreibersite minerals present in those parent bodies (Figure 8.5).

10.3 THE OVERALL PICTURE

We inhabit an extreme universe, characterized by sharp contrasts in many of its basic features. Thus, most of its content adopts an unknown form, referred to as dark matter and dark energy, so that conventional hadronic matter just accounts for about the 5% of the material inventory present in it. If we inspect this relatively small amount of hadronic matter we realize that most of the atomic inventory (up to 90%) is contributed by H and He elements, while the main biogenic elements O, C, and N (the following ones in the elemental abundance ranking) only represent about 0.5% of the atoms present in the universe. The contribution of phosphorus atoms is very much smaller, merely amounting to about 0.01% of the total number of atoms present in the universe (see Table 3.1). Thus, albeit as citizens of our pale blue planet we are used to enjoy beautiful landscapes plenty of all sorts of life forms on earth, this scenery must be regarded as extremely rare in the universe as a whole. In fact, for the moment our planet is the only one where life is known to be actually present, although the widespread presence of carbon atoms in all astrophysical objects ensures that organic chemistry is common elsewhere in the universe (Kwok, 2004).

Albeit hydrogen atoms are overabundant in the universe atomic inventory, H_2 molecules made of two such atoms bonded together cannot be readily formed in the ISM gas phase (Exercise 15). This is due to its very low density, which prevents the assistance of a third body to release the chemical reaction excess energy. This process can only take place on the surface of suitable interstellar grains that catalyze the formation of molecular hydrogen. This constraint nicely illustrates that the high abundance of an atomic species is not a sufficient condition to guarantee the formation of molecules containing it. When those atoms are diluted in an extremely vast volume space, the kinetics of the involved chemical reaction may impose the strongest requirement for molecular formation. On the other hand, the tendency of a given atom to share electrons with other atoms also plays a very important role. Thus, in a cosmos where oxygen is the most abundant chemically reactive element, it is expected that water molecules (made of H_2 and O) become one of the most abundant compounds, along with CO and CO_2 species. Indeed, water is the most abundant molecule in living beings as well, and the simplest carbohydrate, the formaldehyde molecule H_2CO, is among the most abundant molecules in the ISM and comets. Not surprisingly, sugar carbohydrates play fundamental roles in both the structure and the metabolism of most biological species.

When one considers the currently known principal events ultimately leading to the emergence of life on earth through the cosmic history of our unfolding universe one can appreciate some hints supporting the existence of a unifying logic interconnecting different stages of chemical evolution. The general trend is a bottom-up building principle, going from very simple to increasingly more complex material structures, which are progressively spread out throughout the space. In order to properly satisfy the maximum entropy rule, each step towards a higher structural order ("anabolism") is paralleled by processes leading to a substantial degradation of other related matter or energy forms ("catabolism"). Some examples of these processes during the life cycle of stars are provided by stellar winds, external star atmosphere expulsion in the planetary nebula phase, or SN explosions.

According to the fundamental relationship between structure and function, the emergence of new arrangements of matter naturally leads to the establishment of functionally interdependent material systems synergetically interacting with each other as time goes by. This is illustrated in Figure 10.3, which shows the galactic cycle of matter and energy taking place at different astrophysical objects and different moments (compare to Figure 2.2). By zooming in into this picture, we can disclose a number of different relevant networks occurring at different scales. Thus, in the upper left corner, we remind the generation of ^{31}P nuclei due to thermonuclear reactions (Figure 5.3) undergoing at very small volumes inside stellar cores. On the other hand, in the lower right corner, we depict the molecular synthesis networks of several phosphorus bearing compounds taking place on large volumes of ISM space, within molecular clouds (Figure 10.1). In the intermediate size scale, we can find typical biochemical reactions, such as those determining the transfer of electrons among the different molecules involved in metabolic paths occurring within the cells at the micrometer scale.

During the last decades, most people working in the astrobiology field have assumed that prebiotic synthesis routes started from a few basic monomers, which then condensed among them to produce long biopolymers (see Figure 8.1). This is a natural approach, based on an intuitive view, although it also poses some long-standing important drawbacks. For instance, certain necessary precursor monomers, such as some amino acids and nucleobases, along with ribose or deoxyribose sugars, cannot be readily produced abiotically. This is particularly problematic for the formose reaction products. It has been recently suggested that the synthesis of numerous sugar molecules, including ribose, may be possible from ultraviolet (UV) irradiation and thermal treatment of precometary H_2O, CH_3OH, and NH_3 ices deposited onto mantles covering interstellar dust grains (Meinert et al., 2016, 2017). This proposal is based on the assumption that planetesimals (the parent bodies of asteroids, comets, and meteorites) were formed in the solar nebula from the aggregation of icy grains already present in the protosolar nebula and processed at the late molecular cloud stages along with the protoplanetary disk phase, owing to mixing of the protosolar materials. These grains are usually modeled as silicate or carbon cores surrounded by ice mantles subjected to UV photons and/or cosmic ray irradiation leading to complex chemistry in the condensed phase. Interestingly enough, this physico-chemical scenario is in line with the nucleotide abiotic synthesis scheme I proposed in the previous section (Figure 10.2).

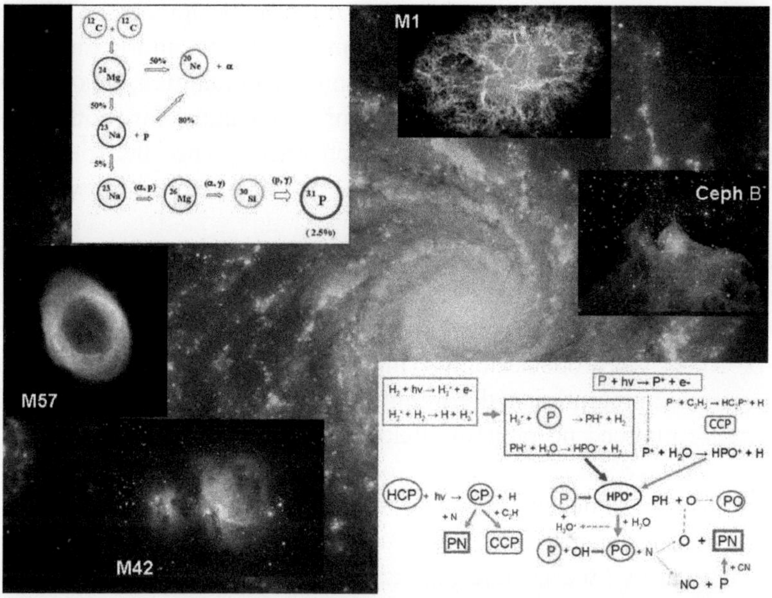

FIGURE 10.3 (See color insert.) The cycling of matter and energy considered at different scales, ranging from thermonuclear reactions taking place at a subatomic scale inside stellar cores (upper left box) to chemical reaction pathways occurring at the atomic and molecular scales (lower right box) in the interstellar medium of spiral galaxies.

By all indications, the phosphorylation problem still stands as one of the major difficulties to be surmounted in prebiotic modeling. A possible solution points toward a concerted action of different chemical reactants of biological interested which contribute to the final synthesis in a synergic way. This route has been recently explored in order to provide a possible answer to the phosphorylation conundrum (see Figure 8.3). In this approach, activated pyrimidine ribonucleotides can be formed in a short sequence that bypasses both free ribose and nucleobases as independent monomers. The starting materials for the synthesis (cyanamide, cyano-acetylene, glycolaldehyde, glyceraldehyde, and phosphate) are plausible prebiotic molecules, and the conditions of the synthesis are consistent with potential early-earth geochemical models (Powner et al., 2009). We note that the presence of different phosphorus compounds all the way long from the initial to the late stages is essential. This result highlights the importance of considering *mixed chemical systems* in which reactants for a particular reaction step can also control other steps in an intertwined way. Thus, chemical evolution is assisted by structural design, hence allowing for additional understanding on the emergence of informative order encoded in an ordered arrangement of matter. In addition, this scenario could help to explain the broad palette of biochemical functions played by phosphate derivatives in current biochemistry, since the role of this compound would have been actively present in different contexts in the very dawn of biopolymers synthesis.

Another aspect to be considered in prebiotic synthesis refers to chemical handling. In terrestrial laboratories, the physicochemical conditions of actual processes taking place in different astrophysical environments can be roughly reproduced only. This has to do with the extremely low temperatures, pressures, and concentrations typical of ISM, along with the intense photon and radiation fields required to properly compare the reaction yields obtained under controlled laboratory experimental conditions with those one knows are present in the astrophysical objects of interest. For instance, the first prebiotic synthesis of adenine was obtained in the laboratory by imposing high molar concentrations of HCN, which certainly are not to be expected on early earth environments. Now, the situation drastically changes if one considers the required pentamer polymer might be obtained in solid-state phase due to UV photons irradiation of an HCN ice layer deposited onto an interstellar grain surface (Figure 10.3). Such a process could likely take place in interstellar grains which were later incorporated inside cometary nuclei, a scenario quite close to the idea inspiring earlier proposals on the role of comets in the origin of life (Oró, 1961).

A related question concerns the timescales involved in actual prebiotic processes occurring in astronomical locations versus the timescales considered in the prebiotic synthesis experiments performed in our laboratories. It is clear that, for practical reasons, our simulation processes must somehow be accelerated in time, and the fundamental question naturally arises concerning if such a procedure substantially affects the ending result (or not). In other words, should we respect the pace observed in naturally occurring chemical evolution processes or are we entitled to speed up such processes in the confidence that in so doing we will obtain essentially the same final outcomes? Perhaps the rhythms of life's metabolic processes may be somehow related to the own cosmic clock rate as the universe unfolds itself…

10.4 SUGGESTIONS FOR FUTURE WORK

Many details of the overall picture just described are to be confirmed and quantitatively assessed in order to be able to achieve a more detailed knowledge of the role of phosphorus and its compounds in chemical evolution and the emergence of life elsewhere in the universe. Keeping this aim in mind, I would like to conclude this last chapter by suggesting some possible research topics well deserving a closer scrutiny in the years to come:

a. Molecules such as PO_2, HC_nP, PH_2CN, CH_3PH_2, PH_2CHO or $CH_3CH_2PH_2$ have complicated rotational levels, and their lines are expected to be weak, so that they are very difficult to be spectroscopically detected. Nonetheless, their systematic search with large enough radiotelescopes would be appealing in order to complete the current P-bearing molecular inventory in the ISM of our Galaxy. Also, the detection of phosphorus compounds in extragalactic sources would be very important in order to understand the primary origin of this biogenic element in the context of chemical evolution at a full-fledged cosmic scale.
b. The detection of P-bearing molecules in the photospheres of late-type stars would be crucial in order gain a better understanding about the nature of the parent molecules of the phosphorus compounds actually observed in the circumstellar envelopes of both C- and O-rich stars.
c. Since the optical constants of the organic refractory grain mantle material are derived from a combination of astronomical and laboratory spectra of residues of UV photoprocessed ices, it would be worthwhile to consider the possible effects of the inclusion of phosphorus compounds, in reduced or oxidized forms, in the matching between astronomical and laboratory obtained infrared

spectra along the research line recently pursued by Turner and co-workers (2016, 2018).

d. Spurred by the recent first detection of phosphorus atoms in comet P67/Churymov-Gerasimenko a continued search for the possible presence of phosphorus oxides and organophosphorus species in the solid phase of cometary nuclei would be most pertinent in future spacecraft missions to comets and asteroids.
e. Detailed analysis, including elemental composition studies, mass spectroscopy, and mineralogical characterization of IDP samples of cometary origin, would be very promising in order to understand the phosphorus chemistry in the earlier stages of the solar system.
f. The search of oxidized phosphorus compounds as a potential guide in the search for extinct or extant life on Mars and other planets should be encouraged. Indeed, the role of the mineral apatite with inclusions of isotopically "light" carbon pointing to possibly the oldest traces of life on earth, although controversial, was reported some time ago (Mojzsis et al., 1996). This interesting result illustrates the wide possibilities that the study of phosphorus compounds in the solar system and other astrophysical environments can provide to the knowledge of the origins of life.
g. The possible detection of P-bearing molecules in the atmospheres of exoplanets would be very interesting, particularly if these planets are located within the habitable zone (HZ) of their planetary system.
h. Similarly, it would be appealing to search for phosphorus compounds in brown dwarf stars.

KEYWORDS

- **biogenic elements enhancement**
- **comets chemical evolution**
- **gas-phase molecular networks**
- **grain chemical processing**
- **mixed chemical systems**
- **phosphorus chemical puzzle**
- **phosphorus nucleosynthesis problem**
- **prebiotic phosphate conundrum**

Bibliography

Abel, T., Bryan, G. L., & Norman, M. L., (2002). The formation of the first star in the Universe. *Science*, 295, 93–98.

Adams, N. G., McIntosh, B. J., & Smith, D., (1990). Production of phosphorus-containing molecules in interstellar clouds. *Astron. Astrophys.*, *232*, 443–446.

Adelberger, E. G., et al., (2011). Solar fusion cross-sections. II. The pp chain and the CBO cycle. *Rev. Mod. Phys.*, *83*, 195–245.

Agúndez, M., & Cernicharo, J., (2006). Oxygen chemistry in the circumstellar envelope of the carbon-rich star IRC +10216. *Astrophys. J.*, *650*, 374–393.

Agúndez, M., Biver, N., Santos-Sanz, P., Bockelée-Morvan, D., & Moreno, R., (2014c). Molecular observations of comets C/2012 S1 (ISON) and C/2013 R1 (Lovejoy): HNC/HCN ratios and upper limits to PH_3. *Astron. Astrophys.*, *564*(L2), p. 5.

Agúndez, M., Cernicharo, J., & Guélin, M., (2007). Discovery of phosphaethyne (HCP) in space: Phosphorus chemistry in circumstellar envelopes. *Astrophys. J.*, *662*, L91–L94.

Agúndez, M., Cernicharo, J., & Guélin, M., (2014b). New molecules in IRC+10216: Confirmation of C_5S and tentative identification of MgCCH, NCCP and SiH_3CN. *Astron. Astrophys.*, *579*, A45–1–9.

Agúndez, M., Cernicharo, J., Decin, L., Encrenaz, P., & Teyssier, T., (2014a). Confirmation of circumstellar phosphine. *Astrophys. J. Lett.*, *790*, L27.

Agúndez, M., Fonfría, J. P., Cernicharo, J., Pardo, J. R., & Guélin, M., (2008). Detection of circumstellar CH_2CHCN, CH_2CN, CH_3CCH, and H_2CS. *Astron. Astrophys.*, *479*, 493–501.

Agúndez, M., Marcelino, N., Cernicharo, J., & Ap, J. L., (2018). Discovery of interstellar isocyanogen (CNCN): Further evidence that dicyanopolynes are abundant in space. *Astrophys. J. Lett.*, *861*(L22), p. 5.

Al Ouahabi, A., Amalian, J. A., Charles, L., & Lutz, J. F., (2017). Mass spectrometry sequencing of long digital polymers facilitated by programmed inter-byte fragmentation. *Nature Commun.*, *8*, 967–975. doi: 10.1038/s41467–017–01104–3.

Altwegg, K., et al., (2016). Prebiotic chemicals – amino acid and phosphorus – in the coma of comet 67P/Churyumov-Gerasimenko. *Sci. Adv.*, *2*, e1600285–e1600289.

Altwegg, K., et al., (2017). Organics in comet 67P – a first comparative analysis of mass spectra from ROSINA – DFMS, COSAC and Ptolemy. *MNRAS*, *469*, S130–S141.

Anders, E., & Grevesse, N., (1989). Abundances of the elements: Meteoritic and solar. *Geochim Cosmochim. Acta*, *53*, 197–214.

Aota, T., & Aikawa, Y., (2012). Phosphorus chemistry in the shocked region L1157 B1. *Astrophys. J.*, *761*, 74–83.

Arndt, P., Bohsung, J., Maetz, M., & Jessberger, E. K., (1996). The elemental abundances in interplanetary dust particles. *Meteor. Planet. Sci.*, *31*, 817–833.

Arrhenius, S., & Sandström, J. W., (1903). *Lehrbuch der Kosmischen Physik*. S. Hirzel: Leipzig.

Asplund, M., Grevesse, N., Sauval, A. J., & Scott, P., (2009). The chemical composition of the Sun. *Annu. Rev. Astron. Astrophys.*, *47*, 481–522.

Aykol, M., Doak, J. W., & Wolverton, C., (2017). Phosphorus allotropes: Stability of black phosphorus re-examined by means of the Van Der Waals inclusive density functional method. *Phys. Rev. B*, *95*, 214115.

Bachiller, R., Forveille, T., Huggins, P. J., & Cox, P., (1997). The chemical evolution of planetary nebulae. *Astron. Astrophys.*, *324*, 1123–1134.

Bada, J., & Lazcano, A., (2002). Origin of life. Some like hot, but not the first biomolecules. *Science*, *296*, 1982–1983.

Baker, G. L., & Blackburn, J. A., (2005). *The Pendulum: A Case Study in Physics*. Oxford University Press: Oxford.

Bakshi, M. S., (2017). Nanotoxicity in systemic circulation and wound healing. *Chem. Res. Toxicol.*, *30*, 1253–1274.

Baltscheffsky, M., Schultz, A., & Baltscheffsky, H., (1999). H+-proton-pumping inorganic pyrophosphatase: A tightly membrane-bound family. *FEBS Lett.*, *452*, 121–127.

Banin, A., Clark, B. C., & Wanke, H., (1992). Surface chemistry and mineralogy. In: Kieffer, H. H., Jakosky, B. M., Snyder, L. W., & Matthews, M. S., (eds.), *Mars* (pp. 594–625). University Arizona Press: Tucson.

Barkana, R., (2006). The first stars in the universe and cosmic reionization. *Science*, *313*, 931–934.

Beck, L. C., (1834). *A Manual of Chemistry: Containing a Condensed View of the Present State of the Science, with Copious References to More Extensive Treatises* (p. 160). Original papers, etc. E.W. & C Skinner.

Belloche, A., Garrod, R. T., Müller, H. S. P., Menten, K. M., Medvedev I., Thomas, J., & Kisiel, Z., (2019). Re-exploring molecular complexity with ALMA (ReMoCa): interstellar detection of urea. *Astron. Astrophys.*, *628*, A10 (68pp).

Benner, S. A., Ricardo, A., & Carrigan, M. A., (2004). Is there a common chemical model for life in the universe? *Current Opin. Chem. Bio.*, *8*, 672–689.

Bentley, M. S., Schmied, R., Mannel, T., Torkar, K., Jeszenszky, H., Romstedt, J., et al., (2016). Aggregate dust particles at comet 67P/Churyumov-Gerasimenko. *Nature*, *537*, 73–76.

Bernal, J. D., (1949). The physical basis of life. *Proc. Phys. Soc. A*, *62*, 537–558.

Bernal, J. D. (1960). The scale of structural units in biopoesis. In: Florkin, M., (ed.), *Aspects of the Origin of Life* (pp. 155–169). Pergamon Press: Oxford.

Bernstein, M. P., Sandford, S. A., Allamandola, L. J., Chang, S., & Scharberg, M. A., (1995). Organic compounds by photolysis of realistic interstellar and cometary ice analogs containing methanol. *Astrophys. J.*, *454*, 327–344.

Bernstein, M. P., Sandford, S. A., Allamandola, L. J., Gillette, J. S., Clemett, S. J., & Zare, R. N., (1999). UV irradiation of polycyclic aromatic hydrocarbons in ices: Production of alcohols, quinones, and ethers. *Science*, *283*, 1135–1138.

Bidelman, W. P., (1960). The unusual spectrum of 3 Centauri. *PASP*, *72*, 24–28.

Binder, A., (1998). Lunar prospector: Overview. *Science*, *281*, 1475–1476.

Blessing, R. H., (1988). New analysis of the neutron diffraction data for anhydrous orthophosphate acid and the structure of the H_3PO_4 molecules in crystals. *Acta Cryst.*, *B44*, 334–340.

Bockelée-Morvan, D., et al., (2000). New molecules found in comet C/1995 O1 (Hale-Bopp). *Astron. Astrophys.*, *353*, 1101–1114.

Borguet, C. J. B., Demonds, D., Arav, N., Benn, Ch., & Chamberlain, C., (2012). BAL phosphorus abundance and evidence for immense ionic column densities in quasar outflows: VLT/X-shooter observations of quasar SDSS J1512+1119. *Astrophys. J., 758*, 69–78.

Bouret, J. C., Lanz, T., & Hillier, D. J., (2005). Lower mass loss rates in O-type stars: Spectral signatures of dense clumps in the wind of two Galactic O4 stars. *Astron. Astrophys., 438*, 301–316.

Bowman, J. D., Rogers, A. E. E., Monsalve, R. A., Mozdzen, T. J., & Mahesh, N., (2018). An absorption profile centered at 78 megahertz in the sky-averaged spectrum. *Nature, 555*, 67–70.

Britvin, S. N., Rudashevsky, N. S., Krivovichev, S. V., Burns, P. C., & Polekhovsky, Y. S., (2002). Allabogdanite, $(Fe,Ni)_2P$, a new mineral from Onello meteorite: The occurrence and crystal structure. *Am. Mineralog., 97*, 1245–1249.

Bromm, V., & Loeb, A., (2003). The formation of the first low-mass stars from gas with low carbon and oxygen abundances. *Nature (London), 425*, 812–814.

Brownlee, D. E., (1994). In: Zolensky, M. E., Wilson, T. L., Rietmeijer, F. J. M., & Flynn, G. J., (eds.), *AIP Conference Proceedings 310: Analysis of Interplanetary Dust* (p. 5). AIP Press, New York.

Bryant, D. E., & Kee, T. P., (2006). Direct evidence for the availability of reactive, water soluble phosphorus on the early earth H-phosphinic acid from the Nantan meteorite. *Chem. Commun.*, 2344–2346.

Burbidge, E. M., Burbidge, G. R., Fowler, W. A., & Hoyle, F., (1957). Synthesis of the elements in stars. *Rev. Mod. Phys., 29*, 547–650.

Burcar, B., Pasek, M., Gull, M., Cafferty, B. J., Velasco, F., Hud, N. V., & Menor-Salván, C., (2016). Darwin's warm little pond: a one-pot reaction for prebiotic phosphorylation and the mobilization of phosphate from minerals in a urea-based solvent. *Angw. Chem. Int. Ed., 55*, 13249–13253.

Buseck, P. R., & Clark, J., (1984). Zaisho – a pallasite containing pyroxene and phosphorus olivine. *Mineral. Mag., 48*, 229–235.

Caffau, E., Andrievsky, S., Korotin, S., Origlia, L., Oliva, E., Sanna, N., Ludwig, H. G., & Bonifacio, P., (2016). GIANO Y-band spectroscopy of dwarf stars: Phosphorus, sulfur, and strontium abundances. *Astron. Astrophys., 585*, A16–A21.

Caffau, E., Bonifacio, P., Faraggiana, R., & Steffen, N., (2011). The galactic evolution of phosphorus. *Astron. Astrophys., 532*, A98–A104.

Cairns-Smith, A. G., (1985). *Seven Clues to the Origin of Life: A Scientific Detective Story.* Cambridge University Press, Cambridge.

Callahan, M. P., Smith, K. E., James, C. II, H., Ruzicka, J., Stern, J. C., Galvin, D. P., House, C. H., & Dworkin, J. P., (2011). Carbonaceous meteorites contain a wide range of extraterrestrial nucleobases. *Proc. Nat. Acad. Sci., 108*, 13995–13998.

Cameron, A. G. W., & Benz, W., (1991). The origin of the moon and the single impact hypothesis IV. *Icarus, 92*, 204.

Caminade, A. M., Turrin, C. O., & Majoral, J. P., (2018). *Phosphorus Dendrimers in Biology and Nanomedicine: Syntheses, Characterization, and Properties*. Pan Stanford Publishing: Singapore.

Canop, R. M., & Asphaug, E., (2001). Origin of the moon in a giant impact near the end of the earth's formation. *Nature, 412*, 708–712.

Cardelli, J. A., (1994). The abundance of heavy elements in interstellar gas. *Science, 265*, 209–213.

Castellanos-Gomez, A., Vicarelli, L., Prada, E., Island, J. O., Narasimha-Acharya, K. L., Blanter, S. I., et al., (2014). Isolation and characterization of few-layer black phosphorus. *2D Materials, 1*(2), 025001.

Cernicharo, J., Kisiel, Z., Tercero, B., Kolesniková, L., Medvedev, I. R., López, A., Frotman, S., Winnewisser, M., De Lucia, F. C., Alonso, J. L., & Guillemin, J. C., (2016). A rigorous detection of interstellar CH_3NCO: An important missing species in astrochemical networks. *Astron. Astrophys., 587*, L4–L47.

Cernicharo, J., Velilla-Prieto, L., Agúndez, M., Pardo, J. R., Fonfría, J. P., Quintana-Lacaci, G., Cabezas, C., Bermúdez, C., Guélin M., (2019). Discovery of the first Ca-bearing molecule in space: CaNC. arXiv:1906.09352 [astro-ph.SR]

Cescutti, G., Matteucci, F., Caffau, E., & François, P., (2012). Chemical evolution of the Milky Way: The origin of phosphorus. *Astron. Astrophys., 540*, A33–A36.

Chandrasekhar, S., (1984). On stars, their evolution and their stability. *Rev. Mod. Phys., 56*, 137–147.

Charnley, S. B., & Millar, T. J., (1994). The chemistry of phosphorus in hot molecular cores. *Mon. Not. R. Astron. Soc., 270*, 570–574.

Chen, Y., Zhao, S., Liu, B., Chen, M., Mao, J., He, H., Zhao, Y., Huang, N., & Wan, G., (2014). Corrosion-controlling and osteo-compatible Mg ion-integrated phytic acid (Mg-PA) coating on magnesium substrate for biodegradable implants application *ACS Appl. Mater. Interfaces, 6*, 19531–19543. doi: 10.1021/am506741.

Chieffi, A., Limongi, M., & Strainiero, O., (1998). The evolution of a 25 M-star from the main sequence up to the onset of the iron core collapse. *Astrophys. J., 502*, 737–762.

Cho, K., Yang, J., & Lu, Y., (2017). Phosphorene: An emerging 2D material. *J. Mater. Res., 32*, 2839–2847.

Chyba, C., & Sagan, C., (1992). Endogenous production, exogenous delivery and impact-shock synthesis of organic molecules: An inventory for the origins of life. *Nature, 355*, 125–132.

Clayton, D. D., (1988). Stellar nucleosynthesis and chemical evolution in the solar neighborhood. In: Kerridge, J. F., & Matthews, M. S., (eds.), *Meteorites and the Early Solar System* (p. 1021). University Arizona Press: Tucson.

Connelly, J. N., Amelin, Y., Krot, A. N., & Bizarro, M., (2008). Chronology of the solar system's oldest solids. *Astrophys. J., 675*, L121–L124.

Cooper, G. W., Onwo, W. M., & Cronin, J. R., (1992). Alkyl phosphonic acids and sulfonic acids in the Murchison meteorite. *Geochim. Cosmochim. Acta, 56*, 4109–4115.

Cooper, G. W., Thiemens, M. H., Jackson, T. L., & Chang, S., (1997). Sulfur and hydrogen isotope anomalies in meteorite sulfonic acids. *Science, 277*, 1072–1074.

Copley, S. D., Smith, E., & Morowitz, H. J., (2007). The origin of the RNA world: Co-evolution of genes and metabolism. *Bioorg. Chem., 35*, 430–433.

Corbridge, D. E. C., & Lowe, E. J., (1952). Structure of white phosphorus: Single crystal x-ray examination. *Nature, 170*, 629.

Cordiner, M. A., Charnley, S. B., Kisiel, Z., McGuire, B. A., & Kuan, Y. J., (2017). Deep K-band observations of TMC-1 with the green bank telescope: Detection of HC_7O, nondetection of $HC_{11}N$, a search for new organic molecules. *Astrophys. J., 850*, 187–193.

Córdoba, A., Hierro-Oliva, M., Pacha-Olivenza, M. A., Fernández-Calderón, M. C., Perelló, J., Isern, B., González-Martín, M. L., Monjo, M., & Ramis, J. M., (2016). Direct covalent grafting of phytate to titanium surfaces through Ti–O–P bonding shows bone stimulating surface properties and decreased bacterial adhesion. *ACS Appl. Mater. Interfaces*, *8*, 11326–11335.

Craig, D. P., (1959). *Aromatic Character in Theoretical Organic Chemistry* (p. 20). Butterworths Scientific Publications: London.

Craig, D. P., & Paddock, N. L., (1958). A novel type of aromaticity. *Nature*, *181*, 1052–1053.

Cronin, J. R., (1989). Origin of organic compounds in carbonaceous chondrites. *Adv. Space Res.*, *9*, 54–64.

Cronin, J. R., & Pizarello, S., (1990). Aliphatic hydrocarbons of the Murchison meteorite. *Geochim. Cosmochim. Acta*, *54*, 2859–2868.

Crovisier, J., Bockelée-Morvan, D., Colom, P., Biver, N., Despois, D., & Lis, D. C., (2004). The composition of ices in comet C/1995 O1 (Hale-Bopp) from radio spectroscopy. *Astron. Astrophys.*, *418*, 1141–1157.

Crowther, P. A., Hillier, D. J., Evans, C. J., Fullerton, A. W., De Marco, O., & Willis, A. J., (2002). Revised stellar temperatures for Magellanic cloud O supergiants from far ultraviolet spectroscopic explorer and very large telescope UV-visual echelle spectrograph spectroscopy. *Astrophys. J.*, *579*, 774–799.

Cummins, C. C., (2014). Phosphorus: From the stars to land & sea. *Daedalus*, *143*, 9–20.

Cunningham et al., (2007). A search for propylene oxide and glycine in Sagittarius B2 (LMH) and Orion. *Mon. Not. R. Astron. Soc.*, *376*, 1201–1210.

Cyburt, R. H., Fields, B. D., Olive, K. A., & Yeh, T. H., (2016). Big bang nucleosynthesis: Present status. *Rev. Mod. Phys.*, *88*, 015004–1–22.

Dai, X., & Guerras, E., (2018). Probing planets in extragalactic galaxies using quasar microlensing. *Astrophys. J. Lett.*, *853*, L27–L31.

Darwin, F., (1959). *The Life and Letters of Charles Darwin* (Vol. II, pp. 202–203). Basic Books, New York, NY, Letter to Joseph Dalton Hooker.

Dass, A. V., Jaber, M., Brack, A., Foucher, F., Kee, T. P., Georgelin, T., & Westall, F., (2018). Potential role of inorganic confined environments in prebiotic phosphorylation. *Life*, *8*, 7–17.

De Beck, E., Kaminski, T., Patel, N. A., Young, K. H., Gottlieb, C. A., Menten, K. M., & Decin, L., (2013). PO and PN in the wind of the oxygen-rich AGB star IK Tauri. *Astron. Astrophys.*, *558*, A132–1–9.

De Graaf, R. M., Visscher, J., & Schwartz, A. W., (1995). A plausibly prebiotic synthesis of phosphonic acids. *Nature*, *378*, 474–477.

Del Rio, E., Barrientos, C., & Largo, A., (1996). Theoretical studies of possible processes for the interstellar production of phosphorus compounds: The reaction of P^+ with C_3H_2. *J. Phys. Chem.*, *100*, 14643–14650.

Delsemme, A. H., (1995). Cometary origin of the biosphere: A progress report. *Adv. Space Research*, *15*, 49–57.

Dobbie, P. D., Barstow, M. A., Hubeny, I., Holberg, J. B., Burleigh, M. R., & Forbes, A. E., (2005). Photospheric phosphorus in the FUSE spectra of GD71 and two similar DA white dwarfs. *Mon. Not. R. Astron. Soc.*, *363*, 763–768.

Domagal-Goldman, S. D., Wright, K. E., et al., (2016). The astrobiology primer v2.0. *Astrobiology*, *8*, 561–653.

Draine, B. T., (2011). *Physics of the Interstellar and Intergalactic Medium*. Princeton University Press: Princeton.

Dufton, P. L., Keenan, F. P., & Hibbert, A., (1986). The abundance of phosphorus in the interstellar medium. *Astron. Astrophys., 164*, 179–183.

Eddington, A. S., (1926). *The Internal Constitution of the Stars*. Cambridge University Press: Cambridge.

Edwards, J. L., & Ziurys, L. M., (2013). The remarkable molecular content of the red spider nebula (NGC 6537). *Astrophys. J. Lett., 770*, L5–L10.

Edwards, J. L., & Ziurys, L. M., (2014). Sulfur and silicon-bearing molecules in planetary nebulae: The case of M2–48. *Astrophys. J. Lett., 794*, L27–L32.

Edwards, J. L., Cox, E. G., & Ziurys, L. M., (2014). Millimeter observations of CS, HCO^+ and CO toward five planetary nebulae: Following molecular abundances with nebular age. *Astrophys. J., 791*, 79–93.

Ehrenfreund, P., (1999). Molecules on a space odyssey. *Science, 283*, 1123–1124.

Ehrenfreund, P., et al., (2002). Astrophysical and astrochemical insights into the origin of life. *Rep. Prog. Phys., 65*, 1427–1487.

Elliot, J. C., (1994). *Structure and Chemistry of the Apatites and Other Calcium Phosphates*. Elsevier: Amsterdam.

Elliott, P., Brugger, J., Caradoc-Davies, T., & Pring, A., (2013). Hylbrownite, $Na_3MgP_3O_{10}.12H_2O$, a new triphosphate mineral from the dome rock mine, South Australia: Description and crystal structure. *Mineral. Mag., 77*, 385–398.

Elser, J. J., (2006). Biological stoichiometry: a chemical bridge between ecosystem ecology and evolutionary biology. *Am. Nat., 168*, S25-S35.

Elsila, J. E., Glavin, D. P., & Dworkin, J. P., (2009). Cometary glycine detected in samples returned by Stardust. *Meteorit. Planet. Sci., 44*, 1323–1330.

Emir, R., Yusof, N., Petermann, I., & Kassim, H. A., (2013). On the nucleosynthesis of phosphorus in massive stars. *AIP Conf. Proc., 1528*, 61–64.

Emsley, J., (1998). *The Elements* (3rd edn.). Clarendon Press: Oxford.

Eschenmoser, A., (1999). Chemical etiology of nucleic acid structure. *Science, 284*, 2118–2124.

Eschenmoser, A., (2007). The search for the chemistry of life's origin. *Tetrahedron, 63*, 12821–12844.

Fayolle, E. C., et al., (2017). Protostellar and cometary detections of organohalogens. *Nature Astron., 1*, 703–708.

Fenton, M. D., (2001). Ferroalloys. *U.S. Geological Survey Minerals Yearbook*, 27.1–27.11. USGS Minerals Publications.

Fernández-García, C., Coggins, A. J., & Powner, M. W., (2017). A chemist's perspective on the role of phosphorus at the origins of life. *Life*, 7, *31*, doi: 10.3390/life7030031.

Ferris, J., et al., (1968). Studies in prebiotic synthesis 3. Synthesis of pyrimidines from cyanoacetylene and cyanate. *J. Molec. Biol., 33*, 643–704.

Fiore, M., & Strazewski, P., (2016). Bringing prebiotic nucleosides and nucleotides down to earth. *Angew. Chem. Int. Ed., 55*, 13930–13933.

Fitzsimmons, A., Snodgrass, C., Rozitis, B., Yang, B., Hyland, M., Seccull, T., Bannister, M. T., Fraser, W. C., Jedicke, R., & Lacerda, P., (2018). Spectroscopy and thermal modeling of the first interstellar object 1I/2017 U1 Oumuamua. *Nature Astronomy, 2*, 133–137.

Flores, E., Ares, J. R., Castellanos-Gómez, A., Barawi, M., Ferrer, I. J., & Sánchez, C., (2015). Thermoelectric power of bulk black-phosphorus. *Appl. Phys. Lett.*, *106*, 022102.

Flynn, G. J., (1994). Interplanetary dust particles collected from the stratosphere: Physical, chemical, and mineralogical properties and implications for their sources. *Planet. Space Sci.*, *42*, 1151–1161.

Flynn, G. J., et al., (2006). Elemental composition of comet 81P/Wild 2 samples collected by Stardust. *Science*, *314*, 1731–1735.

Fontani, F., Rivilla, V. M., Caselli, P., Vasyunin, A., & Palau, A., (2016). Phosphorus-bearing molecules in massive dense cores. *Astrophys. J. Lett.*, *822*, L30–L35.

Fontani, F., Rivilla, V. M., van der Tak, F. F. S., Mininni, C., Beltrán, M. T., & Caselli, P., (2019). Origin of the PN molecule in star-forming regions: the enlarged sample. arXiv: 1908.11280v1 [astro-ph-SR].

Forti, P., Galli, E., Rossi, A., Pint, J., & Pint, S., (2004). Ghar Al Hibashi lava tube: The richest site in Saudi Arabia for cave minerals. *Acta Carsologica*, *33*, 190–206.

Fowler, W. A., (1984). Experimental and theoretical nuclear astrophysics: The quest for the origin of the elements. *Rev. Mod. Phys.*, *56*, 149–179.

Friedman, S. D., Howk, J. C., Andersson, B. G., Sembach, K. R., Ake, T. B., Roth, K., et al., (2000). Far ultraviolet spectroscopic explorer observations of interstellar gas toward the large magallanic cloud star Sk-67°05. *Astrophys. J.*, *538*, L39–L42.

Frost, P. C., Benstead, J. P., Cross, W. F., et al., (2006). Threshold elemental ratios of carbon and phosphorus in aquatic consumers. *Ecology Letters*, *9*, 774–779.

Fuchs, L. H., (1968). The phosphate mineralogy of meteorites. In: Millman, P. M., (ed.), *Meteorite Research, Proceedings of a Symposium on Meteorite Research Held in Vienna* (pp. 7–13, 683–695). Austria. Reidel: Dordrecht.

Gabel, N. W., & Thomas, V., (1971). Evidence for the occurrence and distribution of inorganic polyphosphates in vertebrate tissues. *J. Neurochem.*, *18*, 1229–1242.

George, J. S., et al., (2009). Elemental composition and energy spectra of galactic cosmic rays during solar cycle 23. *Astrophys. J.*, *698*, 1666–1681.

Gerardy, C. L., & Fesen, R. A., (2001). Near-infrared spectroscopy of the Cassiopeia A and Kepler supernova remnants. *Astron. J.*, *121*, 2781–2791.

Gibard, C., Gorrell, I. B., Jiménez, E. I., Kee, T. P., Pasek, M. A., & Krishnamurthy, R., (2019). Geochemical sources and availability of amidophosphates on the early earth. *Angw. Chem. Int. Ed.*, *58*, 8151–8155. doi: 10.1002/anie.201903808.

Gibson Jr., E. K., & Chang, S., (1992). The moon: Biogenic elements. In: Carle, G., Schwartz, D., & Huntington, J., (eds.), *Exobiology in Solar System Exploration* (Vol. 512, p. 29). NASA SP.

Gier, T. E., (1961). HCP, a unique phosphorus compound. *J. Am. Chem. Soc.*, *83*, 1769–1770.

Gilbert, W., (1986). The RNA world. *Nature (London)*, *319*, 618.

Gobrecht, D., Cherchneff, I., Sarangi, A., Plane, J. M. C., & Bromley, S. T., (2016). Dust formation in the oxygen-rich AGB star IK Tauri. *Astron. Astrophys.*, *585*, A6–A20.

Goesmann, F., et al., (2015). Organic compounds on comet 67P/Churyumuv-Gerasimenko revealed by COSAC mass spectrometry. *Science*, *349*, aab0689.

Goldanskii, V. I., Frank-Kamenetskii, M. D., & Barkalov, I. M., (1973). Quantum low-temperature limit of a chemical reaction rate. *Science*, *182*, 1344–1345.

Goodman, N. B., Ley, L. L., & Bullet, D. W., (1983). Valence-band structures of phosphorus allotropes. *Phys. Rev. B*, *27*, 7440–7450.

Gorrell, I. B., Wang, L., Marks, A. J., Bryant, D. E., Bouillot, F., Goddard, A., Heard, D. E., & Kee, T. P., (2006). On the origin of the Murchison meteorite phosphonates. Implications for pre-biotic chemistry. *Chem. Commun., 15*, 1643–1645.

Gras, P., Rey, C., André, G., Charvillat, C., Sarda, S., & Combes, C., (2016). Structure of monoclinic calcium pyrophosphate dihydrate (m-CPPD) involved in inflammatory reactions and osteoarthritis. *Acta Cryst., B72*, 96–101.

Greenstein, J. L., (1961). Stellar evolution and the origin of the chemical elements. *Am. Sci., 49*, 449–473.

Greenwood, N. N., & Earnshaw, A., (1997). *Chemistry of the Elements* (2nd edn.). Butterworth-Heinemann Ltd., Oxford.

Grevesse, N., & Sauval, A. J., (1998). Standard solar composition. *Space Sci. Rev., 85*, 161–174.

Guan, J., Liu, D., Zhu, Z., & Tománek, D., (2016). Two-dimensional phosphorus carbide: Competition between sp^2 and sp^3 bonding. *Nano Lett., 16*, 3247–3252.

Guélin, M., Cernicharo, J., Paubert, G., & Turner, B. E., (1990). Free CP in IRC+10216. *Astron. Astrophys., 230*, L9–L11.

Guélin, M., Muller, S., Cernicharo, J., Apponi, A. J., McCarthy, M. C., Gottlieb, C. A., & Thaddeus, P., (2000). Astronomical detection of the free radical SiCN. *Astron. Astrophys., 363*, L9-L12.

Gulick, A., (1955). Phosphorus as a factor in the origin of life. *Am. Scient., 43*, 479.

Haldane, J. B., (1929). The origin of life. *Rationalist Annual, 148*, 3–10.

Halfen, D. T., Clouthier, D. J., & Ziurys, L. M., (2008). Detection of the CCP radical in IRC +10216: A new interstellar phosphorus-containing molecule. *Astrophys. J., 677*, L101–L104.

Halfen, D. T., Clouthier, D. J., & Ziurys, L. M., (2014). Millimeter/submillimeter spectroscopy of PH_2CN and CH_3PH_2: Probing the complexity of interstellar phosphorus chemistry. *Astrophys. J., 796*, 36–42.

Halfen, D. T., Ilyushin, V. V., & Ziurys, L. M., (2015). Interstellar detection of methyl isocyanate CH_3NCO in Sgr B2(N): A link from molecular clouds to comets. *Astrophys. J. Lett., 812*, L5–L12.

Halmann, M., (1974). Evolution and ecology of phosphorus metabolism. In: Dose, K., Fox, S. W., Deborin, G. A., & Pavlovskaya, T. E., (eds.), *The Origin of Life and Evolutionary Biochemistry* (pp. 169–182). Plenum Press: New York.

Han, X., Han, J., Liu, C., & Sun, J., (2018). Promise and challenge of phosphorus in science, technology, and application. *Adv. Funct. Mater., 28*, 1803471 (56pp). doi: 10.1002/adfm.201803471.

Harrington, J., De Pater, I., Brecht, S. H., Deming, D., Meadows, V., Zahnle, K., & Nicholson, P. D., (2004). Lessons from shoemaker-levy 9 about Jupiter and planetary impacts. In: Bagenal, F., Dowling, T. E., & McKinnon, W. B., (eds.), *Jupiter: The Planet, Satellites and Magnetosphere* (Vol. 1, pp. 159–184). Cambridge University Press, Cambridge Planetary Science.

Haverty, D., Tofail, S. A. M., Stanton, K. T., & McMonagle, J. B., (2005). Structure and stability of hydroxyapatite: Density functional calculation and Rietveld analysis. *Phys. Rev. B, 71*, 094103–1–9.

Hawthorne, F. C., Cooper, M. A., Green, D. I., Starkey, R. E., Roberts, A. C., & Grice, J. D., (1999). Wooldridgeite, $Na_2CaCu_{22}+(P_2O_7)_2(H_2O)_{10}$: A new mineral from Judkins Quarry, Warwickshire, England. *Mineral. Mag., 63*, 13–16.

Hazen, R. M., Papineau, D., Blekeer, W., Down, R. T., Ferry, J. M., McCoy, T. J., Sverjensky, D. A., & Yamg, H., (2008). Mineral evolution. *Am. Mineralog.*, *3*, 1603–1720.

Heckel, W. M., (2002). Molecular self-assembly and the origin of life. In: *Astrobiology: The Quest for the Conditions of Life* (pp. 361–372). Springer, Berlin.

Heger, A., & Woosley, S. E., (2002). The nucleosynthetic signature of population III. *Astrophys. J.*, *567*, 532–543.

Heraclitus, DK B41. Retrieved from: http://www.heraclitusfragments.com/files/ge.html (Accessed on 26 July 2019).

Heritier, K. L., Altwegg, K., Berthelier, J. J., Beth, A., Carr, C. M., De Keyser, J., et al., (2018). On the origin of molecular oxygen in cometary comae. *Nature Commun.*, *9*, 2580. doi: 10.1038/s41467–018–04972–5.

Heritier, K. L., et al., (2017). Ion composition at comet 67P near perihelion: Rosetta observations and model-based interpretation. *Mon. Not. Royal Astron. Soc.*, *469*, S427–S442.

Herschbach, D., (1999). Chemical physics: Molecular clouds, clusters and corrals. *Rev. Mod. Phys.*, *71*, S411–S418.

Herschy, B., Chang, S. J., Blake, R., Lepland, A., Abbott-Lyon, H., Sampson, J., Atlas, Z., Kee, T. P., & Pasek, M. A., (2018). Archean phosphorus liberation induced by iron redox geochemistry. *Nature Comm.*, *9*, 1346–1352.

Hildebrand, R. L., (1983). *The Role of Phosphonates in Living Systems*. CRC Press, Boca Raton, FL.

Hittorf, W., (1865). Zur kenntniss des phosphors. *Ann. Phys. Chem.*, *126*, 193.

Hohsaka, T., & Sisido, S., (2002). Incorporation of non-natural amino acids into proteins. *Current Opinion in Chemical Biology*, *6*, 809–815.

Hollenbach, D. J., & Tielens, A. G. G. M., (1999). Photodissociation regions in the interstellar medium of galaxies. *Rev. Mod. Phys.*, *71*, 173–230.

Hollis, J. M., Lovas, F. J., & Jewell, P. R., (2000). Interstellar glycolaldehyde: The first sugar. *Astrophys. J.*, *540*, L107–L110.

Hollis, J. M., Lovas, F. J., Jewell, P. R., & Coudert, L. H., (2002). Interstellar antifreeze: Ethylene glycol. *Astrophys. J.*, *571*, L59–L62.

Holmström, K. M., Marina, N., Baev, A. Y., Wood, B. W., Gourine, A. V., & Abramov, A. Y., (2013). Signaling properties of inorganic polyphosphate in the mammalian brain. *Nature Comm.*, *4*, 1362, doi: 10.1038/ncomms2364.

Hoyle, F., & Wikramashinghe, C., (1981). *Evolution from Space*. Dover, London.

Hubrig, S., Castelli, F., De Silva, G., González, J. F., Momany, Y., Netopil, M., & Moehler, S., (2009). A high-resolution study of isotopic composition and Chemicals abundances of blue horizontal branch stars in the globular clusters NGC 6397 and NGC 6752. *Astron. Astrophys.*, *499*, 865–878.

Huertas, M. J., & Michán, C., (2012). Indispensable or toxic? The phosphate versus arsenate debate. *Microbial Biotechnology*, *6*, 209–211.

Iafisco, M., & Delgado-López, J. M., (2018). Biomimetic growth of calcium phosphate crystals. *Crystals*, *8*, 5.

Jacobson, H. R., Thanathibodee, T., Frebel, A., Roederer, I. U., Cescutti, G., & Matteucci, F., (2014). The chemical evolution of phosphorus. *Astrophys. J. Lett.*, *796*, L24–L29.

Jarosewich, E., (1990). Chemical analysis of meteorites: A compilation of stony and iron meteorite analysis. *Meteoritics*, *25*, 323–337.

Jasinski, S. M., (2018). *Phosphate Rock* (pp. 122–123). U.S. Geological Survey, Mineral Commodity Summaries.

Jay, E. E., Rushton, M. J. D., & Grimes, R. W., (2012). Migration of fluorine in fluorapatite – a concerted mechanism. *J. Mater. Chem.*, *22*, 6097–6103.

Jenkins, E. B., (2009). A unified representation of gas-phase element depletions in the interstellar medium. *Astrophys. J.*, *700*, 1299–1348.

Jessberger, E. K., Christoforidis, A., & Kissel, J., (1988). Aspects of the major element composition of Halley's dust. *Nature*, *332*, 691–695.

Jeyasingh, P. D., & Weider, L. J., (2007). Fundamental links between genes and elements: Evolutionary implications of ecological stoichiometry. *Molecular Ecology*, *16*, 4649–4661.

Jiménez-Serra, I., Viti, S., Quénard, D., & Holdship, J., (2018). The chemistry of phosphorus-bearing molecules under energetic phenomena. *Astrophys. J.*, *862*, 128–143.

Jing, Y., Ma, Y., Li, Y., & Heine, T., (2017). GeP_3: A small indirect bandgap 2D crystal with high carrier mobility and strong interlayer quantum confinement. *Nano Lett.*, *17*, 1833–1838. doi: 10.1021/acs.nanolett.6b05143.

Jørgensen, G. U., (1997). In: Van Dishoeck, E. F., (ed.), *Molecules in Astrophysics: Probes and Processes* (p. 441). IAU 178, Kluwer Academic Publisher, Dordrecht.

José, J., & Hernanz, M., (1998). Nucleosynthesis in classical novae: CO versus ONe white dwarfs. *Astrophys. J.*, *404*, 680–690.

Joshi, P. R., Ramanathan, N., Sundarajan, K., & Sankaran, K., (2017). Phosphorus bonding in PCl_3:H_2O adducts: A matrix isolation infrared and ab initio computational effects. *J. Mol. Spectr.*, *331*, 44–52.

Jugovic, D., & Uskokovic, D., (2009). A review of recent developments in the synthesis of lithium iron phosphate powders. *J. Power Sour.*, *190*, 538–544.

Jungck, M. H. A., & Niederer, F. R., (2017). From supernova to solar system: Few years only, first solar system components apatite and spinal determined. *Polar Sci.*, *11*, 54–71.

Kafando, I., LeBlanc, F., & Robert, C., (2016). Detailed abundance analysis of five field blue horizontal-branch stars. *Month. Not. Royal Astron. Soc.*, *459*, 871–879.

Kan, S. B. J., Lewis, R. D., Chen, K., & Arnold, F. H., (2016). Directed evolution of cytochrome c for carbon-silicon bond formation: Bringing silicon to life. *Science*, *354*, 1048–1051.

Karki, M., Gibard, C., Bhowmik, S., & Krishamurthy, R., (2017). Nitrogenous derivatives of phosphorus and the origins of life: Plausible prebiotic phosphorylating agents in water. *Life*, *7*, 32, doi: 10.3390/life7030032.

Karlsson, T., Bromm, V., & Bland-Hawthorn, J., (2013). Pregalactic metal enrichment: The chemical signatures of the first stars. *Rev. Mod. Phys.*, *85*, 809–848.

Karttunen, A. J., Linnolahti, M., & Pakkanen, T. A., (2007). Icosahedral and ring-shaped allotropes of phosphorus. *Chem. Eur. J.*, *13*, 5232–5237.

Kasting, J. F., Kopparapu, R., Ramirez, R. M., & Harman, C. E., (2014). Remote life-detection criteria, habitable zone boundaries, and the frequency of earth-like planets around M and late K stars. *Proc. Nat. Acad. Sci.*, *111*, 12641–12646, doi: 10.1073/pnas.1309107110.

Kasting, J. F., Whitmire, D. P., & Reynolds, R. T., (1993). Habitable zones around main-sequence stars. *Icarus*, *101*, 108–128.

Kato, K., & Watanabe, Y., (1996). Atmospheric abundances of light elements in the F-type star procyon. *Publ. Astron. Soc. Japan, 48*, 601–606.

Kaufmann, III, W. J., & Freedman, R. A., (1999). *Universe*, (5th edn.). Freeman, W.H. and Company New York.

Kaur, P., Chakraverty, S., Ganguli, A. K., & Bera, C., (2017). High anisotropic thermoelectric effect in palladium phosphide sulfide. *Phys. Status Solidi B, 254,* 1700021. doi: 10.1002/pssb1700021.

Kawai, J., (2017). Comment on "ribose and related sugars from ultraviolet irradiation of interstellar ice analogs." *Science, 355*, 141a.

Kaye, J. A., & Strobel, D. F., (1983). Phosphine photochemistry in Saturn's atmosphere. *Geophys. Res. Lett., 10*, 957–960.

Keefe, A. D., & Miller, S. L., (1995). Are polyphosphates or phosphate esters prebiotic reagents? *J. Mol. Evol., 41*, 693–702.

Khaoulaf, R., Adhikari, P., Harcharras, M., Brouzi, K., Ez-Zahraouy, H., & Ching, W.-Y., (2019). Atomic-scale understanding of structure and properties of complex pyrophosphate crystals by first-principles calculations. *Appl. Sci. 9,* 840 (16pp). doi: 10.3390/app9050840.

Kikewaga, T., & Iwasaki, H., (1983). An X-ray diffraction study of lattice compression and phase transition of crystalline phosphorus. *Acta Cryst., B39*, 158–164.

Kipp, M. A., & Stüeken, E. E., (2017). Biomass recycling and earth's early phosphorus cycle. *Sci. Adv., 3*, 4795–4800.

Kissel, J., & Krueger, F. R., (1987). The organic component in dust from comet Halley as measured by the PUMA mass spectrometer on board Vega 1. *Nature (London), 326,* 755–760.

Kissel, J., et al., (1986a). Composition of comet Halley dust particles from Vega observations. *Nature (London), 321*, 280–282.

Kissel, J., et al., (1986b). Composition of comet Halley dust particles from Giotto observations. *Nature (London), 321*, 336–337.

Kissel, J., Krueger, F. R., Silén, J., & Clark, B. C., (2004). The cometary and interstellar dust analyzer at comet 81P/Wild 2. *Science, 304*, 1774–1776.

Kitadai, N., & Maruyama, S., (2018). Origins of building blocks of life: A review. *Geoscience Frontiers, 9*, 1117–1153.

Kleine, T., Mezger, K., Palme, H., & Münker, C., (2004). The W isotope evolution of the bulk silicate earth: Constraints on the timing and mechanisms of core formation and accretion. *Earth Plant. Sci. Lett., 228*, 109–123.

Kobayashi, C., Karakas, A. I., & Umeda, H., (2011). The evolution of isotope ratios in the Milky Way galaxy. *Mon. Not. R. Astron. Soc., 414*, 3231–3250.

Kocková-Kratochvílová, A., (1990). *Yeast and Yeast-Like Organisms.* VCH: New York.

Koo, B. C., Lee, Y. H., Moon, D. S., Yoon, S. C., & Raymond, J. C., (2013). Phosphorus in the young supernova remnant Cassiopeia A. *Science, 342*, 1246–1348.

Kopparapu, R. K., Ramirez, R., Kasting, F. J., Eymet, V., Robinson, T. D., et al., (2013). Habitable zones around main-sequence stars: New estimates. *Astrophys. J., 765,* 131–146; *Erratum Astrophys. J., 770*, 82–84.

Krafft, F., (1969). Phosphorus: From elemental light to chemical element. *Angew. Chem. Internat. Edit., 8*, 660–671.

Kreidberg, L., L. Bean, J., Désert, J. M., Line, M. R., Fortney, J. J., Madhusudhan, N., et al., (2014). A precise water abundance measurement for the hot Jupiter WASP -43b. *Astrophys. J. Lett.*, *793*, L27–L32.

Kroto, H. W., (1997). Symmetry, space, stars and C_{60}. *Rev. Mod. Phys.*, *69*, 703–722.

Kroto, H. W., Nixon, J. F., Ohno, K., & Simmons, N. P. C., (1980). The microwave spectrum of phosphaethene, $HP = CH_2$. *JCS Chem. Comm.*, 709a.

Kuan, Y. J., Charnley, S. B., Huang, H. C., Tseng, W. L., & Kisiel, Z., (2003). Interstellar glycine. *Astrophys. J.*, *593*, 848–867.

Kumble, K. D., & Kornberg, A., (1995). Inorganic polyphosphate in mammalian cells and tissues. *J. Bio. Chem.*, *270*, 5818–5822.

Kunde, V. G., et al., (2004). Jupiter's atmospheric composition from the Cassini thermal infrared spectroscopy experiment. *Science*, *305*, 1582–1586.

Kwok, S., (1993). Proto-planetary nebulae. *Ann. Rev. Astron. Astrophys.*, *31*, 63–92.

Kwok, S., (2000). *The Origin and Evolution of Planetary Nebulae.* Cambridge University Press: Cambridge.

Kwok, S., (2004). The synthesis of organic and inorganic compounds in evolved stars. *Nature*, *430*, 985–991.

Kwok, S., (2006). Research highlights journal club. An astronomer is bugged by the scarcity of one of life's vital elements in space. *Nature*, *439*, 637.

Kwok, S., (2007). A new molecular factory. *Nature*, *447*, 1063.

Kwok, S., (2013). *Stardust: The Cosmic Seeds of Life.* Springer Verlag: Heidelberg.

Lambert, J. B., & Gurusamy-Thangavelu, S. A., (2015). Roles of silicon in life on earth and elsewhere. In: Kolb, V. M., (ed.), *Astrobiology: An Evolutionary Approach* (pp. 149–161). CRC Press: Boca Raton.

Lambert, J. B., Gurusamy-Thangavelu, S. A., & Ma, K., (2010). The silicate-mediated formose reaction: Bottom-up synthesis of sugar silicates. *Science*, *327*, 984–986.

Lange, H. C., & Heijnen, J. J., (2001). Statistical reconciliation of the elemental and molecular biomass composition of *Saccharomyces cerevisiae*. *Biotechnol. Bioeng.*, *75*, 334–344.

Larralde, R., Robertson, M. P., & Miller, S. L., (1995). Rates of decomposition of ribose and other sugars: Implications for chemical evolution. *Proc. Nat. Acad. Sci. USA*, *92*, 8158–8160.

Lattanzi, V., Thorwirth, S., DeWayne, T., Halfen, T., Mück, A. L., Ziurys, L. M., Thaddeus, P., Gauss, J., & McCarthy, M. C., (2010). Bonding in the heavy analog of hydrogen cyanide: The curious case of bridged HPSi. *Angew. Chem. Int. Ed.*, *49*, 5661–5664.

Lazcano, A., & Miller, S. L., (1996). The origin and early evolution of life: Prebiotic chemistry, the pre-RNA World, and time. *Cell*, *85*, 793–798.

Lebouteiller, V. K., & Ferlet, R., (2005). Phosphorus in the diffuse interstellar medium. *Astron. Astrophys.*, *443*, 509–517.

Lebouteiller, V., Heap, S., Hubeney, I., & Kunth, D., (2013). Chemical enrichment and physical conditions in I Zw 18. *Astron. Astrophys.*, *553*, A16.

Lebouteiller, V., Kunth, D., Lequeux, J., Désert, J. M., Hébrard, G., Lecavelier des Étangs, A., & Vidal-Madjar, A., (2006). Interstellar abundances in the neutral and ionized gas of NGC 604. *Astron. Astrophys.*, *459*, 161–174.

Lee, C. F., Li, Z. Y., Ho, P. T. P., Hirano, N., Zhang, Q., & Shang, H., (2017). Formation and atmosphere of complex organic molecules of the HH 212 protostellar disk. *Astrophys. J.*, *843*, 27–30.

Lee, Y. H., Koo, B. C., Moon, D. S., Burton, M. G., & Lee, J. J., (2017). Near-infrared knots and dense Fe ejecta in the Cassiopeia A supernova remnant. *Astrophys. J., 837*, 118–133.

Lefloch, B., Vastel, C., Viti, S., Jiménez-Serra, I., Codella, C., Podio, L., Ceccarelli, C., Mendoza, E., Lepine, J. R. D., & Bachiller, R., (2016). Phosphorus-bearing molecules in solar-type star forming regions: First PO detection. *Mon. Not. R. Astron. Soc., 462*, 3937–3944.

Leibniz, G. W., (1710). Historia inventionis phosphori. In: Sumptibus J. C., (ed.), *Miscellanea Berolnensia ad Incrementum Scientiarum I* (pp. 91–98). Papenii: Berlin.

Levasseur-Regourd, A. C., Agarwal, J., Cottin, H., Engrand, C., Flynn, G., Fulle, M., et al., (2018). Cometary dust. *Space Sci. Rev., 214*, 64–120.

Levine, J. S., (1985). *The Photochemistry of Atmospheres: Earth, the Other Planets and Comets*. Academic Press: Orlando.

Levshakov, S. A., Dessauges-Zavadsky, M., D'Odorico, S., & Molaro, P., (2002). Molecular hydrogen, deuterium, and metal abundances in the damped Lyα system at z = 3.025 toward Q0347–3819. *Astrophys. J., 565*, 696–719.

Lewis, J. S., (1969). Observability of spectroscopically active compounds in the atmosphere of Jupiter. *Icarus, 10*, 393.

Li, J., Shen, Z., Wang, J., Chem, X., Li, D., Wu, Y., Dong, J., et al., (2017). Widespread presence of glycolaldehyde and ethylene glycol around Sagittarius B2. *Astrophys. J, 849*, 115–123.

Li, Y., Cai, S., Xu, G., Shen, S., Zhang, M., Zhang, T., & Sun, X., (2015). Synthesis and characterization of phytic acid/mesoporous 45S5 bioglass composite coating on magnesium alloy and degradation behavior. *RSC Adv., 5*, 25708–25716. doi: 10.1039/C5RA00087D.

Liebisch, T. C., Stenger, J., & Ullrich, J., (2019). Understanding the revised SI: Background, consequences, and perspectives. *Ann. Phys. (Berlin), 531*, 1800339 (1 of 11).

Ling, X., Wang, H., Huang, S., Xia, F., & Dresselhaus, M. S., (2015). The renaissance of black phosphorus. *Proc. Nat. Acad. Sci., 112*, 4523–4530.

Lingam, M., & Loeb, A., (2018). Is extraterrestrial life supressed on subsurface ocean worlds due to the paucity of bioessential elements? *Astron. J., 156,* 151 (7pp).

Liszt, H. S., Lucas, R., & Pety, J., (2001). Comparative chemistry of diffuse clouds – V-Ammonia and formaldehyde. *Astron. Astrophys., 448*, 253–259.

Liu, H., Neal, A. T., Zhu, Z., Luo, Z., Xu, X., Tománek, D., & Ye, P. D., (2014). Phosphorene: An unexplored 2D semiconductor with a high hole mobility. *ACS Nano, 8*, 4033–4041.

Liu, Z., Rossi, J.-C., & Pascal, R., (2019). How prebiotic chemistry and early life chose phosphate. *Life, 9,* 26 (16pp). doi: 10.3390/life9010026.

Llorca, J., (2005). Organic matter in comets and cometary dust. *Int. Microbiol., 8*, 5–12.

Lodders, K., (2003). Solar system abundances and condensation temperatures of the elements. *Astrophys. J., 591*, 1220–1247.

Lodders, K., (2004). Revised and updated thermochemical properties of the gases mercapto (HS), disulfur monoxide (Si_2O), thiazyl (NS), and thioxophosphino (PS). *J. Phys. Chem. Ref. Data, 33*, 357–367.

López, S., Reimers, F., D'Odorico, S., & Prochaska, J. X., (2002). Metal abundances and ionization conditions in a possibly dust-free damped Lyα system at z = 2.3. *Astron. Astrophys., 385*, 778–792.

Lowe, C. U., Rees, M. W., & Markham, R., (1963). Synthesis of complex organic compounds from simple precursors: Formation of amino acids, amino-acid polymers, fatty acids, and purines from ammonia cyanide. *Nature*, *199*, 219–222.
Lunine, J. I., (1993). The atmospheres of Uranus and Neptune. *Ann. Rev. Astron. Astrophys.*, *31*, 217–263.
Lyubimkov, K. S., Pokland, D. B., & Rachkovskaya, T. M., (2008). Stratification of phosphorus in the atmosphere of the chemically peculiar B-type star HR 1512. *Astrophysics*, *51*, 197–208.
Maas, Z. G., Pilachowski, C. A., & Cescutti, G., (2017). Phosphorus abundances in FGK stars. *Astrophys. J.*, *841*, 108–116.
Maciá-Barber, E., (2009). *Aperiodic Structures in Condensed Matter: Fundamentals and Applications*. CRC Press: Boca Raton.
Maciá-Barber, E., (2015). *Thermoelectric Materials: Advances and Applications.* Pan Stanford Publishing: Singapore.
Maciá, E., (2005). The role of phosphorus in chemical evolution. *Chem. Soc. Rev.*, *34*, 691–701.
Maciá, E., (2012). Quasicrystals and the quest for next generation thermoelectric materials. *Critical Rev. Solid State Mat. Sci.*, *37*, 215–242.
Maciá, E., Hernández, M. V., & Oró, J., (1997). Primary sources of phosphorus and phosphates in chemical evolution. *Origins Life Evol. Biosphere*, *27*, 459–480.
MacKay, D. D. S., & Charnley, S. B., (2001). Phosphorus in circumstellar envelopes. *Mon. Not. R. Astron. Soc.*, *325*, 545–549.
MacMullan, P., McMahon, G., & McCarthy, G., (2011). Detection of basic calcium phosphate crystals in osteoarthritis. *Joint Bone Spine*, *78*, 358–363.
Madhusuman, N., Agúndez, M., Moses, J. I., & Hu, Y., (2016). Exoplanetary atmospheres – Chemistry, firnation conditions, and habitability. *Space Sci. Rev.*, *205*, 285–348.
Maier, J. P., Lakin, N. M., Walker, G. A. H., & Bohlender, D. A., (2001). Detection of C_3 in diffuse interstellar clouds. *Astrophys. J.*, *553*, 267–273.
Mandel, N. S., (1975). The crystal structure of calcium pyrophosphate dehydrate. *Acta Cryst.*, *B31*, 1730–1734.
Marcano, V., Benitez, P., Campins, J., Matheus, P., Cedeño, X., Falcon, N., & Palacios-Prü, E., (2004). Pyrolisis of phosphorylated molecules and survivability limits during the atmospheric passage in earth-like planets. *Planet. Space Sci.*, *52*, 613–621.
Marcolino, W. L. F., Hillier, D. J., De Araujo, F. X., & Pereira, C. B., (2007). Detailed far-ultraviolet to optical analysis of four [WR] stars. *Astrophys. J.*, *654*, 1068–1086.
Mason, S. F., (1992). *Chemical Evolution.* Clarendon Press: Oxford.
Matsuura, M., Speck, A. K., Smith, M. D., Zijlstra, A. A., et al., (2007). VLT/near-infrared integral field spectrometer observations of molecular hydrogen lines in the knots of the planetary nebula NGC 7293 (the Helix Nebula). *M. N. R. A. S.*, *382*, 1447–1459. doi: 10.1111/j.1365–2966.2007.12496.x.
Matteucci, F., (2016). Introduction to galactic chemical evolution. *J. Phys.: Conference Ser.*, *703*(012004), 1–9.
Mayor, M., & Queloz, D., (1995). A Jupiter-mass companion to a solar-type star. *Nature (London)*, *378*, 355–359.
McCoy, T. J., (2010). Mineralogical evolution of meteorites. *Elements*, *6*, 19–23.

McCoy, T. J., Steele, I. M., Keil, K., Leonard, B. F., & Endress, M., (1993). Chladniite: A new mineral honoring the father of meteoritics. *Meteoritics*, *28*, 394.

McDonald, G. D., (2015). Biochemical pathways as evidence for prebiotic synthesis. In: Kolb, V. M., (ed.), *Astrobiology: An Evolutionary Approach* (pp. 119–147). CRC Press: Boca Raton.

McDonnell, J. A. M., et al., (1986). Density and mass distribution near comet Halley from Giotto observations. *Nature (London)*, *321*, 338–341.

McGuinness, E. T., (2010). Some molecular moments of the Hadean and archaean aeons: A retrospective overview from the interfacing years of the second to third millennia. *Chem. Rev.*, *110*, 5191–5215.

McGuire, B. A., Burkhardt, A. M., Kalenskii, S., Shingledecker, C. N., Remijan, A. J., Herbst, E., & McCarthy, M. C., (2018). Detection of the aromatic molecule benzonitrile (c-C_6H_5CN) in the interstellar medium. *Science*, *359*, 202–205.

McKay, C. P., (1992). In: Carle, G., Schwartz, D., & Huntington, J., (eds.), *Exobiology in Solar System Exploration* (Vol. 512, p. 67). NASA SP.

McSween, Jr., H. Y., (2015). Pretology on mars. *American Mineralogist*, *100*, 2380–2395.

Meiksin, A. A., (2009). The physics of the intergalactic medium. *Rev. Mod. Phys.*, *81*, 1405–1469.

Meinert, C., Myrgorodska, I., De Marcellus, P., Buhse, T., Nahon, L., Hoffmann, S. V., Le Sergeant, D. L., & Meierhenrich, U. J., (2016). Ribose and related sugar from ultraviolet irradiation of interstellar ice analogs. *Science*, *352*, 208–212.

Meinert, C., Myrgorodska, I., De Marcellus, P., Buhse, T., Nahon, L., Hoffmann, S. V., Le Sergeant, D. L., & Meierhenrich, U. J., (2017). Ribose and related sugar from ultraviolet irradiation of interstellar ice analogs. *Science*, *355*, 141, 142.

Meléndez, J., Asplund, M., Gustafsson, B., & Yong, D., (2009). The peculiar solar composition and its possible relation to planet formation. *Astrophys. J.*, *704*, L66–L70.

Mighell, A. D., Smith, J. P., & Brown, W. E., (1969). The crystal structure of phosphoric acid hemihydrate, $H_3PO_4 \cdot ½H_2O$. *Acta Cryst.*, *V25*, 776–781.

Milam, S. N., Halfen, D. T., Tenenbaum, E. D., Apponi, A. J., Woolf, N. J., & Ziurys, L. M., (2008). Constraining phosphorus chemistry in carbon- and oxygen-rich circumstellar envelopes: Observations of PN, HCP and CP. *Astrophys. J.*, *684*, 618–625.

Millar, T. J., Bennett, A., & Herbst, E., (1987). An efficient gas-phase synthesis for interstellar PN. *Mon. Not. R. Astron. Soc.*, *229*, 41p–44p.

Miller, S. L., (1953). A production of amino acids under possible primitive earth conditions. *Science*, *117*, 528–529.

Miller, S. L., (1957). The mechanism of synthesis of amino acids by electric discharges. *Biochim. Biophys. Acta*, *23*, 480–489.

Miller, S. L., (1997). Peptide nucleic acids and prebiotic chemistry. *Nature Struct. Biol.*, *4*, 167–169.

Miller, S. L., (1998). The endogenous synthesis of organic compounds. In: Brak, A., (ed.), *The Molecular Origins of Life: Assembling Pieces of the Puzzle*. Cambridge University Press: Cambridge.

Miller, S. L., & Orgel, L. E., (1974). *The Origins of Life on Earth.* Prentice-Hall: New Jersey.

Miller, S. L., & Parris, M., (1964). Synthesis of pyrophosphate under primitive earth conditions. *Nature*, *204*, 1248–1249.

Miller, S. L., & Schlesinger, G., (1984). Carbon and energy yields in prebiotic syntheses using atmospheres containing CH_4, CO and CO_2. *Orig. Life*, *14*, 83–90.

Mininni, C., Fontanic, F., Rivilla, V. M., Beltrán, M. T., Caselli, P., & Asyunin, A., (2018). On the origin of phosphorus nitride in star-forming regions. *Mon. Not. Roy. Astron. Soc.*, *476*, L39–L44.

Mo, H., Van Den Bosch, F. C., & White, S., (2010). *Galaxy Formation and Evolution.* Cambridge University Press: Cambridge.

Mojzsis, S., Arrhenius, G., McKeegan, K., Harrison, K. D., Nutman, A. P., & Friend, C. R. L., (1996). Evidence for life on earth before 3,800 million years ago. *Nature*, *384*, 55–58.

Molaro, P., Levshakov, S. A., D'Odorico, S., Bonifacio, P., & Centurión, M., (2001). UVES observations of QSO 0000 -2620: Argon and phosphorus abundances in the dust-free damped Lyα system at z = 3.3901. *Astrophys. J.*, *549*, 90–99.

Morbidelli, A., Chambers, J., Lunine, J. I., Petit, J. M., Robert, F., Valsecchi, G. B., & Cyr, K. E., (2000). Source regions and timescales for the delivery of water to the earth. *Meteorit. Planet. Sci.*, *35*, 1309–1320.

Morton, S. C., & Edwards, M., (2005). Reduced phosphorus compounds in the environment. *Critical Rev. Environ. Sci. Tech.*, *35*, 333–364.

Müller, U., (2007). *Inorganic Structural Chemistry* (2nd edn.). John Wiley & Sons: Chichester.

Muñoz-Caro, G. M., Meierhenrich, U. J., Schutte, W. A., Barbier, B., Arcones-Segovia, A., Rosenbauer, H., Thiemann, W. H. P., Brack, A., & Greenberg, J. M., (2002). Amino acids from ultraviolet irradiation of interstellar ices analogs. *Nature*, *416*, 403–406.

Murphy, D. M., Thomson, D. S., & Mahoney, M. J., (1998). In situ measurements of organics, meteoritic material, mercury, and other elements in aerosols at 5 to 19 kilometers. *Science*, *282*, 1664–1669.

Nash, W. P., (1984). Phosphate minerals in terrestrial igneous and metamorphic rocks. In: Nriagu, J. O., & Moore, P. B., (eds.), *Phosphate Minerals* (pp. 215–241). Springer-Verlag: Berlin.

Natta, G., & Passerini, L., (1930). The crystal structure of white phosphorus. *Nature*, *125*, 707.

Nazarov, M. A., Kurat, G., Brandstaetter, F., Ntaflos, T., Chaussidon, M., & Hoppe, P., (2009). Phosphorus-bearing sulfides and their associations in CM chondrites. *Petrology*, *17*, 101–123.

Nicolau, D. V., Lard, M., Korten, T., Falco, C. M., Van Delft, J. M., Persson, M., et al., (2016). Parallel computation with molecular-motor-propelled agents in nanofabricated networks. *Proc. Nac. Acad. Sci.*, *113*, 2591–2596.

Noble, D., (2008). *The Music of Life*. Oxford University Press, Oxford.

Norman, P. D., (1993). Models of the meson bond structures of the most abundant products of stellar nucleosynthesis. *Eur. J. Phys.*, *14*, 36–42.

Nriagu, J. O., (1984). Phosphate minerals: Their properties and general modes of occurrence. In: Nriagu, J. O., & Moore, P. B., (eds.), *Phosphate Minerals* (pp.1–136). Springer-Verlag: Berlin.

O'Dell, C. R., Balick, B., Hajian, A. R., Henney, W. J., et al., (2002). Knots in nearby planetary nebulae. *The Astronomical Journal*, *123*, 3329–3347. doi: 10.1086/340726.

Oberhummer, H., Csoto, A., & Schlattl, H., (2000). Stellar production rates of carbon and its abundance in the universe. *Science*, *289*, 88–90.

Ohishi, M., Yamamoto, S., Saito, S., Kawaguchi, K., Suzuki, H., Kaifu, N., Ishikawa, S. I., Takano, S., & Tsuji, T., (1988). The laboratory spectrum of the PS radical and related astronomical search. *Astrophys. J., 329*, 511–516.

Ohk, R. G., Chayer, P., & Moos, H. W., (2000). Photospheric metals in the far ultraviolet spectroscopic explorer spectrum of the subdwarf B star PG 0749+658. *Astrophys. J., 538*, L95–L98.

Okudera, H., Dinnebier, R. E., & Simon, A., (2005). The crystal structure of γ-P_4, a low temperature modification of white phosphorus. *Z. Kristallogr., 220*, 259–264.

Oliva, E., Marconi, A., Maiolino, R., et al., (2001). NICS-TNG infrared spectroscopy of NGC 1068: The first extragalactic measurement of [PII] and a new tool to constrain the origin of [FeII] line emission in galaxies. *Astron. Astrophys., 369*, L5–L8.

Oliva, E., Moorwood, A. F. M., Draptz, S., Lutz, D., & Sturm, E., (1999). Infrared spectroscopy of young supernova remnants heavily interacting with the interstellar medium. *Astron. Astrophys., 343*, 943–952.

Oparin, A. I., (1924). *The Origin of Life*. Moscow Worker Publisher: Moscow.

Örberg, K. I., Guzmán, V. V., Furuya, K., Qi, C., Aikawa, Y., Andrews, S. M., Loomis, R., & Wilner, D. J., (2015). The comet-like composition of a protoplanetary disk as revealed by complex cyanides. *Nature (London), 520*, 198–201.

Orgel, L. E., (1989). Was RNA the first genetic polymer? *Evolutionary Tinkering in Gene Expression* (pp. 215–224). Springer US.

Orgel, L. E., & Lahrmann, R., (1974). Prebiotic chemistry and nucleic acid replication. *Acc. Cham. Res., 7*, 368–377.

Oró, J., (1960). Synthesis of adenine from ammonium cyanide. *Biochem. Biophys. Res. Commun., 2*, 407–412.

Oró, J., (1961). Comets and the formation of biochemical compounds on the primitive earth. *Nature, 190*, 389–390.

Oró, J., (1963). Studies in experimental organic cosmochemistry. *Ann. N. Y. Acad. Sci., 108*, 464–481.

Oró, J., (1965). Stages and mechanisms of prebiological organic synthesis. In: Fox S. W., (ed.), *Origins of Prebiological Systems* (pp. 137–171). Academic Press Inc.

Oró, J., (1995). From cosmochemistry to life and man. In: Poglazov, B., et al., (eds.), *Evolutionary Biochemistry and Related Areas of Physicochemical Biology.* Bach Institute of Biochemistry: Moscow.

Oró, J., & Kamat, S. S., (1961). Amino acids synthesis from hydrogen cyanide under possible primitive earth conditions. *Nature, 590*, 442–443.

Oró, J., Lazcano, A., & Ehrenfreund, P., (2006). Comets and the origin and evolution of life. In: Thomas, P. J., Hicks, R. D., Chyba, C. F., & McKay, C. P., (eds.), *Comets and the Origin and Evolution of Life* (2nd edn.). Springer: Berlin.

Oró, J., Miller, S. L., & Lazcano, A., (1990). The origin and early evolution of life on earth *Ann. Rev. Earth Planet. Sci., 18*, 317–356.

Oró, J., Miller, S. L., & Lazcano, A., (1992). Comets and the formation of biochemical compounds on the primitive earth. *Origins Life Evol. Biosphere, 21*, 267–277.

Oró, J., Sherwood, E., Elchberg, J., & Epps, D., (1978). Formation of phospholipids under primitive earth conditions and the role of membranes in prebiological evolution. In: Deamer, D., (ed.), *Light Transducing Membranes, Structure, Function and Evolution* (pp. 1–21). Academic Press: New York.

Owen, T., Bar-nun, A., & Kleinfled, I., (1992). Possible cometary origin of heavy noble gases in the atmospheres of Venus, Earth and Mars. *Nature (London)*, *358*, 43.

Oxtoby, D. W., Gillis, H. P., & Butler, L. J., (2015). *Principles of Modern Chemistry* (8th edn., p. 934). Cengage Learning: Boston, MA.

Palache, C., Berman, H., & Frondel, C., (1951), *The System of Mineralogy of James Dwight Dana and Edward Salisbury Dana, Yale University 1837–1892* (7th edn., Vol. II, p. 684). John Wiley and Sons, Inc., New York.

Papike, J. J., Taylor, L. A., & Simons, S. B., (1991). Lunar minerals. In: *Lunar Sourcebook: A User's Guide to the Moon* (pp. 121–181). Cambridge University Press: Cambridge.

Pasek, M. A., (2008). Rethinking early earth phosphorus geochemistry. *Proc. Nat. Acad. Sci.*, *105*, 853–858.

Pasek, M. A., (2015a). Phosphorus as a lunar volatile. *Icarus*, *255*, 18–23.

Pasek, M. A., (2015b). Role of phosphorus in prebiotic chemistry. In: Kolb, V. M., (ed.), *Astrobiology: An Evolutionary Approach* (pp. 257–270). CRC Press: Boca Raton.

Pasek, M. A., & Kee, T. P., (2011). On the origin of phosphorylated biomolecules. In: Egel, R., et al., (eds.), *Origin of Life: The Primal Self-Organization* (pp. 57–84). Springer-Verlag: Berlin.

Pasek, M. A., & Lauretta, D. S., (2005). Aqueous corrosion of phosphide minerals from iron meteorites: A highly reactive source of prebiotic phosphorus on the surface of the early earth. *Astrobiology*, *5*, 515–535.

Pasek, M. A., Dworkin, J. P., & Lauretta, D. S., (2007). A radical pathway for organic phosphorylation during schreibersite corrosion with implications for the origin of life. *Geochim. Cosmochim.*, *71*, 1721–1736.

Pasek, M. A., Gull, M., & Herschy, B., (2017). Phosphorylation on the early earth. *Chem. Geo.*, *475*, 149–170.

Pasek, M. A., Harnmeijer, J. P., Buick, R., Gull, M., & Atlas, Z., (2013). Evidence for reactive reduced phosphorus species in the early Archean ocean. *Proc. Nat. Acad. Sci.*, *110*, 10089–10094.

Peacor, D. R., Dunn, P. J., Simmons, W. B., & Wicks, F. J., (1985). Canaphite, a new sodium-calcium phosphate hydrate from the Paterson area, New Jersey. *Mineral. Theatr. Record*, 16, 467–468.

Penrose, R., (2006). *The Road to Reality: A Complete Guide to the Laws of the Universe* (p. xix). Alfred, A. Knopf: New York.

Piccirillo, J., (1980). Interpretation of the gross molecular spectra of S stars. *Mon. Not. R. Astr. Soc.*, *190*, 441–457.

Pila-Díez, B., (2015). *Structure and Substructure in the Stellar Halo of the Milky Way.* Netherlands' Research School for Astronomy, Leiden University, PhD Dissertation.

Pirim, C., Pasek, M. A., Sokolov, D. A., Sidorov, A. N., Gann, R. D., & Orlando, T. M., (2014). Investigation of schreibersite and intrinsic oxidation products from Sikhote-Alin, Seymchan, and Odessa meteorites and Fe_3P and Fe_2NiP synthetic surrogates. *Geochim. Cosmochim. Acta*, *140*, 259–274.

Popova, V. I., Popov, V. A., Sokolova, E. V., Ferraris, G., & Chukanov, N. V., (2002). Kanonerovite, $MnNa_3P_3O_{10}\cdot 12H_2O$, first triphosphate mineral (Kazennitsa pegmatite, Middle Urals, Russia). *Neues Jahrbuch für Mineralogie-Monatshefte*, *3*, 117–127.

Powner, M. W., Gerland, B., & Sutherland, J. D., (2009). Synthesis of activated pyrimidine ribonucleotides in prebiotically plausible conditions. *Nature*, *459*, 239–242.

Prialnik, D., (2011). *An Introduction to the Theory of Stellar Structure and Evolution* (Chapter 2, 2nd edn.). Cambridge University Press: Cambridge.

Pritekel, C., (2015). The crystal structure of meteoritic schreibersite: Refinement of the absolute crystal structure. *Geological Sciences Departmental Honors Thesis*. University of Colorado, Boulder. PhD Dissertation.

Radicati Di, B. F., Bunch, T., & Chang, S., (1986). Laser microprobe study of carbon in interplanetary dust particles (IDP). *Origins Life Evol. Biosphere, 16*, 236–237.

Ralchenko, Y., Kramida, A. E., & Reader, J., (2010). NIST ASD Team. *NIST Atomic Spectra Database Version 4.0.0*. Gaithersburg, MD: National Institute of Standards and Technology, http://www.nist.gov/physlab/data/asd.cfm (Accessed on 27 July 2019).

Rauchfuss, H., (2008). *Chemical Evolution and Origin of Life*. Springer-Verlag: Berlin.

Raunier, S., Chiavassa, T., Duvernay, F., Borget, F., Aycard, J. P., Dartois, E., & D'Hendecourt, L., (2004). Tentative identification of urea and formamide in ISO-SWS infrared spectra of interstellar ices *Astron. Astrophys., 416*, 165–169.

Reach, W. T., & Rho, J., (1998). Detection of far-infra-red water vapor, hydroxyl, and carbon monoxide emissions from the supernova remnant 3C 391. *Astrophys. J., 507*, L93–L97.

Redfield, A. C., (1958). The biological control of chemical factors in the environment. *Am. Sci., 46*, 205–221.

Remijan, A. J., Snyder, L. E., McGuire, B. A., Kuo, H. L., Looney, L. W., Friedel, D. N., Golubiatnikov, G. Y., Lovas, F. J., Ilyushin, V. V., & Alekseev, E. A., (2014). Observational results of a multi-telescope campaign in search of interstellar urea [$(NH_2)_2CO$]. *Astrophys. J., 783*, 77–92.

Rice, K., (2015). Origin of elements and formation of solar system, planets and exoplanets. In: Kolb, V. M., (ed.), *Astrobiology: An Evolutionary Approach* (p. 27). CRC Press: Boca Raton.

Richards, A. M. S., Yates, J. A., & Cohen, R. J., (1998). MERLIN observations of water maser proper motions in VY Canis Majoris. *Mon. Not. R. Astron. Soc., 299*, 319–331.

Rieder, R., Economou, T., Wänke, H., Turkevich, A., Crisp, J., Brückner, J., Dreibus, G., McSween, Jr. H. Y., (1997). The chemical composition of Martian soil and rocks returned by the mobile Alpha proton x-ray spectrometer: Preliminary results from the x-ray mode. *Science, 278*, 1771–1774.

Ritchey, A. M., Federman, S. R., & Lambert, P. L., (2018). Abundances and depletions of neutron-capture elements in the interstellar medium. *Astrophys. J. Suppl. Ser., 236*, 36–75.

Rivilla, V. M., Fontani, F., Beltrán, M. T., Vasyunin, A., Caselli, P., Martín-Pintado, J., & Cesaroni, R., (2016). The first detection of the key prebiotic molecule PO in star-forming regions. *Astrophys. J., 826*, 161–168.

Rivilla, V. M., Jiménez-Serra, I., Zeng, S., Martín, S., Martín-Pintado, J., Armijos-Abendaño, J., et al., (2018). Phosphorus-bearing molecules in the galactic center. *Month. Not. Roy. Astron. Soc., 475*, L30–L34.

Roberge, A., Feldman, P. D., Weinberger, A. J., Deleuil, M., & Bouret, J. C., (2006). Stabilization of the disk around β Pictoris by extremely carbon-rich gas. *Nature (London), 441*, 724–726.

Roederer, I. U., Jacobson, H. R., Thanathibodee, T., Frebel, A., & Toller, E., (2014). Detection of neutral phosphorus in the near-ultraviolet spectra of late-type stars. *Astrophys. J., 797*, 69–80.

Roederer, I. U., Placco, V. M., & Beers, T. C., (2016). Detection of phosphorus, sulfur, and zinc in the carbon-enhanced metal-poor star BD+44°493. *Astrophys. J. Lett., 824*, L19–L22.

Rolfs, C. E., & Rodney, W. S., (1988). *Cauldrons in the Cosmos*. University of Chicago Press: Chicago.

Rothschild, L. J., & Mancinelli, R. L., (2001). Life in extreme environments. *Nature, 409*, 1092–1101.

Ruck, M., Hoppe, D., Wahl, B., Simon, P., Wang, Y., & Seifert, G., (2005). Fibrous red phosphorus. *Angew. Chem. Int. Ed., 44*, 7616–7619.

Sánchez-Lavega, A., (2011). *An Introduction to Planetary Atmospheres; Chapter 3.* Taylor & Francis, CRC Press: Boca Raton.

Sauval, A. J., (1978). Predicted presence and tentative identification of new molecules in the pure S star R CYG Astron. *Astrophys., 62*, 295–298.

Savage, B. D., & Sembach, K. R., (1996). Interstellar abundances from absorption-line observations with the Hubble space telescope. *Ann. Rev. Astron. Astrophys., 34*, 279–330.

Scharf, C., Storrie-Lombardi, M., & McDonald, G. (2011). *The Biomass Capacity of the Galaxy*. (unpublished).

Scheer, M., Balázs, G., & Seitz, A., (2010). P_4 activation by main group elements and compounds. *Chem. Rev., 110*, 4236–4256.

Schink, B., & Friedrich, M., (2000). Bacterial metabolism: Phosphite oxidation by sulfate reduction. *Nature, 406*, 37–38.

Schipper, W. J., Klapwijk, A., Potjer, B., Rulkens, W. H., Temmenik, B. G., Kiestra, F. D. G., & Lijnbach, A. C. M., (2001). Phosphorus recycling in the phosphorus industry. *Environ. Technol., 22*, 1337–1345.

Schlesinger, M. E., (2002). The thermodynamic properties of phosphorus and solid binary phosphides. *Chem. Rev., 102*, 4267–4301.

Schmidt, D. R., & Ziurys, L. M., (2017). New identifications of the CCH radical in planetary nebulae: A connection to C60? *Astroph. J., 850*, 123–132.

Schowanek, D., & Verstraete, W., (1990). Phosphonate utilization by bacterial cultures and enrichments from environmental samples. *Appl. Environ. Microbiol., 56*, 895–903.

Schrödinger, E., (1944). *What is Life? The Physical Aspect of the Living Cell*. Cambridge University Press: Cambridge.

Schutte, W. A., Allamandola, L. J., & Sandford, S. A., (1993a). Formaldehyde and organic molecule production in astrophysical ices at cryogenic temperatures. *Science, 259*, 1143–1145.

Schutte, W. A., Allamandola, L. J., & Sandford, S. A., (1993b). An experimental study of the organic molecules produced in cometary and interstellar ice analogs by thermal formaldehyde reactions. *Icarus, 104*, 118–137.

Scott, P., Grevesse, N., Asplund, M., Sauval, A. J., Lind, K., Takeda, Y., Collet, R., Trampedach, R., & Hayek, W., (2015). The elemental composition of the Sun I. The intermediate mass elements Na to Ca. *Astron. Astrophys., 573*(A25), 1–19.

Seaquist, E. R., Frayer, D. T., & Bell, M. B., (1998). HCO^+ in the starburst galaxy M82. *Astrophys. J., 507*, 745–758.

Sedaghati, E., Boffin, H. M. J., MacDonald, R. J., Gandhi, S., Madhusudhan, N., Gibson, N. P., Oshagh, M., Claret, A., & Rauer, H., (2017). Detection of titanium oxide in the atmosphere of a hot Jupiter. *Nature (London), 549*, 238–241.

Sevic, C., & Sevincli, H., (2016). Promising thermoelectric properties of phosphorenes. *Nanotechnology*, *27*, 355705.

Shaw, A. M., (2006). *Astrochemistry: From Astronomy to Astrobiology.* John Wiley & Sons: West Sussex.

Shepard, C. U., (1878). On a new mineral, pyrophosphorite, and anhydrous pyrophosphate of lime from the West Indies. *Am. J. Sci.*, *15*, 49–51.

Shirley, V. S., (1996). *Table of Isotopes* (Vol. I, p. 8). Wiley, New York.

Simon, A., Borrmann, H., & Craubner, H., (1987). Crystal structure of ordered white phosphorus (β – P). *Phosphorus and Sulfur*, *30*, 507–510.

Simon, A., Borrmann, H., & Horakh, J., (1997). On the polymorphism of white phosphorus. *Eur. J. Inorg. Chem.*, *130*, 1235–1240.

Sion, E. W., & Sparks, W. M., (2014). On the effect of explosive thermonuclear burning on the accreted envelopes of white dwarfs in cataclysmic variables. *Astrophys. J.*, *796*, L10–L13.

Sion, E. W., Cheng, D. H., Sparks, W. M., Szkody, P., Huang, M., & Hubeny, I., (1997). Evidence of a thermonuclear runaway and proton-capture material on a white dwarf in a dwarf nova. *Astrophys. J.*, *480*, L17–L20.

Snow, T. P., & Witt, A. N., (1996). Interstellar depletions updated: Where all the atoms went. *Astrophys. J.*, *468*, L65–L68.

Soderberg, T., (2016). http://chem.libretexts.org/Textbook_Maps/Organic_Chemistry_Textbook_Maps/Map%3A_Organic_Chemistry_With_a_Biological_Emphasis_(Soderberg)/10%3A_Phosphoryl_transfer_reactions/10.1%3A_Overview_of_phosphates_and_phosphoryl_transfer_reactions (Accessed on 27 July 2019).

Souhassou, M., Espinosa, E., Lecomte, C., & Blessing, R. H., (1995). Experimental electron density in crystalline H_3PO_4. *Acta Cryst.*, *B51*, 661–668.

Srinivasan, V., & Morowitz, H. J., (2004). *Intermediary Metabolism in the Early Assembly of Cellular Life*. Abstracts of papers, 228th ACS National Meeting.

Stachel, D., (1995). Phosphorus pentoxide at 233 K. *Acta Cryst.*, *C51*, 1049–1050.

Staggs, S., Dunkley, J., & Page, L., (2018). Recent discoveries from the cosmic microwave background: A review of recent progress. *Rep. Prog. Phys.*, *81*, 044901–044934.

Standford, S. A., et al., (2016). Organics captured from comet 81P/Wild 2 by the stardust spacecraft. *Science*, *314*, 1720–1724.

Stanghellini, L., Shaw, R. A., Balick, B., & Blades, J. C., (2000). Large magellanic cloud planetary nebula morphology: Probing stellar populations and evolution. *Astrophys. J. Lett.*, *534*, L167–L171.

Sterenberg, B. T., Scoles, L., & Carty, A. J., (2002). Synthesis, structure, bonding and reactivity in clusters of the lower phosphorus oxides. *Coordination Chem. Rev.*, *231*, 183–197.

Sterner, R. W., & Elser, J. J., (2002). *Ecological Stoichiometry: The Biology of Elements from Molecules to the Biosphere.* Princeton University Press: Princeton.

Struve, O., (1930). Phosphorus in stellar spectra. *Astrophys. J.*, *71*, 150–152.

Subramanian, H., & Gatenby, R. A., (2018). Chiral monomers ensure orientational specificity of monomer binding during polymer self-replication. *J. Molecular Evol.* https://doi.org/10.1007/s00239–018–9845–9 (Accessed on 27 July 2019).

Summers, M., & Trefil, J., (2017). *Exoplanets.* Smithsonian Books: Washington DC.

Sun, X., (2002a). Pentacoordinated AB_5 – type main group molecules favorably adopt sp^2 hybridization in the central atom: Bonding without d-orbital participation. *Chem. Educator*, *7*, 11–14.

Sun, X., (2002b). The three-center, four-electron bond in hexacoordinated AB_6-type main group molecules: An alternative model of bonding without d-orbital participation in the central atom. *Chem. Educator*, *7*, 261–264.

Suzuya, K., Price, D. L., Loong, Ch. K., & Martin, S. W., (1998). Structure of vitreous P_2O_5 and alkali phosphate glasses. *J. Non-Cryst. Solids*, *232–234*, 650–657.

Svara, J., Weferling, S. J., N., & Hofmann, T., (2008). Phosphorus compounds, organic. In: *Ullmann's Encyclopedia of Industrial Chemistry.* Wiley-VCH: Weinheim. doi: 10.1002/14356007.a19_545.pub2.

Tashiro, T., Ishida, A., Hori, M., Igisu, M., Koike, M., Méjean, P., Takahara, N., Sano, Y., & Komiya, T., (2017). Early trace of life from 3.95 Ga sedimentary rocks in Labrador, Canada. *Nature (London)*, *549*, 516–519.

Tenenbaum, E. D., & Ziurys, L. M., (2008). PH_3 detected in the proto-planetary nebula CRL 2688. *Astrophys. J. Lett.*, *680*, L121–L124.

Tenenbaum, E. D., Dodd, J. L., Milam, S. N., Woolf, N. J., & Ziurys, L. M., (2010). Comparative spectra of oxygen-rich versus carbon-rich circumstellar shells: VY Canis Majoris and IRC +10216 at 215–285 GHz. *Astrophys. J. Lett.*, *720*, L102–L107.

Tenenbaum, E. D., Milam, S. N., Woolf, N. J., & Ziurys, L. M., (2009). Molecular survival in evolved planetary nebulae: Detection of H_2CO, c-C_3H_2 and C_2H in the Helix. *Astrophys. J. Lett.*, *704*, L108–L112.

Tenenbaum, E. D., Woolf, N. J., & Ziurys, L. M., (2007). Identification of phosphorus monoxide in VY Canis Majoris: Detection of the first P-O bond in space. *Astrophys. J.*, *666*, L29–L32.

Tennyson, J., (2003). Molecules in space. In: Wilson, S., (ed.), *Handbook of Molecular Physics and Quantum Chemistry: Molecules in the Physico-Chemical Environment: Spectroscopy, Dynamics and Bulk Properties* (Vol. 3, pp. 356–369). John Wiley & Sons, Ltd.

Thomas, K. L., Blandford, G. E., Kellier, L. P., Klock, W., & McKay, D. S., (1993). Carbon abundance and silicate mineralogy of anhydrous interplanetary dust particles. *Geochim. Cosmochim. Acta*, *57*, 1551–1566.

Thomson, D. W., (1992). *On Growth and Form: The Complete Revised Edition.* Dover Books on Biology.

Thorne, L. R., Anicich, V. G., Prasad, S. S., & Huntress, W. T., (1984). The chemistry of phosphorus in dense interstellar clouds. *Astrophys J.*, *280*, 139–143.

Thurn, H., & Krebs, H., (1969). Über struktur und eigenshcaften der halbmetalle. XXII. Die kristallstruktur des Hittorfschen phosphors. *Acta Crystallogr. Sect. B*, *25*, 125–135.

Tielens, A. G. G. M., (2013). The molecular universe. *Rev. Mod. Phys.*, *85*(3), 1021–1081.

Tobin, W., & Kaufmann, J. P., (1984). Analysis of the three high-velocity B stars HD 125824, 165955 and CPD -72°1184. *Mon. Not. R. Soc.*, *207*, 369–392.

Trimble, V., (1975). The origin and abundances of the chemical elements. *Rev. Mod. Phys.*, *47*, 877–976.

Trimble, V., (1982). Supernovae. Part I: The events. *Rev. Mod. Phys.*, *54*, 1183–1224.

Trimble, V., (1983). Supernovae. Part II: The aftermath. *Rev. Mod. Phys.*, *55*, 511–563.

Tsubota, G., (1955). Phosphate reduction in the paddy field. *Soil Plant Food, 10*, 1959.

Tsuji, T., (1973). Molecular abundances in stellar atmospheres. II. *Astron. Astrophys.*, *23*, 411–431.

Turner, A. M., Abplanalp, M. J., & Kaiser, R. I., (2016). Probing the carbon-phosphorus bond coupling in low-temperature phosphine (PH_3) – methane (CH_4) interstellar ice analogs. *Astrophys. J.*, *819*, 97–106.

Turner, A. M., Abplanalp, M. J., Blair, T. J., Dayuha, R., & Kaiser, R. I., (2018). An infrared spectroscopy study toward the formation of alkylphosphonic acids and their precursors in extraterrestrial environments. *Astrophys. J. Suppl. Ser.*, 234–256.

Turner, B. E., & Bally, J., (1987). Detection of interstellar PN – the first identified phosphorus compound in the interstellar medium. *Astrophys J.*, *321*, L75-L79.

Turner, B. E., Tsuji T., Bally, J., Guelin, M., & Cernicharo, J., (1990). Phosphorus in the dense interstellar medium. *Astrophys J.*, *365*, 565–585.

Umeda, H., & Nomoto, K., (2002). Nucleosynthesis of zinc and iron peak elements in population III type II supernovae: Comparison with abundances in very metal-poor halo stars. *Astrophys. J.*, *565*, 385–404.

Vallet-Regí, M., Salinas, A. J., Ramírez-Castellanos, J., & González-Calbet, J. M., (2005). Nanostructure of bioactive sol-gel glasses and organic-inorganic hybrids. *Chem. Mater.*, *17*, 1874–1879.

Van Cappellen, P., & Ingall, E. D., (1996). Redox stabilization of the atmosphere and oceans by phosphorus-limited marine productivity. *Science*, *271*, 493–496.

Van Dishoeck, E. F., & Black, J. H., (1986). Comprehensive models of diffuse interstellar clouds: physical conditions and molecular abundances. *Astrophys. J. Sup. Ser.*, *62*, 109–145.

Verschuur, G. L., (1992). Interstellar molecules. *Sky & Telescope*, 379–384.

Viana, R. B., & Pimentel, A. S., (2007). New insights in the formation of thioxophosphine: A quantum chemical study. *J. Chem Phys.*, *127*, 204306.

Vidal-Madjar, A., Désert, J. M., Lecavelier, D. E. A., Hébrard, G., Ballester, G. E., Ehrenreich, D., Ferlet, R., McConnell, J. C., Mayor, M., & Parkinson, C. D., (2004). Detection of oxygen and carbon in the hydrodynamically escaping atmosphere of the extrasolar planet HD 209458B. *Astrophys. J.*, *604*, L69-L72.

Visscher, C., Lodders, K., & Fegley, Jr., B., (2006). Atmospheric chemistry in giant planets, brown dwarfs, and low-mass dwarf stars. II Sulfur and phosphorus. *Astrophys J.*, *648*, 1181–1195.

Wallerstain, G., et al., (1997). Synthesis of the elements in stars: Forty years of progress. *Rev. Mod. Phys.*, *69*, 995–1084.

Weinberg, S. W., (1977). *The First Three Minutes.* A. Deutsch: London.

Westheimer, F. H., (1987). Why nature chose phosphates. *Science*, *235*, 1173–1178.

White, S. D., & Rees, M. J., (1978). Core condensation in heavy halos – a two-stage theory for galaxy formation and clustering. *Month. Not. Royal Astr. Soc.*, *183*, 341–358.

Whittet, D. C.B., & Chiar, J. E., (1993). Cosmic evolution of the biogenic elements and compounds. *The Astron. Rev.*, *5*, 1–35.

Willacy, K., & Millar, T. J., (1997). Chemistry in oxygen-rich circumstellar envelopes. *Astron. Astrophys.*, *324*, 237–248.

Wischer, M. C., & Langanke, K., (2015). Light: Cosmic messages from the past. *Europhysics News*, *46*(4), 27–31.

Woese, C., Kandler, O., & Wheelis, M., (1990). Towards a natural system of organisms: Proposal for the domains Archaea, Bacteria, and Eucarya. *Proc. Natl. Acad. Sci. USA*, *87*, 4576–4579.

Wolfe-Simon, F., Blum, J. S., Kulp, T. R., Gordon, G. W., Hoeft, S. E., Pett-Ridge, J., et al., (2010). A bacterium that can grow by using arsenic instead of phosphorus. *Science*, 1197258, doi: 10.1126/science.1197258.

Wolfe-Simon, F., Blum, J. S., Kulp, T. R., Gordon, G. W., Hoeft, S. E., & Pett-Ridge, J., (2011). A bacterium that can grow by using arsenic instead of phosphorus. *Science, 332*, 1163–1166.

Wolfe-Simon, F., Davies, P. C. W., & Anbar, A. D., (2009). Did nature also choose arsenic? *Inter. J. Astrobiol., 8*, 69–74.

Wolszczan, A., & Frail, D., (1992). A planetary system around the millisecond pulsar PSR1257 + 12. *Nature, 355*, 145–147.

Wood, J. A., & Chang, S., (1985). The cosmic history of the biogenic elements and compounds. *NASA SP, 476*, Washington D. C.

Woosley, S. E., Heger, A., & Weaver, T. A., (2002). The evolution and explosion of massive stars. *Rev. Mod. Phys., 74*, 1015–1071.

Wright, I. P., Sheridan, S., Barber, S. J., Morgan, G. H., Andrews, D. J., & Morse, A. D., (2015). CHO-bearing organic compounds at the surface of 67P/Churyumov-Gerasimenko revealed by Ptolemy. *Science, 349*, aab0673–1–3.

Wu, J., Li, Q., Liu, J., Xue, C., Yang, S., Shao, C., Tu, L., Hu, Z., & Luo, J., (2019). Progress in precise measurements of the gravitational constant. *Ann. Phys. (Berlin), 531*, 1900013 (1 of 14).

Xu, J. P., Zhang, J. Q., Tain, H., Ho, W., & Xie, M., (2017). One-dimensional phosphorus chain and blue phosphorene grown on Au(111) by molecular-beam epitaxy. *Phys. Rev. Mat., 1*, 061002(R).

Yamaguchi, T., Takano, S., Sakai, N., Sakai, T., Liu, S. Y., Su, Y. N., et al., (2011). Detection of phosphorus nitride in the Lynds 1157 B1 shocked region. *Publ. Soc. Japan, 63*, L37–L41.

Yan, Z., Han, W., Peñuelas, J., Sardans, J., Elser, J. J., Du, E., Reich, P. B., & Fang, J., (2016). Phosphorus accumulates faster than nitrogen globally in freshwater ecosystems under antropogenic impacts. *Ecology Lett., 19*, 1237–1246.

Yang, X., Liu, G., Shi, Y., Huang, W., Shao, J., & Dong, X., (2018). Nano-black phosphorus for combined cancer phototherapy: Recent advances and prospects. *Nanotechnology, 29*, 222001–222014.

Yang, Z., Sismou, A. M., Sheng, P., Puskar, N. L., & Benner, S. A., (2007). Enzymatic incorporation of a third nucleobase pair. *Nucleic Acids Research, 35*, 4238–4249.

York, D. G., (1994). The elemental composition of interstellar dust. *Science, 265*, 191–192.

Zahnle, K., Schaefer, L., & Fegley, B., (2010). Earth's earliest atmospheres. *Cold Spring Harb. Perspec. Biol., 2*, a004895. doi: 10.1101/cshperspect.a004895.

Zatovsky, I. V., Strutynska, N. Y., Hizhnyi, Y. A., Baumer, V. N., Ogorodnyk, I. V., Slobodyanik, N. S., Odynets, I. V., & Klyui, N. I., (2018). New complex phosphates $Cs_3MBi(P_2O_7)_2$ (M – Ca, Sr and Pb): Synthesis, characterization, crystal and electronic structure. *Dalton Trans., 47*, 2274–2284.

Zeng, S., Jiménez-Serra, I., Rivilla, V. M., Martín, S., Martín-Pintado, J., Requena-Torres, M. A., Armijos-Abendaño, J., Riquelme, D., & Aladro, R., (2018). Complex organic molecules in the galactic center: The N-bearing family. *Month. Not. Roy. Astron. Soc., 478*, 2962–2975.

Zhang, Y., Kwok, S., Nakashima, J. I., & Trung, D. V., (2008). A spectral line survey of NGC 7027 at millimeter wavelengths. *Astrophys. J.*, *678*, 328–346.

Zhang, Y., Kwok, S., Nakashima, J. I., Chau, W., & Trung, D. V., (2013). A molecular line survey of the carbon-rich protoplanetary nebula AFGL 2688 in the 3 mm and 1.3 mm windows. *Astrophys. J.*, *773*, 71–90.

Ziurys, L. M., (1987). Detection of interstellar PN: The first phosphorus bearing species observed in molecular clouds. *Astrophys J.*, *321*, L81–L85.

Ziurys, L. M., (2008). Organic chemistry in circumstellar envelopes: Setting the stage for prebiotic synthesis. In: Kwok, S., & Sandford, S., (eds.), *Organic Matter in Space* (Vol. 251, pp. 147–156). Proceedings IAU Symposium.

Ziurys, L. M., Adande, G. R., Edwards, J. L., Schmidt, D. R., Halfen, D. T., & Woolf, N. J., (2015). Prebiotic chemical evolution in the astrophysical context. *Orig. Life Evol. Biosph.*, *45*, 275–288.

Ziurys, L. M., Milam, S. N., Apponi, A. J., & Woolf, N. J., (2007). Chemical complexity in the winds of the oxygen-rich supergiant star VY Canis Majoris. *Nature*, *447*, 1094–1097.

Ziurys, L. M., Schmidt, D. R., & Bernal, J. J., (2018). New circumstellar sources of PO and PN: The increasing role of phosphorus chemistry in oxygen-rich stars. *Astrophys. J.*, *856*, 169–179.

Exercises with Solutions

1. *Making use of the physical constants values listed in the Appendix determine the relative intensities of the gravitational and electrostatic forces between the proton and the electron in the Bohr's model of the hydrogen atom.*

According to the Newton gravitation law, the force between a proton of mass m_p and an electron of mass m_e is given by:

$$F_g = G\frac{m_p m_e}{a_0^2},$$

where, G is the gravitational constant and a_0 is the Bohr radius, which measures the size of the hydrogen atom. On the other hand, the Coulomb law for electrostatics reads:

$$F_e = \frac{1}{4\pi\varepsilon_0}\frac{e^2}{a_0^2},$$

where, $(4\pi\varepsilon_0)^{-1} \cong 9\times10^9$ JmC^{-2}, and e is the electron charge. By dividing both expressions, we get the ratio:

$$\frac{F_g}{F_e} = 4\pi\varepsilon_0 G\frac{m_p m_e}{e^2},$$

which depends on fundamental constants only. Plugging their values, we obtain a dimensionless number indicating that the relative intensity between both forces is really huge. Not surprisingly then, gravitational interaction, which rules the formation of structures at the astronomical scale, plays a negligible role at macroscopic scales, where electrostatic interactions take over the scene.

$$\frac{F_g}{F_e} = \frac{10^{-9}}{9} \text{J}^{-1}\text{m}^{-1}\text{C}^2 \times 6.673\times10^{-11} \, Nm^2kg^{-2} \frac{1.673\times10^{-27} \, kg \times 9.109\times10^{-31} \, kg}{(1.602\times10^{-19} \, C)^2} = 4.403\times10^{-40},$$

2. *Determine the neutron to proton ratio at the time when the neutron decay meantime becomes lower than the universe age.*

Assuming that the universe is in thermal equilibrium at the considered time, the neutron to proton ratio is given by the Boltzmann distribution:

$$\frac{n}{p} = \exp\left(-\frac{m_n - m_p}{T^*}\right) = e^{-1.6345} = 0.195,$$

where, $T^* = 9.5\times10^9$ K (see Table 2.1), m_n = 939.550 MeV/c², and m_p = 938.256 MeV/c², and we have used the conversion factors given in the Appendix. Accordingly, p/n = 5.1 ~ 5 at this time.

3. *Determine the average density of matter in the Galaxy.*

Making use of the data indicated in the caption of Figure 2.5 for the Galaxy size, which we approximate to the shape of a flattened cylinder of radius *R* and thickness *h*, we get the volume:

$$V = \pi R^2 h = \pi(7.71\times10^{20})^2 \, m^2 \times 1.89\times10^{19} \, m = 3.53\times10^{61} \, m^3$$

Assuming a Galaxy mass of about 1.4 × 10¹¹ solar masses, we obtain the average density value:

$$\rho = \frac{M}{V} = \frac{1.4\times10^{11} \times 1.989\times10^{30} \, kg}{3.53\times10^{61} \, m^3} = 7.89\times10^{-21} \, kgm^{-3} = 7.89\times10^{-24} \, gcm^{-3}$$

Dividing the above figure by the mass of the hydrogen atom given in the Appendix, we can express the Galaxy mass density as the approximate atomic density:

$$n = \frac{\rho}{m_H} = \frac{7.89\times10^{-24} \, gcm^{-3}}{1.673\times10^{-24} \, g/at} = 4.7 \text{ atoms per cubic centimeter.}$$

4. *(a) Estimate the time required for the star IRC + 10216 to expel an amount of mass equivalent to the Earth mass. (b) Determine the atomic density reached at the outer envelope located at about one light-year from the star.*

(a) Making use of the Earth-mass given in the Appendix and the rate mass indicated in the text, we obtain:

$$t = \frac{M_E}{\dot{M}} = \frac{5.974\times10^{24}\,kg}{2\times10^{-5}\,M_O\,yr^{-1}\times1.989\times10^{30}\,kg} = 0.15\,yr,$$

That is, about 55 days.

(b) In the text, we indicate that the atomic density in the inner shell is $n = 10^{11}$ atoms per cubic centimeter. This shell extends to about $r_* = 40$ AU $= 5.98\times10^{12}$ m from the star, whereas the outer envelope is located at about $r_\infty = 9.46\times10^{15}$ m. Therefore, the atomic density at the outer region can be roughly estimated as:

$$n_\infty = \frac{nV_*}{V_\infty} = n\left(\frac{r_*}{r_\infty}\right)^3 = 10^{11}\left(\frac{5.98\times10^{12}\,m}{9.46\times10^{15}\,m}\right)^3 = 25.3 \text{ atoms per cubic}$$

centimeter.

This figure is about five times the average atomic density of the Galaxy obtained in the previous exercise.

5. *Assuming that the baryonic mass of the Galaxy is $M_x = 1.8\times10^{10}\,M_\odot$, the hydrogen mass fraction in the Galaxy is $\mu = 0.77$, a typical protoplanetary disk mass is $M_d = 10^{-7}\,M_\odot$, the planets to Sun mass ratio is $\eta = 1.5\times10^{-3}$, the Earth biomass is $M_E = 2\times10^{15}$ kg, the phosphorus content of the Earth biomass is 0.6% by mass, and adopting a phosphorus elemental abundance [P] = 5.41 (cf. Table 3.1), a) determine the total amount of available phosphorus in the Galaxy; b) estimate the possible number of Earth-like biomasses in the Galaxy by assuming that about 10% of stars are endowed with a planetary system.*

(a) The available amount of phosphorus in the Galaxy is given by

$$M_P = \mu M_x \frac{P}{H} = 0.77\times1.8\times10^{10}\times1.989\times10^{30}\,kg\times2.56\times10^{-7} = 7.06\times10^{33}\,kg$$

where, $\frac{P}{H} = 10^{5.41-12} = 2.56 \times 10^{-7}$. This represents a large P budget, amounting about 3,550 $M_{\odot}$.

(b) If all the available phosphorus was condensed to form Earth-like planets, we would obtain the following number of possible biomasses in the Galaxy leading to the amusing ratio of about 6×10^{9} biomasses per star!

$$N = \frac{M_P}{0.006 M_E} = \frac{7.1 \times 10^{33} kg}{6 \times 10^{-3} \times 2 \times 10^{15} kg} = 5.9 \times 10^{20},$$

Now, we know that not all the available P atoms in the Galaxy will be condensed to form Earth-like planets, but most of them remain bounded in stars, dense stellar remnants, interstellar gas, and dust, or comet nuclei, so that only a minor phosphorus quantity would be effectively incorporated in biological systems. In order to obtain a suitable upper bond to this quantity let us assume that all the P atoms present in a typical protoplanetary disk are eventually fixed in the resulting planets biomasses. Then, we have

$$\tilde{M}_P = \mu M_d \eta \frac{P}{H} = 0.77 \times 10^{-7} \times 1.989 \times 10^{30} kg \times 1.5 \times 10^{-3} \times 2.56 \times 10^{-7} = 5.9 \times 10^{13} kg$$

where, as a first approximation, we have assumed that the solar system planets to Sun ratio are applicable to other planetary systems as well. Thus, the number of possible Earth-like biomasses in a planetary system will be"

$$N = \frac{\tilde{M}_P}{0.006 M_E} = \frac{5.9 \times 10^{13} kg}{6 \times 10^{-3} \times 2 \times 10^{15} kg} = 4.9 \cong 5,$$

And assuming that only 10% of the stars are endowed with a planetary system (a conservative value) the figure above gives a ratio of about 0.5 biomasses per star in the Galaxy.

6. *Brand's original method used about 5,500 liters of urine to produce just 120 grams of white phosphorus with a density* $\rho = 1.828\ gcm^{-3}$. *Taking into account that 1 liter of adult human urine contains about 1.4 g of sodium ammonium phosphorus salt* $(NH_4)NaHPO_4$, *estimate the maximum possible yield of pure white phosphorus and compare it with that obtained by Brand.*

Each pure white phosphorus crystal unit cell contains 58×4 = 232 P atoms (see Section 4.2.1 and Figure 4.5a), so that the mass of given amount of white phosphorus can be expressed as $m_w = \rho N_u V_u$, where N_u is the number of unit cells contained in a given volume of material and $V_u = a^3$, where a = 1.88 nm is the lattice parameter of the cubic unit cell. In turn, N_u = N/232, where N is the number of phosphorus atoms contained in a given mass, *m*, of the phosphorus ammonium salt, which can be obtained as:

$$N = \frac{m}{M} N_A,$$

where, N_A is the Avogadro's number, M = 137.0079 is the atomic mass of the $(NH_4)NaHPO_4$ molecule, and we take into account that each phosphate molecule just contributes one P atom. Therefore, in a liter of urine, we have:

$$m_W = \rho \frac{m N_A}{232 M} a^3 = 1.828\, g\, cm^{-3} \frac{1.4\, g \times 6.022 \times 10^{23}}{232 \times 137.0079} (1.88 \times 10^{-7})^3 cm^3 = 0.322 \text{ g}$$

Thus, the maximum amount of white phosphorus which could be obtained would be 0.322 g/l × 5,500 l = 1771 g. Thus, we conclude that Brand's process renders a very low yield, of less than 7%. Indeed, it was subsequently realized that the salt part he originally discarded contained most of the available phosphate.

7. *A typical human being uses 6 moles of ATP per hour. Estimate her/his power consumption rate in a day.*

The power consumed is given by the ratio:

$$P_h = \frac{\Delta E}{\Delta t} = \frac{n N_A E_{ATP}}{\Delta t},$$

where, n is the number of moles, N_A is the Avogadro's number (see the Appendix), and E_{ATP} = 0.31 eV is the energy released in the hydrolysis of an ATP molecule. By plugging the corresponding numerical values we obtain:

$$P_h = \frac{6 \times 6.022 \times 10^{23} \times 0.31 eV \times 1.602 \times 10^{-19} J(eV)^{-1}}{3600\, s} = 49.8\, W.$$

This figure is similar to that of a medium-sized old-fashioned light bulb or that delivered by 20 AA batteries (with voltage 4.9 V rendering 0.5 A). Hence, the power consumed by a human body in a day is $P_d = 24 \times 50$ W = 1,200 W. This amount is somewhat less than the typical power delivered by the Sun on a square meter of Earth's surface (the so-called solar constant flux value of 1,360 Wm^{-2}).

8. *Derive expression (5.6).*

It can be reasonably assumed that the density of a star is a monotonous function, which increases from the surface to the center, where it takes the value ρ_c. Let us now consider the case of a star having the uniform density $\rho(r) = \rho_c$, which corresponds to the denser star one may think of. The mass enclosed within a surface of radius r from the center can be expressed as $M(r) = \frac{4\pi}{3}\rho_c r^3$. By integrating the hydrostatic equilibrium Eq. (5.1) from the center (r = 0) to the surface (r = R), where pressure vanishes (otherwise the surface will expand outwards), we get:

$$P_c = \int_0^R \frac{GM(r)}{r^2}\rho(r)dr = \frac{4\pi}{3}G\rho_c^2 \int_0^R r dr = \frac{2\pi}{3}G\rho_c^2 R^2.$$

Now, the total mass of the considered star is given by $M = \frac{4\pi}{3}\rho_c R^3$, so that $R = \left(\frac{3M}{4\pi\rho_c}\right)^{1/3}$, and the expression above can be rewritten in the form:

$$P_c = \frac{G}{2}\left(\frac{4\pi}{3}\right)^{1/3} \rho_c^{4/3} M^{2/3},$$

which provides the upper bond to any star's central pressure.

9. *Making use of Eq. (5.8) determine the radiation pressure upper limit at the center of a star with* $M = M_\odot$.

According to Eq. (5.8) the radiation pressure upper bond at the center of a star with a solar mass is given by:

$$0.183 = \left(\frac{\beta_c^{1/4}}{1-\beta_c} \right)^2,$$

which, leads to the quartic equation $3.35\times10^{-2}(1-\beta_c)^4 - \beta_c = 0$, whose numerically obtained real solution reads $\beta_c = 0.0296$, that is, the radiation pressure amounts to 2.96 % of the total pressure (the other real solution $\beta_c = 4.382$ has no physical meaning, since $\beta_c \leq 1$ by definition).

10. *Making use of Eq. (5.13) determine the threshold distance d_* for the following nuclear reactions: (a) p-p chain ($Z_i = Z_j = 1$, ignition temperature $T = 4\times10^6$ K), (b) triple alpha process ($Z_i = Z_j = 2$, ignition temperature $T = 1\times10^8$ K), and C-burning ($Z_i = Z_j = 6$, ignition temperature $T = 7\times10^8$ K).*

Making use of Eq. (5.13) we can write Eq. (5.12) in the form:

$$d_*[m] = 1.115\times10^{-5}\frac{Z_iZ_j}{T[K]},$$

where, we have used $(4\pi\varepsilon_0)^{-1} \cong 9\times10^9$ JmC^{-2}, and $e^2/k_B \cong 1.859\times10^{-15}$ C^{2}J^{-1}K (see the Appendix). Plugging the values of nuclear charge and ignition temperatures indicated above, we get $d_* \cong 2.8\times10^{-12}$ m, for the *p-p* chain, $d_* \cong 4.5\times10^{-13}$ m, for the 3α process, and $d_* \cong 5.7\times10^{-13}$ m, for the C-burning process. As we see, all the obtained threshold values are significantly larger than the Fermi scale (10^{-15} m) required for the strong nuclear force to be of some significance.

11. *The last nuclear reaction leading to the formation of ^{31}P nuclide (see Figure 5.3) involves the collision of a proton ($Z = 1$) with a ^{30}Si nuclide ($Z = 14$). a) Making use of Eq. (5.12) determine the threshold temperature required for the colliding particles to reach the critical distance $d_* = 10^{-15}$ m. b) Making use of Eq. (5.14) obtain the temperature corresponding to the optimal ratio for this nuclear reaction. Discuss the obtained results.*

(a) Making use of the expression obtained in the solution to Exercise 10, we have:

$$T[K] = 1.115\times10^{-5}\frac{Z_iZ_j}{d^*[m]} = 1.115\times10^{-5}\frac{14\times1}{10^{-15}} \cong 1.56\times10^{11}\text{ K}$$

(b) Taking the derivative of Eq. (5.14) with respect to the temperature and equating it to zero, we obtain the optimal condition:

$$T^* = \frac{27\mu}{64k_B}\left(\frac{\pi Z_i Z_j e^2}{h\varepsilon_0}\right)^2 = \frac{27\pi^4}{4}(4\pi\varepsilon_0)^{-2}\frac{e^4}{h^2 k_B}(Z_i Z_j)^2\mu = 1.82\times10^{12}\text{ K},$$

where, we have used the fundamental constants values given in the Appendix, along with:

$$(4\pi\varepsilon_0)^{-1} \cong 9\times10^9 \text{ JmC}^{-2},\ Z_H = 1,\ Z_{Si} = 14,$$
$$\mu = \frac{30m_H^2}{m_H + 30m_H} = \frac{30}{31}\times1.66\times10^{-27}\text{ kg}.$$

As we see, the optimum temperature is about one order of magnitude higher than the threshold temperature one. Since the temperatures reached in the explosive Ne-burning stage of a type-II supernova are in the range $T = 2-3\times10^9$ K (well below the above-mentioned figures), we understand the low yield of the nuclear reaction producing ^{31}P nucleus according to the pathway indicated in Figure 5.3.

12. *The resonance energy relative to the 3α reaction threshold is ε = 380 keV. Making use of Eq. (5.15) determine the change in this resonant reaction rate if one increases the resonance energy value by a 5%, keeping the values of all the remaining parameters the same. Discuss the possible implications such a change may have for the nucleosynthesis of carbon at a cosmic scale.*

Making use of Eq. (5.15) we have:

$$\frac{r'_{3\alpha}}{r_{3\alpha}} = e^{-\frac{\Delta\varepsilon}{k_B T_{3\alpha}}} = \exp\left(-\frac{0.05\times380\times10^3 eV}{8.617\times10^{-5} eVK^{-1}10^8 K}\right) = e^{-2.205} \cong 0.11,$$

where, $\Delta\varepsilon \equiv \varepsilon' - \varepsilon$, and we used $\Delta\varepsilon/\varepsilon = 0.05$ and $T_{3\alpha} = 10^8$ K. As we see, the reaction rate is reduced by almost an order of magnitude, in this case, hence indicating that the carbon nucleosynthesis is very sensitive to minute changes in the resonance energy value. Such dependence may give rise to a paramount reduction in the stellar production of carbon (thereby of oxygen as well), significantly lowering their abundances in the universe, where the strong nuclear force strength just changed by a mere 0.5% from its actual value (Oberhummer et al., 2000).

13. *Making use of the P-bearing molecules abundances data listed in Table 1.5 and the P/H = 2.57×10^{-7} cosmic value, estimate the amount of phosphorus atoms available for condensed matter forms in the different astrophysical objects.*

From the abundance data given in Table 1.3, we obtain the below $\eta \equiv (\sum P_j/2)/(P/H)$ ratios listed in Table 1.3, where P_j stands for the abundance (relative to hydrogen) of any P-bearing molecule detected in the considered source.

Star	**η**	**Gas Phase P (%)**
CRC 2688	0.399	40
R Cas	0.233	23
NML Cyg	0.142	14
TX Cam	0.126	13
VY CMa	0.111	11
IK Tau	0.107	11
IRC +10216	0.100	10

These results indicate that most phosphorus mainly resides in a condensed form in all the considered sources, although there are also major gas-phase carriers of phosphorus in the protoplanetary CRC 2688 (namely HCP), and to a lesser extent in the AGB star R Cas (namely, PO and PN).

14. *Estimate the collision rate value in typical (a) diffuse and (b) dense ISM clouds by assuming the reaction rate $k \approx 1 \times 10^{-9}$ cm^{3}s^{-1} as a representative figure.*

(a) The typical density of a diffuse cloud is n ~ 10^2 cm^{-3}, so that we get the collision rate k = 10^2 cm^{-3}$\times 1 \times 10^{-9}$ cm^{3}s^{-1} = 10^{-7} s^{-1}. Taking into account that one day contains 86,400 s, the obtained frequency can be conveniently expressed as:

$$\frac{86400 s/d}{10^7 s} = 8.6 \times 10^{-3} d^{-1} \times 120\, d \cong 1.04,$$

That is, about one collision every four months.

(b) The typical density of a diffuse cloud is n ~ 10^5 cm^{-3}, so that we get the collision rate k = 10^5 cm^{-3}×1 × 10^{-9} cm^{3}s^{-1} = 10^{-4} s^{-1}. Taking into account that one day contains 86400 s, the obtained frequency can be conveniently expressed as:

$$\frac{86400 s / d}{10^4 s} = 8.6 d^{-1},$$

That is, about one collision every 3 hours approximately.

15. *Let us assume that hydrogen molecules were formed in the ISM through the three-body collision processes H + H +H → H_2 + H, with a rate constant k ≈ 1 × 10^{-32} cm^{6}s^{-1}. Assuming that the resulting H_2 molecules are not subsequently destroyed: (a) obtain the evolution of the fractional abundance of H_2 with time, (b) determine the amount of H_2 produced in a typical molecular cloud, where n_H = 10^5 cm^{-3} is the total number of H nuclei in the cloud, during a time t* = 1.4 × 10^{10} years (comparable to the age of the universe, see Table 2.1).*

(a) The differential formation rate of H_2 molecules is given by $dn(H_2) = kn(H)^3 dt$ cm^{-3}. The fractional abundance of H_2 is given by the ratio $f = 2n(H_2)/n_H$, where $n_H \equiv 2n(H_2) + n(H)$, measures the total H nuclei content of the cloud in cm^{-3}, so that the differential variation of this ratio in time can be expressed as:

$$df = \frac{2dn(H_2)}{n_H} = \frac{2k}{n_H}\left(n_H - 2n(H_2)\right)^3 dt = 2kn_H^2 (1-f)^3 dt,$$

a relation that can be readily integrated to obtain:

$$2kn_H^2 t = \int_{f(0)}^{f} \frac{df'}{(1-f')^3} = \frac{1}{2}\left[(1-f)^{-2} - 1\right],$$

where, we have assumed $f(0) = 0$. Therefore, we finally get:

$$f = 1 - \frac{1}{\sqrt{1 + 4kn_H^2 t}} \qquad (15.1)$$

a curve describing a continuous increase of H_2 molecules.

(b) Evaluating Eq. (15.1) at time $t^* = 1.4 \times 10^{10}$ yr $= 4.4\times10^{17}$ s,with $n_H = 10^5$ cm^{-3}, we obtain $f(t^*) = 8.8\times10^{-5}$, so that the density of hydrogen molecules is given by $n(H_2) = f(t^*)n_H/2 = 8.8\times10^{-5}\times10^5$ cm^{-3}/2 = 4.4 cm^{-3}. This result clearly indicates that the three-body collision process is highly inefficient in order to get H_2 molecules in the ISM.

16. *Phosphorylation reactions may work better if polyphosphates are used in place of phosphate. Determine as to whether the glucose-6-phosphate reaction may turn to be spontaneous by using pyrophosphate molecules, which spontaneously hydrolyze according to the reaction $H_2P_2O_7^{-2} + H_2O \rightarrow 2H_2PO_4^-$, with a Gibbs free energy* $\Delta G = -19.2$ kJ mol^{-1}.

Let us consider a general chemical reaction written in the form

$$aA + bB \rightarrow cC + dD \tag{16.1}$$

Its reaction constant is given by the ratio

$$K = \frac{[C]^c[D]^d}{[A]^a[B]^b}, \tag{16.2}$$

which is related to the corresponding Gibbs free energy by the relationship

$$\Delta G = -RT \ln K, \tag{16.3}$$

where R = 8.314 Jmol^{-1}K^{-1} is the perfect gas constant. Accordingly, if we multiply all the terms in (16.1) by a given number α, Eq. (16.2) will read K^α and, after Eq. (16.3), the corresponding Gibbs energy becomes αΔG. Now, multiplying Eq. (8.1) by two and adding the pyrophosphate hydrolysis reaction above, we have the net reaction

$$2C_6H_{12}O_6 + H_2P_2O_7^{-2} \rightarrow 2C_6H_{12}O_6PO_3^- + H_2O, \tag{16.4}$$

describing the phosphorylation of glucose by using pyrophosphate molecules. The Gibbs free energy of this reaction is $\Delta G = 2 \times 13.8 - 19.2 = +8.4$ kJmol^{-1}. Plugging this value in Eq. (16.3) we obtain:

$$K = e^{-\frac{\Delta G}{RT}} = \exp\left(-\frac{8.4 \times 10^3}{8.314 \times 298}\right) = 0.034. \tag{16.5}$$

By comparing with the reaction constant value given by Eq. (8.2) we see that (16.5) is almost an order of magnitude larger, hence indicating that the pyrophosphate phosphorylation reaction, albeit it does not occur spontaneously, proceeds more readily than the phosphate one.

Glossary

A

Albedo: is the measure of the diffuse reflection of solar radiation out of the total solar radiation received by an astronomical body (e.g., a planet like Earth). It is dimensionless and measured on a scale from 0 (corresponding to a black body that absorbs all incident radiation) to 1 (corresponding to a body that reflects all incident radiation).

Alcohol: any organic compound in which the hydroxyl functional group (–OH) is bound to a saturated carbon atom. The suffix -ol appears in the IUPAC chemical name of all substances where the hydroxyl group is the functional group with the highest priority; in substances where a higher priority group is present the prefix hydroxy- will appear instead. The term alcohol originally referred to the primary alcohol ethanol (ethyl alcohol).

Aldehyde: or alkanol is an organic compound containing a functional group with the structure −CHO, consisting of a carbonyl center (a carbon double-bonded to oxygen) with the carbon atom also bonded to hydrogen and to an R group, which is any generic alkyl or side chain. The group—without R—is the aldehyde group, also known as the formyl group.

Allotrope: any of two or more physical forms in which an element can exist. For instance, diamond and graphite are allotropes of carbon

Amide: a compound with the functional group $R_nE(O)_xNR'2$ (R and R′ refer to H atoms or organic groups). Most common are carboxamides (organic amides) (n = 1, E = C, x = 1), but many other important types of amides are known, including phosphoramides (n = 2, E = P, x = 1 and many related formulas) and sulfonamides (E = S, x = 2). The simplest amides are derivatives of ammonia wherein one hydrogen atom has been replaced by an acyl group. The ensemble is generally represented as $RC(O)NH_2$ and is described as a primary amide. Closely related and even more numerous are secondary amides which can be derived from primary amines ($R'NH_2$) and have the formula RC(O)NHR′. Tertiary amides are commonly derived from secondary amines (R′R″NH) and have the general structure RC(O)NR′R″. Amides are usually regarded as derivatives of carboxylic acids in which the hydroxyl group has been replaced by an amine or ammonia.

Amine: in organic chemistry are compounds and functional groups that contain a basic nitrogen atom with a lone pair. Amines are formally derivatives of ammonia, wherein one or more hydrogen atoms have been replaced by a substituent such as an alkyl or aryl group. Important amines include amino acids.

Amino acids: the primary building block of proteins, are molecules that contain an amine group (NH_2), a carboxylic group (COOH) and a side chain.

Anabolism: is the use of certain organic products for the synthesis of biomolecules such as amino acids, nucleic acid bases, and lipids. Anabolic pathways generally take a small number of building blocks, many of them products of catabolism, and assemble them.

Anaerobic: any organism whose redox metabolism does not depend on free oxygen.

Archea: (also referred to as Archeobacteria) One of the three domains of life characterized by single-celled microorganisms with no cell nucleus, most of which live in extreme environments.

Archean: The archean era (4.55 – 2.5 Ga) is generally divided into three mean intervals: the Hadean, which comprises the initial period after the formation of Earth (4.55 – 3.9 Ga), the early Archean (3.9 – 2.9 Ga), and the late Archean (2.9 – 2.5 Ga).

Astronomical unit (AU): The average distance from the Sun to the Earth (approximately 150 million km).

Asymptotic giant branch (AGB): A region on the Herstprung-Russell diagram occupied by stars that have ascended the red giant branch for the second and final red-giant phase, once the helium in their cores has been completely exhausted.

Autotrophs: Organisms that use the energy captured to reduce CO_2, creating organic carbon molecules, such as sugars.

B

Bacteria: One of the three domains of life characterized by single-celled microorganisms with no cell nucleus, most of which living in common environments.

Baryonic matter: matter that is composed of protons and neutrons.

Biogenic elements: Those atoms needed in large quantities to make living organisms, which include H, O, C, N, P, and S.

Bioglass: is a glass specifically composed of 45 wt% SiO_2, 24.5 wt% CaO, 24.5 wt% Na_2O, and 6.0 wt% P_2O_5. Glasses are non-crystalline amorphous solids that are commonly composed of silica-based materials with other minor additives. Compared to soda-lime glass (commonly used, as in windows or bottles), bioglass contains less silica and higher amounts of calcium and phosphorus. This high ratio of calcium to phosphorus promotes the formation of apatite crystals; calcium and silica ions can act as crystallization nuclei.

Biomass: different biopolymers, namely, proteins, sugars, lipids, ribonucleic acids (RNA), and deoxyribonucleic acids (DNA) present in living organisms, together with smaller molecules, such as water, phosphates, sulfates, and a few metallic ions.

Biosphere: also known as the ecosphere (from Greek οἶκος oîkos "environment" and σφαῖρα), is the worldwide sum of all ecosystems. It can also be termed the zone of life on Earth, a closed system (apart from solar and cosmic radiation and heat from the interior of the Earth), and largely self-regulating.

Black hole: A black hole is a region of spacetime exhibiting such strong gravitational effects that nothing—not even particles and electromagnetic radiation such as light—can escape from inside it. The theory of general relativity predicts that a sufficiently compact mass can deform spacetime to form a black hole. The boundary of the region from which no escape is possible is called the event horizon.

Blue compact galaxy: These galaxies are often low mass, low metallicity, dust-free objects. Because they are dust-free and contain a large number of hot, young stars, they are often blue in optical and ultraviolet colors.

Brown dwarf: A stellar-like object that is not massive enough to sustain nuclear reactions in its core. They have masses between 13–80 Jupiter mass and surface temperatures below 2500 K, defining three new spectral classes: L (1500–1300 K), T (1300–700 K), and Y (> 600 K). The coolest known brown dwarf is about room temperature. Observations are finding about one brown dwarf for every six regular stars.

C

Calchophile elements: those metals and heavier nonmetals that have a low affinity for oxygen and preferably bond with sulfur as highly insoluble sulfides. The chalcophile elements include: Ag, As, Bi, Cd, Cu, Ga, Ge, Hg, In, Pb, S, Sb, Se, Sn, Te, Tl and Zn

Carbene: in chemistry, a carbene is a molecule containing a neutral carbon atom with a valence of two and two unshared valence electrons. The general formula is R-(C:)-R' or R=C: where the R represent substituents or hydrogen atoms.

Carbonaceous chondrite: A type of meteorite that has a high abundance of carbon-containing species and volatile compounds which are characterized by the presence of once-molten globules of rock called chondrules.

Carboxylic acid: any of a class of organic compounds in which a carbon (C) atom is bonded to an oxygen (O) atom by a double bond and to a hydroxyl group (−OH) by a single bond. A fourth bond links the carbon atom to a hydrogen (H)

atom or to some other univalent combining group. The carboxyl (COOH) group is so-named because of the carbonyl group (C=O) and hydroxyl group.

Catabolism: is the breakdown of certain molecules into a relatively small number of intermediate products, in order to convert chemical energy into a form that can be used for biological processes.

Cepheid variable: a yellow supergiant, pulsating star that exhibits a periodic variation (within the interval 1–50 days) in its luminosity with a direct relation between the luminosity value and its period which was discovered by Henrietta Leavitt in 1908. These stars are important distance markers and are named after δ Cephei, identified as a variable star by John Goodricke in 1784. There are two different families of Cepheid variables, with different period-luminosity relationships.

Chalcogenide: A chalcogenide is a chemical compound consisting of at least one chalcogen anion and at least one more electropositive element. Although all group 16 elements of the periodic table are defined as chalcogens, the term chalcogenide is more commonly reserved for sulfides, selenides, tellurides, and polonides, rather than oxides.

Chandrasekhar limit: is the maximum mass of a stable white dwarf star. The currently accepted value of the Chandrasekhar limit is about 1.4 $M_{\odot}$ (2.765×10^{30} kg).

Chelation: is a type of bonding of ions and molecules to metal ions. It involves the formation or presence of two or more separate coordinate bonds between a multiple bonded ligand and a single central atom. Chelation is useful to remove toxic metals from the body, in manufacturing using homogeneous catalysts, in chemical water treatment to assist in the removal of metals, and in fertilizers.

Chemical evolution: a trend of matter to go from simple to complex structures as time goes on.

Chemically peculiar stars: are the main sequence A and B type stars in the spectra of which lines of a number of elements appear abnormally strong or weak with respect to the bulk of normal A and B stars of the same temperature. Altogether, these classes of stars include the so-called He-rich, He-weak, HgMn, Si, SrCrEu and Am families.

Chemisorption: is a kind of adsorption which involves a chemical reaction between the surface and the adsorbate, so that new chemical bonds are generated at the adsorbent surface. Examples include macroscopic phenomena that can be very obvious, like corrosion, and subtler effects associated with heterogeneous catalysis.

Chemotrophs: Organisms that derive their energy from any chemical energy source (including organic carbon compounds).

Chondrite: Chondrites are stony (non-metallic) meteorites that have not been modified due to melting or differentiation of the parent body. They are formed when various types of dust and small grains that were present in the early solar system accreted to form primitive asteroids. They are the most common type of meteorite that falls to Earth with estimates for the proportion of the total fall that they represent is about 86%.

Circumstellar: Space close to a star external atmosphere.

Classical nova: a transient astronomical event that involves an interaction between two stars that causes a flare-up that is perceived as a new entity that is much brighter than the stars involved. The main sub-classes of novae are classical novae, recurrent novae, and dwarf novae.

Cosmic-ray: A highly energetic, fast-moving particle emitted from a star.

D

Damped Lyman α systems: are concentrations of neutral hydrogen gas, detected in the spectra of quasars. The observed spectra consist of neutral hydrogen Lyman alpha absorption lines, which are broadened by radiation damping. They are defined as systems where the column density (density projected along the line of sight to the quasar) of hydrogen is larger than 2×10^{20} atoms cm^{-2}. These systems can be observed in quantity at relatively high redshifts of 2–4, so that they span the whole epoch of galaxy formation and are generally considered the progenitors of the present-day galaxies.

Dark energy: an unknown form of energy which is hypothesized to permeate all of space, tending to accelerate the expansion of the universe. Dark energy is the most accepted hypothesis to explain the observations since the 1990s, indicating that the universe is expanding at an accelerating rate.

Dark matter: Any form of matter that does not emit nor absorbs electromagnetic radiation at any wavelength.

Direct gap: when in a semiconductor material the conduction band minimum occurs at the same point in reciprocal space as the valence band maximum then it is said we have a direct energy gap, because in this case the energy gap value can be directly determined from optical absorption processes involving photons with an energy $\hbar\omega \geq E_g$. On the contrary, as is often the case, the minimum and maximum occur at different points in k-space, then for crystal momentum, $p = \hbar k$ to be conserved a phonon must also participate in the process, which is then known as an indirect transition. Accordingly, in that case, it is said we have an indirect gap.

DNA: Deoxyribonucleic acid. The genetic information-bearing molecule in most (but not all) viruses and cells.

Dwarf Nova: is a type of cataclysmic variable star consisting of a close binary star system in which one of the components is a white dwarf that accretes matter from its companion.

Dwarf Planet: is a planetary-mass object that is neither a planet nor a natural satellite. That is, it is in direct orbit of a star, and is massive enough for its gravity to crush it into a hydrostatically equilibrious shape (usually a spheroid), but has not cleared the neighborhood of other material around its orbit.

E

Eccentricity: an orbital parameter that measures the ellipticity of a planet, moon or comet orbit. It is comprised in the interval $0 < e \leq 1$, where $e = 0$ stands for a perfectly circular closed orbit and e = 1 for an open parabolic one. Hyperbolic orbits correspond to $e > 1$ values.

Enantiomorphic: either of the two crystal forms of a substance that are mirror images of each other.

Endergonic: (also known as endothermic) in chemical thermodynamics, an endergonic reaction is a chemical reaction in which the standard change in free energy is positive, so that energy is absorbed to proceed. In metabolism, an endergonic process is anabolic, meaning that energy is stored; in many such anabolic processes energy is supplied by coupling the reaction to adenosine triphosphate (ATP) and consequently resulting in a high energy, negatively charged organic phosphate and positive adenosine diphosphate.

Energy gap: In the band diagrams the Fermi energy level E_F (which is usually set to zero) indicates the energy of the electrons located at the topmost occupied band in the fundamental state of the system. The charge carriers located at the energy levels closer to E_F play the more significant roles in the transport properties of the solid. In metallic systems, like lead, the Fermi level crosses through the bands, so that electrons can readily increase their energy in response to outside influences, such as an electric field or temperature gradient. It is this availability to electrons of easily accessible vacant, excited levels in an unfilled band that is at the heart of metallic behavior. On the contrary, insulators and semiconducting materials are characterized by the presence of an appreciable energy gap separating the top occupied band (referred to as the valence band) from the lowest unoccupied band (referred to as the conduction band). The main difference between both kinds of materials is related to the magnitude of the energy bandgap, E_g, which ranges from a few tenths of eV for narrow gap semiconductors to about 1–2 eV for broadband semiconductors, up to 5–10 eV for typical insulating materials.

Ester: is a chemical compound derived from an acid (organic or inorganic) in which at least one –OH (hydroxyl) group is replaced by an –O–alkyl group.

Usually, esters are derived from a carboxylic acid and an alcohol. Glycerides, which are fatty acid esters of glycerol, are important esters in biology, being one of the main classes of lipids, and making up the bulk of animal fats and vegetable oils.

Ether: a class of organic compounds that contain an oxygen atom connected to two alkyl or aryl groups. They have the general formula R–O–R′, where R and R′ represent the alkyl or aryl groups. Ethers can again be classified into two varieties: if the alkyl groups are the same on both sides of the oxygen atom, then it is simple or symmetrical ether, whereas if they are different, the ethers are called mixed or unsymmetrical ethers. A typical example of the first group is the solvent and anesthetic diethyl ether, commonly referred to simply as "ether" (CH_3–CH_2–O–CH_2–CH_3).

Eukarya: One of the three domains of life characterized by single- or multicellular organisms with a cell nucleus.

Eutrophication: is when a body of water becomes overly enriched with minerals and nutrients that induce excessive growth of plants and algae. This process may result in oxygen depletion of the water body. One example is the "bloom" or great increase of phytoplankton in a water body as a response to increased levels of nutrients. Eutrophication is almost always induced by the discharge of nitrate or phosphate-containing detergents, fertilizers, or sewage into an aquatic system.

Exergonic: (also known as exothermic) a chemical reaction producing energy and therefore occurring spontaneously. (*Cf.* endergonic reaction).

Exoplanet: (also referred to as Extrasolar planet) A planet orbiting a star other than the Sun.

Extremophile: A bacterium that has adapted to life in an extreme environment of temperature (thermophiles), pressure (barophile), acidity (acidophiles), salinity (halophiles) or ionizing radiation (radioresistant).

F

Ferroelectrics: a class of materials exhibiting spontaneous electrical polarization below the so-called ferroelectric Curie temperature. The polarization direction can be modified by the applied electric field. At temperatures above Curie temperature, the crystals are nonpolar and no longer ferroelectric and behave like normal dielectrics. Ferroelectrics belong to a wider class of materials called piezoelectric class. All ferroelectric materials are also piezoelectric, but the opposite is not true. For example, quartz and ZnO are piezoelectric, but not ferroelectric.

Ferrophosphorus: Metallurgical additive that contains 50–60 wt% iron and 18–28 wt% phosphorus (Fe_2P plus a small amount of Fe_3P and FeP). This additive is produced during the thermal processing of phosphate rock.

Fischer-Tropsch reaction: is a collection of chemical reactions that converts a mixture of carbon monoxide and hydrogen into liquid hydrocarbons. These reactions occur in the presence of certain metal catalysts, typically at temperatures of 150–300°C and pressures of one to several tens of atmospheres.

Formose reaction: A possible prebiotic reaction producing many kinds of different sugars (pentoses and hexoses), although little ribose and negligible amounts of deoxyribose.

G

Genotype: the part of the genetic makeup of a cell, and therefore of an organism or individual, which determines its characteristics (the so-called phenotype).

Globular cluster: a large spherical ensemble of stars, typically found in the outer regions of a galaxy. A typical globular cluster contains up to 1 million stars, each with an average mass of about 1 $M_{\odot}$, in a volume less than 100 pc across.

Glycolysis: is the metabolic pathway that converts glucose, $C_6H_{12}O_6$, into pyruvate, $CH_3COCOO^- + H^+$. The free energy released in this process is used to form the high-energy molecules ATP (adenosine triphosphate) and NADH (reduced nicotinamide adenine dinucleotide).

H

Habitable zone: is the region around a star where a planet, with sufficient atmospheric pressure, could support liquid water on its surface.

Hadean: the first geological eon in Earth's history which spans the period from the end of the accretion to the beginning of the Archean eon at 3.8 billion years ago.

Hadron: is a composite particle made of quarks held together by the strong force in a similar way as molecules are held together by the electromagnetic force. Hadrons are categorized into two families: baryons, made of three quarks, and mesons, made of one quark and one antiquark. Protons and neutrons are examples of baryons; pions are examples of mesons.

Halo: a spherical distribution of globular clusters and Population II stars that surround a spiral galaxy center.

Heliosphere: is the bubble-like region of space dominated by the Sun, which extends far beyond the orbit of Pluto. Plasma blown out from the Sun, known as the solar wind, creates and maintains this bubble against the outside pressure of the interstellar medium that permeates the Galaxy.

Helium flash: A rapid burst of nuclear reactions indicating the beginning of the helium-burning in the core of a red giant star.

Hertztsprung-Russell diagram: A graphical plot of luminosity (or absolute magnitude) of stars against their surface temperature (or spectral type). According to this classification scheme, most stars can be grouped into the main sequence, subgiants, giants, supergiants, and white dwarfs classes.

Heterotrophs: Organisms that take organic carbon molecules from other organisms.

HII region: gas phase of the ISM where the hydrogen atoms have been ionized by the UV radiation from hot stars.

Homeostasis: any self-regulating process by which biological systems tend to maintain stability while adjusting to conditions that are optimal for survival. If homeostasis is successful, life continues; if unsuccessful, disaster or death ensues. The stability attained is actually a dynamic equilibrium, in which continuous change occurs yet relatively uniform conditions prevail.

Horizontal branch: is a stage of stellar evolution that immediately follows the red giant branch in stars whose masses are similar to the Sun's. Horizontal-branch stars are powered by helium fusion in the core (via the triple-alpha process) and by hydrogen fusion (via the CNO cycle) in a shell surrounding the core and have roughly constant luminosity. The so-called blue horizontal branch stars are those hotter (~ 11,500 K) than objects in the RR Lyrae instability strip, that display abundance anomalies in their atmospheres such as an under abundance of He and an overabundance of several metals.

Hot-Jupiters: are gas giant exoplanets with sizes like that of Jupiter but much shorter orbital periods. Due to its close location to the star the planet receives a great amount of radiation, so that the planet's effective temperature can be over 2,000 K.

Hubble time: a fairly accurate estimate of the age of the universe given by the reciprocal of the Hubble constant H = 73 km s^{-1} Mpc^{-1}. This estimate assumes that the expansion of the universe occurred at the same rate since its origin, although we know that the expansion rate was somewhat lower in the past.

Hypernova: supernovae explosions of larger than 20 $M_{\odot}$ stars releasing about an order of magnitude more energy (~ 10^{45} J) than standard type-II supernovae.

Hypertermophile: An extremophile bacterium that has adapted to an environment of very high temperature, up to 120°C.

I

Inflationary epoch: an extremely rapid expansion episode, by a factor of about 10^{50}, experienced by the early universe shortly after the Planck time (10^{-35}–10^{-33} s).

Intergalactic medium (IGM): the tenuous medium containing gas among the galaxies.

Interplanetary dust particle (IDP): micron to millimeter-sized particles orbiting the Sun around planets coming from meteorite impacts on asteroids and cometary coma particles left over.

Interplanetary medium: is the material which fills the solar system, and through which all the larger solar system bodies, such as planets, dwarf planets, asteroids, and comets, move. It includes interplanetary dust, cosmic rays, and hot plasma from the solar wind.

Interstellar medium (ISM): the tenuous medium containing gas and dust among stars with atomic densities as low as 1 molecule cm^{-3} in diffuse clouds rising to 10^6 molecules cm^{-3} in giant molecular clouds. Temperatures may be as low as 10–40 K.

Interstellar objects: cometary or asteroidal bodies ejected during the formation of planetary systems other than the solar system. The first known object, dubbed Oumuamua, was discovered by the Pan-STARRS1 telescope in October 2017.

K

Kerogen: organic material occurring in oil shales and derived from the decaying of living organisms such as algae and other low plant forms. This material is usually found around minerals and chondrules in carbonaceous chondrites.

Ketone: is an organic compound with the structure RC(=O)R', where R and R' can be a variety of carbon-containing substituents. Ketones and aldehydes are simple compounds that contain a carbonyl group (a carbon-oxygen double bond). Examples include many sugars (ketoses) and the industrial solvent acetone, which is the smallest ketone.

Kuiper belt objects: minor bodies located beyond Neptune orbit at a heliocentric distance between 30 and 50 AU.

L

Lagrangian points: five points in the orbital plane of two bodies revolving about each other in near circular orbits where the third object of negligible mass can remain in equilibrium.

Lepton: an elementary particle of half-integer spin (spin 1/2) that does not undergo strong interactions. Two main classes of leptons exist: charged leptons (also known as the electron-like leptons), and neutral leptons (better known as neutrinos). Charged leptons can combine with other particles to form various composite particles such as atoms or positronium, while neutrinos rarely interact with anything, and are consequently rarely observed. The best known of all leptons is the electron.

Liposome: an ensemble of molecules with hydrophobic tails and hydrophilic heads which spontaneously aggregate together to form a spherical structure in aqueous solution.

Lithophile elements: mainly consist of the highly reactive metals of the s- and f-blocks. They also include a small number of reactive nonmetals, and the more reactive metals of the d-block such as titanium, zirconium, and vanadium

Lithotrophs: organisms that derive their metabolic energy from chemical species, other than organic carbon, instead of light.

Low-mass star: stars with initial masses on the zero-age-main-sequence comprised between 1 and 8 $M_{\odot}$.

M

Magnetocaloric effect: the reversible change of temperature accompanying the change of magnetization of a ferromagnetic or paramagnetic material. This change in temperature may be of the order of 1°C.

Magnetoelastic effect: the change of the magnetic susceptibility of a material when subjected to a mechanical stress.

Mass spectroscopy: A mass spectrometer is an analytical instrument routinely used by chemists that ionize samples using techniques such as electron bombardment and produces a signature of the chemical structure showing the mass-to-charge ratio of the elements in the sample. The shorter segments produced are easier to analyze than longer ones.

Massive compact halo objects (MACHOs): dim brown or white dwarfs stars or low-mass black holes that may constitute part of the unseen dark matter in our Galaxy.

Metabolism: a set of chemical reactions, usually occurring in living cells, that allows organisms to harness chemical energy and synthesize necessary biomolecules. In general metabolic pathways function primarily in either catabolism or anabolism modes, although some of them such as the tricarboxylic acid cycle are amphibolic, meaning they have both catabolic and anabolic functions.

Metallicity: Abundance of elements heavier than helium in a given astrophysical object. It is generally denoted by the letter Z.

Meteor: A meteoroid entering a planet's atmosphere at speed greater than 70 kms^{-1} that burns up completely before it hits the planet's surface.

Meteorite: A meteoroid entering the atmosphere of a planet that survives the journey fall to land on the ground.

Meteoroid: A particle of interplanetary debris that can enter the atmosphere of a planet to become either a meteor or a meteorite.

Mira variable: named after the prototype star Mira (o Cet), are a class of pulsating variable stars characterized by very red colors, pulsation periods longer than 100 days, and amplitudes greater than one magnitude in infrared and 2.5 magnitudes at visual wavelengths. They are red giants in the very late stages of stellar evolution, on the asymptotic giant branch.

N

Nanomaterials: materials exhibiting a nanometric scale in one of their dimensions at least.

Nanoparticle: an aggregate of atoms whose size is in the 10^{-9} m range.

Near-Earth Object (NEO): is any small solar system body whose orbit can bring it into proximity with Earth. By definition, a solar system body is a NEO if its closest approach to the Sun (perihelion) is less than 1.3 AU. If a NEO's orbit crosses the Earth's and the object is larger than 140 meters across, it is considered a potentially hazardous object.

Neutron star: a dense, highly packed neutron sphere a few kilometers in diameter which results from high mass stars' core-collapse after supernova explosion event.

Nucleosynthesis: The process of building up complex nuclei from simpler nuclei, protons, and neutrons.

O

Oligoelements: chemical elements which are required in minor quantities for most living beings, including K, Na, Mg, Ca, Fe, Mn, Cu, and Zn.

Oort cloud: is a theoretical cloud of predominantly icy planetesimals proposed to surround the Sun at distances ranging from 50,000 to 200,000 AU (0.8 and 3.2 ly). It is divided into two regions: a disk-shaped inner Oort cloud and a spherical outer Oort cloud. Both regions lie beyond the heliosphere and in interstellar space.

Open cluster: a loose gravitational association of young stars in the disk of our Galaxy.

Organophosphorus compounds: are organic compounds containing phosphorus, and organophosphorus chemistry is the corresponding science of the properties and reactivity of organophosphorus compounds.

P

Pallasite: a class of stony-iron meteorite. Its main feature consists of centimeter-sized olivine crystals in an iron-nickel matrix.

Phenotype: (see Genotype).

Phosphoester bond: bond formed between a phosphate group and a hydroxyl group. A phosphodiester bond contains one phosphate group participating in two phosphoester bonds, as it occurs in DNA and RNA.

Phosphorylation: is a reaction that produces an organophosphate compound by adding a phosphate group (PO_4^{-3}) to an organic molecule through an oxygen atom.

Photon: is a type of elementary particle, the quantum of the electromagnetic field including electromagnetic radiation such as light, and the force carrier for the electromagnetic force. The photon has zero mass and always moves at the speed of light within a vacuum.

Photosynthesis: is a process used by plants and other organisms to convert light energy into chemical energy that can later be released to fuel the organisms' activities (metabolism). This chemical energy is stored in carbohydrate molecules, such as sugars, which are synthesized from carbon dioxide and water. In most cases, oxygen is also released as a waste product. Photosynthesis is largely responsible for producing and maintaining the oxygen content of the Earth's atmosphere, and supplies all of the organic compounds and most of the energy necessary for life on Earth

Phototrophs (photoautotrophs): organisms performing photosynthesis, which include cyanobacteria and most algae and plants.

Physisorption: also called physical adsorption, is a process in which the electronic structure of the atom or molecule is barely perturbed upon adsorption. The fundamental interacting force of physisorption is caused by van der Waals force.

Piezoelectricity: is the ability of a material to generate an electric potential in response to applied mechanical stress (and vice versa).

Planck time: a fundamental interval of time defined in terms of the fundamental constant of gravitation G, the Planck's constant, h, and the speed of light c, as $t_p = (Gh/c^5)^{1/2} = 1.35\times10^{-43}$ s.

Planetary nebula: a luminous shell of glowing, ionized gas ejected from an old, low-mass star as it begins to die. Despite their name, planetary nebulae have nothing to do with planets. This misleading name was introduced in the nineteenth century because these glowing objects looked like little, fuzzy planetary disks when viewed through the small telescopes then available.

Planetesimal: a small body of primordial durst and ice from which the planets and comets formed.

Polycyclic aromatic hydrocarbon (PAH): organic molecules containing several fused six-carbon rings building up from benzene (C_6H_6). These compounds may be the most abundant organic molecules in space, making up 20% of the total cosmic carbon. They contain several hundred atoms and are very stable due to the delocalization of electrons over their carbon skeleton. PAHs are readily synthesized during combustion processes of organic substances and are responsible for the formation of soot particles.

Population I star: a star whose spectrum exhibits spectral lines of many elements heavier than helium (metal-rich star).

Population II star: a star whose spectrum exhibits comparatively few spectral lines of elements heavier than helium (metal-poor star)

Population III star: hypothetical very massive stars made only of hydrogen and helium which formed and exploded during the early stages of the universe.

Prebiotic chemistry: the natural process by which life arises from non-living matter, such as simple organic compounds, where one assumes that the transition from non-living to living entities was not a single event, but a gradual process of increasing complexity.

Preplanetary disk: A disk of material encircling a protostar or a newborn star.

Prokaryote: is a unicellular organism that lacks a membrane-bound nucleus, mitochondria, or any other membrane-bound organelle. Prokaryotes are divided into two domains, Archaea and Bacteria.

Proplyd: a syllabic abbreviation of an ionized protoplanetary disk, is an externally illuminated photoevaporating disk around a young star. Nearly 180 proplyds have been discovered in the Orion Nebula. Images of proplyds in other star-forming regions are rare, while Orion is the only region with a large known sample due to its relative proximity to Earth.

Proteome: is the entire set of proteins that is, or can be, expressed by a genome, cell, tissue, or organism at a certain time. It is the set of expressed proteins in a given type of cell or organism, at a given time, under defined conditions. Proteomics is the study of the proteome.

Proterozoic: is a geological eon representing the time just before the proliferation of complex life on Earth. The name Proterozoic comes from Greek and means "earlier life." The Proterozoic Eon extended from 2500 to 540 million years ago, and is the most recent part of the Precambrian Supereon. It can be also described as the time range between the appearance of oxygen in Earth's atmosphere and the appearance of first complex life forms (like trilobites or corals).

Protogalaxy: is a cloud of gas which is forming into a galaxy. It is believed that the rate of star formation during this period of galactic evolution will determine whether a galaxy is a spiral or elliptical galaxy; a slower star formation tends to produce a spiral galaxy. The smaller clumps of gas in a protogalaxy form into stars.

Protoplanet: an object intermediate in size between a planetesimal and a planet; an intermediate stage in the formation of a planet.

Protoplanetary nebula: is a rapid (~ 1,000 years) transition phase between the asymptotic giant branch and full-fledged planetary nebula phases

Protostar: a star in its earliest stages of formation, shortly after the initial gravitational collapse.

Purines: one of the two types of nitrogen-heterocyclic bases (adenine and guanine) present in the DNA and RNA nucleic acids.

Pyrimidines: one of the three types of nitrogen-heterocyclic bases present in the DNA (cytosine and thymine) and RNA (uracil) nucleic acids.

Q

Quarks: particles thought to be the internal constituents of heavier subatomic particles, such as protons and neutrons.

Quasar: abbreviation of a quasi-stellar object: a very luminous object with a very large redshift and a star-like appearance.

Quinones: a class of organic compounds that are formally derived from aromatic compounds, such as benzene or naphthalene, by conversion of an even number of –CH= groups into –C(=O)– groups, with any necessary rearrangement of double bonds, resulting in a fully conjugated cyclic structure. The class includes some heterocyclic compounds.

R

Racemic: a racemic mixture, or racemate, is one that has equal amounts of left- and right-handed enantiomers of a chiral molecule. The first known racemic mixture was racemic acid, which Louis Pasteur found to be a mixture of the two enantiomeric isomers of tartaric acid. A sample with only a single enantiomer is an enantiomerically pure, enantiopure or homochiral compound.

Recombination era: refers to the epoch at which charged electrons and protons first became bound to form electrically neutral hydrogen atoms. Recombination occurred about 378,000 years after the origin of the universe (at a redshift of $z = 1{,}100$).

Red giant: is a luminous giant star of low or intermediate mass (roughly 0.3–8 solar masses) in a late phase of stellar evolution. The outer atmosphere is inflated and tenuous, making the radius large and the surface temperature around 5,000 K or lower. The appearance of the red giant is from yellow-orange to red, including the spectral types K and M, but also class S stars and most carbon stars.

Redox reactions: in these processes reduction entails becoming more electron-rich (i.e., gaining electrons), whereas oxidation refers to becoming more electron-deficient (i.e., losing electrons).

Redshift: The Doppler shifting to longer wavelengths of the light from remote galaxies and quasars receding sources.

Reflux: heating a solution at its boiling point. This ensures heating at a constant temperature. A typical setup includes a flask with the solution with an attached condenser that is placed in a vertical position. This guarantees that the boiled liquid is not lost, but condenses and returns to the flask.

Regolith: is a layer of loose, heterogeneous superficial deposits covering solid rock. It includes dust, soil, broken rock, and other related materials and is present on Earth, the Moon, Mars, some asteroids, and other terrestrial planets and moons.

Ribose: the sugar that attaches to the DNA bases to yield a nucleoside.

Ribozymes: ribonucleic acids which are capable of acting as catalysts. RNA enzymes can catalyze a wide variety of reactions, encompassing all types of transformations necessary for the protocell activity.

RNA: Nucleic acid analog to DNA by replacing deoxyribose by ribose and thymine by uracil.

Roche lobe: a teardrop-shaped volume surrounding a star in a binary system inside which gases are gravitationally bound to that star.

Rogue planet: planets which are not orbiting around a star but wander in the galaxies orbiting around the galactic center.

RR Lyra star: a type of variable pulsating star with a period of less than one day.

S

Seebeck coefficient: also referred to as thermopower or thermoelectric power, measures the magnitude of an induced thermoelectric voltage in response to a temperature difference across the material. Its magnitude (usually comprised within the range from μVK^{-1} to mVK^{-1}) generally depends on the temperature of the junction and its sign is determined by the materials composing the circuit.

Seebeck effect: describes the conversion of thermal energy into electrical energy in the form of an electrical current. The magnitude of this effect can be expressed in terms of the Seebeck voltage related to the electromotive force set up under open-circuit conditions. For not too large temperature differences between the junctions, this voltage is found to be proportional to their temperature difference,

$\Delta V = S\ \Delta T$, where the coefficient of proportionality S(T) is a temperature-dependent property of the junction materials called the Seebeck coefficient, and it is expressed in VK^{-1} units.

Semiconductor: materials characterized by the presence of an appreciable energy gap separating the top occupied band (referred to as the valence band) from the lowest unoccupied band which ranges from a few tenths of eV for narrow gap semiconductors, to about 1–2 eV for broadband semiconductors. With increasing temperature, electrons in the valence band begin to be thermally excited. Once the electrons are excited into the conduction band, an equal amount of holes is left behind in the valence band. The material of this type is called an intrinsic semiconductor.

Seyfert galaxy: a spiral galaxy with a bright nucleus whose spectrum exhibits emission lines.

Siderophile: chemical elements such as iridium or gold that tend to bond with metallic iron.

Snow line: is the particular distance in the preplanetary nebula from the central protostar where it is cold enough for volatile compounds such as water, ammonia, methane, carbon dioxide or carbon monoxide to condense into solid ice grains. This condensation temperature depends on the volatile substance and the partial pressure of the vapor in the protostar nebula.

Solar analog: dwarf stars belonging to the spectral classes G0 – G5.

Solar twin: stars almost identical to the Sun. To date, no solar twin that exactly matches the Sun has been found. However, there are some stars that come very close to being identical to that of the Sun, and are such considered solar twins by members of the astronomical community. An exact solar twin would be a G2V star with a 5,778K temperature, be 4.6 billion years old, with the correct metallicity and a 0.1% solar luminosity variation.

Strecker synthesis: a possible prebiotic reaction to produce amino acids.

Stromatolites: layered bio-chemical accretionary structures formed in shallow water by the trapping, binding and cementation of sedimentary grains by biofilms (microbial mats) of microorganisms, especially cyanobacteria. Fossilized stromatolites provide ancient records of life on Earth by these remains.

S-type star: a cool giant star with approximately equal quantities of carbon and oxygen in its atmosphere. The presence of zirconium monoxide (ZrO) molecular bands in their spectra are a defining feature of the S stars.

Subdwarf B star: is a kind subdwarf star with spectral type B (notation sdB) being much hotter and brighter than a typical subdwarf, so that they are prominent on ultraviolet images and some of them pulsate with typical periods of 100–300 s. Masses of these stars are around 0.5 solar masses, and they contain only about

1% hydrogen, with the rest being helium. Their radius is from 0.15 to 0.25 solar radii, and their temperature is from 20,000 to 40,000K. These stars represent a late stage in the evolution of some stars, caused when a red giant star loses its outer hydrogen layers before the core begins to fuse helium. Accordingly, they are situated at the "extreme horizontal branch" of the Herzsprung-Rusell diagram and are progenitors of low-mass white dwarfs.

Supercluster: a collection of clusters of galaxies.

Super-Earths: exoplanets with masses between 2–8 times that of the Earth.

Supernovae: are classified by their spectroscopic appearance and are sorted in two main classes: type I, which lack hydrogen lines in the spectrum of the emitted light, and type II, which does show hydrogen lines. Type Ia supernova is a gigantic thermonuclear explosion of a white dwarf star that is triggered by accretion material from a star companion in a binary system. Types II, Ib, and Ic are referred to as core-collapse supernovae, and they represent the violent death of a single massive star as a consequence of the gravitational collapse of the stellar core.

Superphosporic acid: is a blend of orthophosphoric acid and polyphosphoric acid. Polyphosphoric acid is composed of linear polyphosphate species, which include pyrophosphate, tripolyphosphate, tetrapolyphosphate, and longer chains.

Synodic period: the time interval between successive occurrences of the same orbital configuration of a planet or moon.

T

Troposphere: is the lowest layer of Earth's atmosphere, and is also where nearly all weather conditions take place. It contains approximately 75% of the atmosphere's mass and 99% of the total mass of water vapor and aerosols. The average depths of the troposphere are 20 km in the tropics, 17 km in the mid-latitudes, and 7 km in the polar regions.

T-Tauri star: a class of variable stars associated with young stars whose ages are less than about ten million years old. This class is named after the prototype, T-Tauri, a young star in the Taurus star-forming region. They are found near molecular clouds and identified by their optical variability and strong chromospheric lines. T-Tauri stars are pre-main-sequence stars in the process of contracting to the main sequence along the so-called Hayashi track, a luminosity–temperature relationship obeyed by infant stars of less than three solar masses in the pre-main-sequence phase of stellar evolution.

V

Van Der Waals interaction: distance-dependent force between atoms or molecules. Unlike ionic or covalent bonds, these attractions are not a result of any chemical electronic bond, and they are comparatively weak and more susceptible to being perturbed. Van der Waals forces quickly vanish at longer distances between interacting molecules. Van der Waals forces play a fundamental role in fields as diverse as supramolecular chemistry, structural biology, polymer science, nanotechnology, surface science, and condensed matter physics. Van der Waals forces also define many properties of organic compounds and molecular solids, including their solubility in polar and non-polar media.

Volatile element: in planetary science, volatiles are the group of chemical elements and chemical compounds with low boiling points that are associated with a planet's or moon's crust or atmosphere. Examples include nitrogen, water, carbon dioxide, ammonia, hydrogen, methane, and sulfur dioxide. In astrogeology, these compounds, in their solid-state, often comprise large proportions of the crusts of moons and dwarf planets. In contrast with volatiles, elements and compounds with high boiling points are known as refractory substances.

W

Weakly interacting massive particles (WIMPs): hypothetical massive ($10 - 10^4$ times greater than a proton) elementary particles that may make up part of the unseen dark matter in the universe. These particles do not emit or absorb electromagnetic radiation but interact with regular matter through the weak nuclear force.

White dwarf: A hot, compact sphere of helium, carbon, and oxygen, resulting from the core of low- to intermediate-mass stars undergoing the ejection of their outer layers during the planetary nebula phase. The material in a white dwarf no longer undergoes fusion reactions, so the star has no source of energy. As a result, it cannot support itself by the heat generated by fusion against gravitational collapse, but is supported only by electron degeneracy pressure, causing it to be extremely dense: its mass is comparable to that of the Sun, while its volume is comparable to that of Earth. The compositions of the photospheres of white dwarfs are observed to be dominated by the lightest elements H and He, since the high surface gravities of these objects cause heavier elements to settle out on time-scales of mere days. Nevertheless, UV observations have revealed trace quantities of metals in the photospheres of a large number of hot enough (30,000–40,000 K) white dwarfs, whose presence is attributed to an interplay between radiative levitation and gravitational settling. White dwarfs spectral classification includes DA-type, which have hydrogen-dominated atmospheres and make up the majority (approximately 80%) of all observed white dwarfs. The next class in number is of DB-type (approximately 16%), which have helium-dominated atmospheres. The hot (above 15,000 K) DQ class (roughly 0.1%) has carbon-dominated atmospheres.

Wolf-Rayet star: are a rare heterogeneous set of stars with unusual spectra showing prominent broad emission lines of highly ionized helium and nitrogen or carbon. The spectra indicate a very high surface enhancement of heavy elements, depletion of hydrogen, and strong stellar winds. Their surface temperatures range from 30,000 K to around 200,000 K, hotter than almost all other stars. Classic (or Population I) Wolf–Rayet stars are evolved, massive stars that have completely lost their outer hydrogen and are fusing helium or heavier elements in the core. A subset of the population I WR stars show hydrogen lines in their spectra and are known as WNh stars; they are young extremely massive stars still fusing hydrogen at the core, with helium and nitrogen exposed at the surface by strong mixing and radiation-driven mass loss. A separate group of stars with WR spectra are the central stars of planetary nebulae (CSPNe), post-Asymptotic Giant Branch stars that were similar to the Sun while on the main sequence, but have now ceased fusion and shed their atmospheres to reveal a bare carbon-oxygen core.

Z

Zodiacal light: is a faint, diffuse white glow that is visible in the night sky and appears to extend from the Sun's direction and along the zodiac, straddling the ecliptic. Sunlight scattered by interplanetary dust causes this phenomenon. Zodiacal light is best seen during twilight after sunset in spring and before sunrise in autumn, when the zodiac is at a steep angle to the horizon.

Appendix

Some Important Astronomical Quantities

Astronomical unit	1 AU	1.496×10^{11} m
Light-year	1 ly	9.460×10^{15} m
Parsec	1 pc	3.086×10^{16} m
Solar mass	1 M	1.989×10^{30} kg
Jupiter mass	10^{-3} M	1.899×10^{27} kg
Earth-mass	1 M	5.974×10^{24} kg
Earth radius	1 R	6.378×10^{6} m
Solar radius	1 R	6.960×10^{8} m
Solar luminosity	1 L	3.900×10^{26} W
Galaxy radius	25 kpc	7.714×10^{20} m

Some Important Physical Constants

Speed of light	c	2.9979×10^{8} ms^{-1}
Gravitational constant	G	6.6740×10^{-11} Jm kg^{-2}
Planck's constant	h	6.6261×10^{-34} J s
		4.1357×10^{-15} eV s
Boltzmann constant	k	1.3807×10^{-23}J/K
		8.6174×10^{-5} eV/K
Stefan-Boltzmann constant	σ	5.6705×10^{-8} J m^{-3} K^{-4}
Mass of electron	m_e	9.1094×10^{-31} kg
Mass of proton	m_p	1.6726×10^{-27} kg
Mass of neutron	m_n	1.6749×10^{-27} kg
Mass of hydrogen atom	m_H	1.6735×10^{-27} kg
Atomic mass unit	amu	1.6605×10^{-27} kg

Rydberg constant	R	1.0968×10^{7} m^{-1}
Bohr radius	a_0	5×10^{-11} m
Avogadro number	N_A	6.022×10^{23} mol^{-1}
Perfect gas constant	R	8.314 $Jmol^{-1}K^{-1}$
Hubble constant	H	73.52 $kms^{-1}Mpc$
Vacuum dielectric constant	ε_0	8.854×10^{-12} Fm^{-1}
Vacuum permeability	μ_0	1.257×10^{-6} NA^{-2}
Electron charge	e	1.602×10^{-19} C

(Updated values taken from Liebisch et al., 2019; Wu et al., 2019).

Conversion Factors

1 eV	1.6022×10^{-19} J
	1.2×10^{4} Kc^{2}
1 kJ mol^{-1}	10.35×10^{-3} eV
1 pc	3.26 ly
1 Å	10^{-10} m
$(4\pi\varepsilon_0)^{-1}$	9×10^{9} JmC^{-2}
1 cal	4.186 J

Index

D

Q

R

S

T